T0235174

Lecture Notes in Computer Science 10556

Commenced Publication in 1973
Founding and Former Series Editors:
Gerhard Goos, Juris Hartmanis, and Jan van Leeuwen

More information about this series at http://www.springer.com/series/7407

Roberto Battiti · Dmitri E. Kvasov
Yaroslav D. Sergeyev (Eds.)

Learning and Intelligent Optimization

11th International Conference, LION 11
Nizhny Novgorod, Russia, June 19–21, 2017
Revised Selected Papers

 Springer

Editors
Roberto Battiti ⓘ
University of Trento
Trento
Italy

and

Lobachevsky University of Nizhny
  Novgorod
Nizhny Novgorod
Russia

Dmitri E. Kvasov ⓘ
University of Calabria
Rende
Italy

and

Lobachevsky University of Nizhny
  Novgorod
Nizhny Novgorod
Russia

Yaroslav D. Sergeyev ⓘ
University of Calabria
Rende
Italy

and

Lobachevsky University of Nizhny
  Novgorod
Nizhny Novgorod
Russia

ISSN 0302-9743   ISSN 1611-3349 (electronic)
Lecture Notes in Computer Science
ISBN 978-3-319-69403-0   ISBN 978-3-319-69404-7 (eBook)
https://doi.org/10.1007/978-3-319-69404-7

Library of Congress Control Number: 2017956726

LNCS Sublibrary: SL1 – Theoretical Computer Science and General Issues

Printed on acid-free paper

This Springer imprint is published by Springer Nature
The registered company is Springer International Publishing AG
The registered company address is: Gewerbestrasse 11, 6330 Cham, Switzerland

Preface

This volume edited by R. Battiti, D.E. Kvasov, and Y.D. Sergeyev contains peer-reviewed papers from the 11th Learning and Intelligent International Optimization conference (LION-11) held in Nizhny Novgorod, Russia, during June 19–21, 2017. The LION-11 conference has continued the successful series of the constantly expanding and worldwide recognized LION events (LION-1: Andalo, Italy, 2007; LION-2 and LION-3: Trento, Italy, 2008 and 2009; LION-4: Venice, Italy, 2010; LION-5: Rome, Italy, 2011; LION-6: Paris, France, 2012; LION-7: Catania, Italy, 2013; LION-8: Gainesville, USA, 2014; LION-9: Lille, France, 2015; LION-10: Ischia, Italy, 2016). This edition was organized by the Lobachevsky University of Nizhny Novgorod, Russia, as one of the key events of the Russian Science Foundation project No. 15-11-30022 "Global Optimization, Supercomputing Computations, and Applications." Like its predecessors, the LION-11 international meeting explored advanced research developments in such interconnected fields as mathematical programming, global optimization, machine learning, and artificial intelligence. Russia has a long tradition in optimization theory, computational mathematics, and "intelligent learning techniques" (in particular, cybernetics and statistics), therefore, the location of LION-11 in Nizhny Novgorod was an excellent occasion to meet researchers and consolidate research and personal links.

More than 60 participants from 15 countries (Austria, Belgium, France, Germany, Hungary, Italy, Lithuania, Portugal, Russia, Serbia, Switzerland, Taiwan, Turkey, UK, and USA) took part in the LION-11 conference. Four plenary lecturers shared their current research directions with the LION-11 participants:

Renato De Leone, Camerino, Italy: "The Use of Grossone in Optimization: A Survey and Some Recent Results"

Nenad Mladenovic, Belgrade, Serbia: "Less Is More Approach in Heuristic Optimization"

Panos Pardalos, Gainesville, USA: "Quantification of Network Dissimilarities and Its Practical Implications"

Julius Žilinskas, Vilnius, Lithuania: "Deterministic Algorithms for Black Box Global Optimization"

Moreover, three tutorials were also presented during the conference:

Adil Erzin, Novosibirsk, Russia: "Some Optimization Problems in the Wireless Sensor Networks"

Mario Guarracino, Naples, Italy: "Laplacian-Based Semi-supervised Learning"

Yaroslav Sergeyev, University of Calabria, Italy, and Lobachevsky University of Nizhny Novgorod, Russia: "Numerical Computations with Infinities and Infinitesimals"

A total of 20 long papers and 15 short papers were accepted for publication in this LNCS volume after thorough peer reviewing (up to three review rounds for some manuscripts) by the members of the LION-11 Program Committee and independent reviewers. These papers describe advanced ideas, technologies, methods, and applications in optimization and machine learning. This volume also contains the paper of the winner (Francesco Romito, Rome, Italy) of the second edition of the Generalization-Based Contest in Global Optimization (GENOPT: http://genopt.org).

The editors thank all the participants for their dedication to the success of LION-11 and are grateful to the reviewers for their valuable work. The support of the Springer LNCS editorial staff is greatly appreciated.

The editors express their gratitude to the organizers and sponsors of the LION-11 international conference: Lobachevsky University of Nizhny Novgorod, Russia; Russian Science Foundation; EnginSoft Company, Italy; NTP Truboprovod, Russia; and the International Society of Global Optimization. Their support was essential for the success of this event.

August 2017

Roberto Battiti
Dmitri E. Kvasov
Yaroslav D. Sergeyev

Organization

General Chair

Yaroslav Sergeyev   University of Calabria, Italy and Lobachevsky
University of Nizhny Novgorod, Russia

Steering Committee

Roberto Battiti (Head)	University of Trento, Italy and Lobachevsky University of Nizhny Novgorod, Russia
Holger Hoos	University of British Columbia, Canada
Youssef Hamadi	École Polytechnique, France
Mauro Brunato	University of Trento, Italy
Thomas Stützle	Université Libre de Bruxelles, Belgium
Christian Blum	Spanish National Research Council
Martin Golumbic	University of Haifa, Israel
Marc Schoenauer	Inria Saclay, Île-de-France
Xin Yao	University of Birmingham, UK
Benjamin Wah	The Chinese University of Hong Kong and University of Illinois, USA
Yaroslav Sergeyev	University of Calabria, Italy and Lobachevsky University of Nizhny Novgorod, Russia
Panos Pardalos	University of Florida, USA

Technical Program Committee

Annabella Astorino	ICAR-CNR, Italy
Roberto Battiti	University of Trento, Italy and Lobachevsky University of Nizhny Novgorod, Russia
Bernd Bischl	Ludwig Maximilians University Munich, Germany
Christian Blum	Spanish National Research Council
Mauro Brunato	University of Trento, Italy
Sonia Cafieri	École Nationale de l'Aviation Civile, France
Andre de Carvalho	University of São Paulo, Brazil
John Chinneck	Carleton University, Canada
Andre Cire	University of Toronto Scarborough, Canada
Renato De Leone	University of Camerino, Italy
Luca Di Gaspero	University of Udine, Italy
Bistra Dilkina	Georgia Institute of Technology, USA
Adil Erzin	Sobolev Institute of Mathematics SB RAS, Russia
Giovanni Fasano	University Ca'Foscari of Venice, Italy
Paola Festa	University of Naples Federico II, Italy

Antonio Fuduli	University of Calabria, Italy
David Gao	Federation University, Australia
Martin Golumbic	University of Haifa, Israel
Vladimir Grishagin	Lobachevsky University of Nizhny Novgorod, Russia
Tias Guns	Vrije Universiteit Brussel, Belgium
Youssef Hamadi	École Polytechnique, France
Frank Hutter	Albert-Ludwigs-Universität Freiburg, Germany
George Katsirelos	MIAT-INRA, France
Michael Khachay	Krasovsky Institute of Mathematics and Mechanics UB RAS, Russia
Oleg Khamisov	Melentiev Institute of Energy Systems SB RAS, Russia
Yury Kochetov	Sobolev Institute of Mathematics SB RAS, Russia
Lars Kotthoff	University of British Columbia, Canada
Dmitri Kvasov	University of Calabria, Italy and Lobachevsky University of Nizhny Novgorod, Russia
Dario Landa-Silva	The University of Nottingham, UK
Hoai An Le Thi	Université de Lorraine, France
Daniela Lera	University of Cagliari, Italy
Vasily Malozemov	Saint Petersburg State University, Russia
Marie-Eléonore Marmion	CRISTAL/Inria/Lille University, France
Silvano Martello	University of Bologna, Italy
Kaisa Miettinen	University of Jyväskylä, Finland
Nenad Mladenovic	Mathematical Institute SANU, Serbia
Evgeni Nurminski	Far Eastern Federal University, Russia
Barry O'Sullivan	University College Cork, Ireland
Panos Pardalos	University of Florida, USA
Remigijus Paulavičius	Imperial College London, UK and Vilnius University, Lithuania
Thomas Pock	Graz University of Technology, Austria
Mikhail Posypkin	Dorodnicyn Computing Centre, FRC CSC RAS, Russia
Oleg Prokopyev	University of Pittsburgh, USA
Helena Ramalhinho Lourenço	Universitat Pompeu Fabra, Spain
Massimo Roma	Sapienza University of Rome, Italy
Francesca Rossi	University of Padova, Italy and Harvard University, USA
Valeria Ruggiero	University of Ferrara, Italy
Horst Samulowitz	IBM T.J. Watson Research Center, USA
Marc Schoenauer	Inria Saclay, Île-de-France
Meinolf Sellmann	General Electric Global Research, USA
Yaroslav Sergeyev (Chair)	University of Calabria, Italy and Lobachevsky University of Nizhny Novgorod, Russia
Carlos Soares	University of Porto, Portugal
Alexander Strekalovskiy	Matrosov Institute for System Dynamics and Control Theory SB RAS, Russia
Thomas Stützle	Université Libre de Bruxelles, Belgium

Éric Taillard   University of Applied Science of Western Switzerland
Tatiana Tchemisova   University of Aveiro, Portugal
Gerardo Toraldo   University of Naples Federico II, Italy
Michael Trick   Carnegie Mellon University, USA
Daniele Vigo   University of Bologna, Italy
Petr Vilím   IBM Czech, Czech Republic
Luca Zanni   University of Modena and Reggio Emilia, Italy
Anatoly Zhigljavsky   Cardiff University, UK
Antanas Žilinskas   Vilnius University, Lithuania
Julius Žilinskas   Vilnius University, Lithuania

Additional Reviewers

Steven Adriaensen   Marat Mukhametzhanov
Yair Censor   Duy Nhat Phan
Elena Chistyakova   Soumyendu Raha
Ivan Davydov   Ivan Takhonov
Alessandra De Rossi   Ider Tseveendorj
Giuseppe Fedele   Vo Xuanthanh
Jonathan Gillard

Local Organization Committee

Roman Strongin (Chair)   Lobachevsky University of Nizhny Novgorod, Russia
Victor Kasantsev   Lobachevsky University of Nizhny Novgorod, Russia
  (Vice-chair)
Vadim Saygin   Lobachevsky University of Nizhny Novgorod, Russia
  (Vice-chair)
Konstantin Barkalov   Lobachevsky University of Nizhny Novgorod, Russia
Lev Afraimovich   Lobachevsky University of Nizhny Novgorod, Russia
Dmitri Balandin   Lobachevsky University of Nizhny Novgorod, Russia
Victor Gergel   Lobachevsky University of Nizhny Novgorod, Russia
Vladimir Grishagin   Lobachevsky University of Nizhny Novgorod, Russia
Dmitri Kvasov   Lobachevsky University of Nizhny Novgorod, Russia
Iosif Meerov   Lobachevsky University of Nizhny Novgorod, Russia
Grigory Osipov   Lobachevsky University of Nizhny Novgorod, Russia
Mihail Prilutskii   Lobachevsky University of Nizhny Novgorod, Russia
Alexadner Sysoyev   Lobachevsky University of Nizhny Novgorod, Russia
Dmitri Shaposhnikov   Lobachevsky University of Nizhny Novgorod, Russia
Ekaterina Goldinova   Lobachevsky University of Nizhny Novgorod, Russia

Acknowledgements. The organization of the LION-11 conference was supported by the Russian Science Foundation, project No. 15-11-30022 "Global optimization, supercomputing computations, and applications". The organizers also gratefully acknowledge the support of the following sponsors: EnginSoft company, Italy; NTP Truboprovod, Russia; and the International Society of Global Optimization.

Contents

GENOPT Paper

Short Papers

Long Papers

An Importance Sampling Approach to the Estimation of Algorithm Performance in Automated Algorithm Design

Steven Adriaensen$^{(\boxtimes)}$, Filip Moons, and Ann Nowé

Vrije Universiteit Brussel, Pleinlaan 2, 1050 Brussels, Belgium
steven.adriaensen@vub.be

Abstract. In this paper we consider the problem of estimating the relative performance of a given set of related algorithms. The predominant, general approach of doing so involves executing each algorithm instance multiple times, and computing independent estimates based on the performance observations made for each of them. A single execution might be expensive, making this a time-consuming process. We show how an algorithm in general can be viewed as a distribution over executions; and its performance as the expectation of some measure of desirability of an execution, over this distribution. Subsequently, we describe how Importance Sampling can be used to generalize performance observations across algorithms with partially overlapping distributions, amortizing the cost of obtaining them. Finally, we implement the proposed approach as a Proof of Concept and validate it experimentally.

Keywords: Performance evaluation · Algorithms · Importance Sampling · Programming by Optimization · Automated algorithm synthesis

1 Introduction

Often, there are many ways to solve a given problem. However, not all of these are equally "good". In the Algorithm Design Problem [2,14] (ADP) we are to find the "best" way to solve a given problem; e.g. using the least computational resources (time, memory etc.), and/or maximizing the quality of the solutions obtained. In a sense it is the problem of "how to best solve a given problem".

To date, algorithms for many real-world problems are most commonly designed following a manual, ad-hoc, trial & error approach, making algorithm design a tedious and costly process, often leading to mediocre results. Recently, Programming by Optimization [8] (PbO) was proposed as an alternative design paradigm. In PbO, difficult choices are deliberately left open at design time, thus programming a family of algorithms (*design space*), rather than a single algorithm. Subsequently, optimization methods are applied to automatically determine the best algorithm instance (*design*) for a specific use-case. Often, the latter

Steven Adriaensen is funded by a Ph.D. grant of the Research Foundation Flanders (FWO).

R. Battiti et al. (Eds.): LION 2017, LNCS 10556, pp. 3–17, 2017.
https://doi.org/10.1007/978-3-319-69404-7_1

requires evaluating each design numerous times, as we are typically interested in performance across multiple inputs and observations thereof might be noisy. In turn, a single evaluation can be expensive as it involves algorithm execution, followed by some sort of quantification of its desirability. As the design space considered is often huge, this quickly becomes a time-consuming process.

In this paper we propose a novel way of estimating the relative performance of a given set of related algorithms, targeted at reducing the number of evaluations needed to solve a given ADP, i.e. solve it faster. This approach is based on the observation that different algorithm instances in the design space might, with some likelihood, generate the exact same execution, i.e. perform the exact same sequence of instructions. Some motivating examples are the following:

- *Conditional parameters in algorithm configuration* [10]: Often not all parameters are used during every execution of an algorithm, i.e. the likelihood of an execution does only depend on the values of those parameters that were actually used.
- *Selecting a cooling schedule for Simulated Annealing* [15] (SA): A cooling schedule, determines the likelihood of accepting worsening proposals in SA at each point in time. Most schedules, at each iteration i, accept a worsening proposals with a probability $0 < p_i < 1$, i.e. can generate any possible execution.

The crux is that even though an execution e might have been obtained using some design c, it might as well (with some likelihood) have been generated using a different design c'. As such, the observed desirability of e does not only provide information about the performance of c, but also about that of c'. In this work we describe how Importance Sampling (IS) can be used to combine all performance observations relevant to an algorithm, into a consistent estimator of its performance, assuming we are able to compute the (relative) likelihood of generating a given execution, using a given design.

The remainder of this paper is structured as follows. First, in Sect. 2, we formally define the ADP and related concepts such as the *design space*, *desirability of an execution* and *algorithm performance*. Subsequently, in Sect. 3.1, we summarize contemporary approaches to performance estimation in PbO; before introducing our IS approach in Sect. 3.2, discussing its benefits in Sect. 4 and examining its theoretical feasibility in Sect. 5. Finally, we discuss some challenges encountered when trying to implement IS in practice, and how we have addressed these in the implementation of a Proof of Concept in Sect. 6, which we validate experimentally in Sect. 7, before concluding in Sect. 8.

2   The Algorithm Design Problem (ADP)

Informally, we refer to the ADP as any attempt to (automatically) decide how to best solve a given problem. In the following, we formalize this notion and introduce concepts and notation used throughout this article.

Let C be the set of alternative algorithm instances considered, i.e. the *design space*. Let $\mathcal{D}$ denote a distribution over X, the set of possible inputs (e.g. problem instances to be solved and budgets available for doing so). Let E be the *execution space*, i.e. the set of all possible executions of any $c \in C$, on any $x \in X$. Let $Pr(e|c)$ denote the likelihood that executing algorithm instance c, on an input $x \sim \mathcal{D}$, results in an execution e, and $f : E \to \mathbb{R}$ a function quantifying the *desirability of an execution*. We define *algorithm performance* as a function $o : C \to \mathbb{R}$:

$$o(c) = \sum_{e \in E} Pr(e|c)f(e). \tag{1}$$

i.e. every algorithm instance in the design space can be viewed as a distribution over executions, whose performance corresponds to the expectation of f, over this distribution (as illustrated in Fig. 2). Remark that executing an algorithm instance (on $x \sim \mathcal{D}$) corresponds to sampling from its corresponding distribution. The objective in the ADP is to find $c^* = \arg \max_{c \in C} o(c)$.

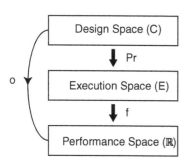

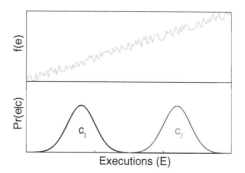

Fig. 1. A diagram showing how Pr and f relate designs to their performance.

Fig. 2. Distribution perspective

Thus far, optimizers of choice for PbO have been algorithm configurators, i.e. "Programming by Configuration" (PbC). Here, alternative designs are represented as configurations and configurators search the space of configurations C for one maximizing performance. As such, the ADP is treated as an Algorithm Configuration Problem [10] (ACP). Our formulation above is very similar and can in fact be seen as a simplified, specialization of the ACP. The most relevant difference, in context of this paper, lies in our definition of o. As argued in [2], PbC treats algorithm evaluation as a (stochastic) black box function $C \to \mathbb{R}$. Above, we define this mapping as a consequence of algorithm execution (see Fig. 1), i.e. our choice of design c affects execution e in a particular way (Pr), which in turn relates to its observed desirability (f). The performance estimation technique proposed in this paper exploits this feature, assuming Pr (and f) to be computable (black box) functions. In Sect. 5 we argue why this assumption is reasonable in context of PbO.

3   Performance Estimation in PbO

3.1   Prior Art

Computing $o(c)$ exactly, e.g. using (1), is often intractable due to the size of E. As such, general configuration frameworks rely on estimates $\tilde{o}(c)$. Most (e.g. [3,10,12]) maintain an independent sample average as performance estimate for each design c. They obtain a set of executions E_c' by repeatedly evaluating c (see Figs. 3 and 4) and estimate $o(c)$ as $\tilde{o}(c) = \bar{o}(c) = \frac{1}{|E_c'|} \sum_{e \in E_c'} f(e)$. Here, so called Sequential Model-Based Optimization (SMBO) frameworks are a notable exception (e.g. SMAC [9]). SMBO frameworks maintain a regression model $M : C \to \mathbb{R}$ (a.k.a. response surface model) trained using all previous performance observations. As such M gives us a performance prediction for any design. Note that these predictions are, to the best of our knowledge, only used to guide the search (identify unexplored, potentially interesting areas in the configuration space), i.e. to decide on which design to evaluate next. The choice of incumbent, i.e. which design to return as solution at any time, is based solely on performance observations obtained using the design itself.

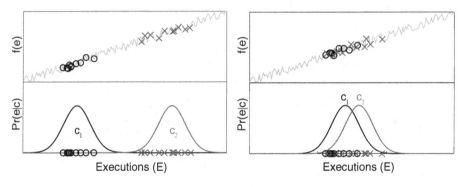

Fig. 3. Two different designs

Fig. 4. Two similar designs (color figure online)

3.2   An Importance Sampling Approach

The use of independent estimates is reasonable if the distributions in the design space are disjoint, as in Fig. 3. However, in what follows, we will argue that we can do better if two or more distributions overlap, as in Fig. 4, i.e. if there exists an execution that can be generated by at least two different algorithm instances: $\exists e \in E \land c_i, c_j \in C : Pr(e|c_i) * Pr(e|c_j) > 0$. When distributions overlap, evaluations of one design, potentially provide information about another. As evaluations can be extremely expensive, we would like to maximally utilize the information they provide; e.g. in Fig. 4, also use the red crosses to estimate the expectation over the blue distribution and vice versa.

Importance Sampling (IS) is a general technique for estimating properties of one distribution, using samples generated by another. IS originates from the field of rare event sampling [6], where some events might strongly affect some statistic, yet only occur rarely under the nominal distribution $\mathcal{P}$. To reduce the estimation error on the sample average, it is beneficial to draw samples from a different distribution $\mathcal{G}$, called the importance or generating distribution, which increases the likelihood of sampling these *important* events. To account for having sampled from this other distribution, we have to adjust our estimate by weighing observations by the likelihood ratio ($\frac{\mathcal{P}}{\mathcal{G}}$). In this paper, we propose a different use of IS, i.e. to combine all performance observations $f(e)$ relevant for a design c (i.e. $Pr(e|c) > 0$) into a consistent estimator of its performance. Concretely, let E' be the sample of executions obtained thus far, with $E' \sim \mathcal{G}$ (e.g. see Fig. 5), the IS estimate of the performance of any c is then given by

$$\tilde{o}(c) = \hat{o}(c) = \frac{\sum_{e \in E'} w_c(e) * f(e)}{\sum_{e \in E'} w_c(e)} \qquad \text{with} \qquad w_c(e) = \frac{Pr(e|c)}{\mathcal{G}(e)}. \qquad (2)$$

where $\mathcal{G}(e)$ is the likelihood of generating e using $\mathcal{G}$. While the estimate of each design c is based on the same E'; the weight function w_c will be different, weighing performance observations according to their relevance to c; e.g. in Fig. 6, observations on the left and right hand side are more relevant for c_1 and c_2 respectively. $\hat{o}(c)$ is a consistent estimate of $o(c)$, as long as $w_c(e)$ is bounded, $\forall\, e \in E$. In practice, it is also important that we can actually compute $\mathcal{G}(e)$ for any e. We will discuss the choice of $\mathcal{G}$ in more detail in Sect. 6.1, but for now it should be clear that both conditions are met if $\mathcal{G}$ is some mixture of all $c \in C$, i.e.

$$\mathcal{G}(e) = \sum_{c \in C} \lambda(c) * Pr(e|c). \qquad (3)$$

where $\lambda(c) > 0$ is the likelihood that c is used. Note that $w_c(e) \leq \frac{1}{\lambda(c)}$ holds. Furthermore, remark that if distributions do not overlap, $w_c(e) = \frac{1}{\lambda(c)}$ and $\bar{o} = \hat{o}$.

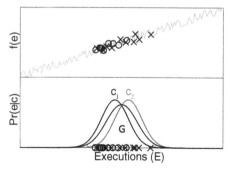

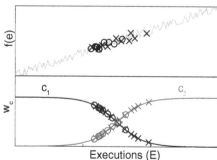

Fig. 5. $\mathcal{G}$ as equal mixture   **Fig. 6.** Design specific weights

4   Envisioned Benefits

In this section we will discuss the benefits of using IS (2) to estimate algorithm performance in an ADP setting. In addition, we will illustrate some of these experimentally for the abstract setup shown in Figs. 2, 3, 4, 5 and 6. Here, our design space consists of 2 normal distributions $c_1 = \mathcal{N}(\mu_1, 1)$ and $c_2 = \mathcal{N}(\mu_2, 1)$. Our objective is to determine which of these has the greatest mean, based on samples E' generated by alternately sampling each. To make this more challenging we add uniform white noise in $[-1, 1]$ to these observations. Results shown are averages of 1000 independent runs, generating 1000 samples each. Obviously, this particular instance is not representative for the ADP. Nonetheless, we would like to argue that the observations in this section generalize to the full ADP setting; e.g. computing Pr requires us to actually know μ, i.e. o, this however is a peculiarity of this simple setup and is definitely not the case in general (see also Sect. 5). As IS treats Pr as a black box, this fact did not affect the generality of our results. For the critical reader, an analogous argument is made in [1] using a somewhat more realistic ADP, "benchmark 1" (see Sect. 7.1), as a running example.

First, *IS increases data-efficiency*. By using a single performance observation in the estimation of many designs, we amortize the cost of obtaining it, and will need less evaluations to obtain similarly accurate estimates. Figure 7 illustrates this, comparing estimation errors $|\bar{o}(c_2) - o(c_2)|$ (dashed line) and $|\hat{o}(c_2) - o(c_2)|$ (full lines) respectively, after x evaluations, for multiple setups with different $\Delta = \mu_2 - \mu_1$. Clearly, if Δ is large, the distributions for c_1 and c_2 do not overlap, i.e. samples from c_1 are not relevant for c_2 and $\bar{o} \simeq \hat{o}$. However, the lower Δ, the greater the overlap. As Δ approaches 0, observations become equally relevant for both designs and only half the evaluations are needed to obtain similarly accurate estimates.

Related, yet arguably more important in an optimization setting, is that *we can determine more quickly which of 2 similar designs performs better*, i.e. have a more reliable gradient to guide the search. As similar designs share performance observations, their estimation errors will be correlated, i.e. the error on relative performance will be smaller. In the extreme case where distributions fully

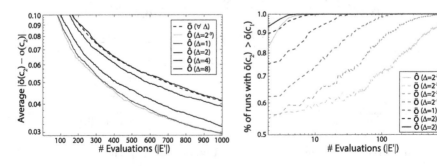

Fig. 7. Absolute estimation error   **Fig. 8.** Gradient accuracy

overlap (e.g. $\Delta = 0$), performance estimates are the same. Figure 8 illustrates this, comparing the fraction of the runs for which $\bar{o}(c_1) < \bar{o}(c_2)$ (dashed lines) and $\hat{o}(c_1) < \hat{o}(c_2)$ (full lines) holds after x evaluations, for different Δ. For high Δ values both perform good, as $o(c_1) \ll o(c_2)$. However, for Δ approaching 0, designs become more similar, and the independent estimate of c_1 is frequently better, even after many evaluations. However, $\hat{o}(c_1) < \hat{o}(c_2)$ holds, even for small Δ, after only a few evaluations. In summary, using IS we generally expect to need less evaluations to solve a given ADP, where at least two or more designs overlap.

Thus far we have discussed why one would use IS estimates, as opposed to independent sample averages. But how about using regression model predictions instead? Similar to IS, these allow one to generalize observations beyond the design used to obtain them, as such improving data-efficiency. The main difference is that *IS is model-free*. By choosing a regression model, one introduces model-bias, i.e. makes prior assumptions about what the fitness landscape (most likely) looks like. Clearly, the specific assumptions are model-specific. As parametric models (e.g. linear, quadratic) typically impose very strict constraints, mainly nonparametric models (e.g. Random Forests [5], Gaussian Processes [13]) are used in general ADP settings. While nonparametric models are more flexible, they nonetheless hinge on the assumption that the performance of similar designs is correlated (i.e. smoothness), which is in essence reasonable, however the key issue is that this similarity measure is defined in representation space. Small changes in representation (e.g. a configuration) might result in large performance differences, while large changes may not.

Using IS estimates, similar designs will also have similar estimates, but rather than being based on similarity in representation space, they are based on similarity in execution space. In fact, *we can derive similarity measures for designs* based on the overlap of their corresponding distributions of executions. This is interesting in analysis, but can also be used in solving the ADP; e.g. to maintain the diversity in a population of designs. One way to measure overlap would be by computing the Bhattacharyya Coefficient [4].

To conclude this section, the use of the proposed IS approach does not exclude the use of regression models. Both approaches are complementary, as different executions might have a similar desirability, something one could learn based on correlations in performance observations.

5   Theoretical Feasibility

In what follows we will examine whether it is at all possible to compute $\hat{o}$, what information is required for doing so, and whether this information can be reasonably assumed to be available in a PbO context.

Clearly, we can compute $\hat{o}$ if we are able to calculate the weights $w_c(e)$ for any $e \in E'$ and $c \in C$. From (2) and (3) it follows that the ability to compute $Pr(e|c)$, i.e. the likelihood of generating an execution e, using configuration c, on

$x \sim \mathcal{D}$, is a sufficient condition.[1] Here, Pr can be viewed as encoding how design decisions affected algorithm execution. In [2] it was argued that this information is available in a PbO context, and can therefore be exploited without loss of generality. In the remainder of this section we will briefly sketch how $Pr(e|c)$ could be computed in general. Please, bear in mind that the IS approach described in this paper does not require Pr to be computed as described below, and in specific scenarios it can often be done more time/space/information efficiently.

In [2] algorithm design is viewed as a sequential decision problem. Intuitively, we start execution, leaving design choices open. We continue execution as long as the next instruction does not depend on the decisions made for any of these. If it does, a choice point is reached and we must decide which of the possible instructions (A) to execute next. This process continues until termination. Our objective is to make these decisions as to maximize the desirability of the execution f. Any candidate design $c \in C$ corresponds to a policy of the form $\pi_c : \Omega \times A \rightarrow [0,1]$, where $\pi_c(\omega, a)$ indicates the likelihood of making decision a in execution context ω using design c. Here, ω "encodes" the features of the input and execution so far, on which the decision made by any of the designs of interest (i.e. C) depends. For instance, in our conditional parameter example from Sect. 1, choice points correspond to the (first) use of a parameter, A to its possible values and each $\omega \in \Omega$ encodes the parameter used. In our SA example, choice points correspond to the acceptance decisions faced every iteration, $A = \{accept, reject\}$ and each $\omega \in \Omega$ encodes the magnitude of worsening proposed and the time elapsed/remaining.

Let $\Psi(e)$ be the likelihood of any stochastic events[2] that occurred during an execution e. The likelihood of generating e using c with $x \sim \mathcal{D}$ is given by

$$Pr(e|c) = \mathcal{D}(x)\Psi(e) \prod_{i=1}^{n} \pi_c(\omega_i, a_i). \qquad (4)$$

where x is the input used, a_i the decision made in the ith choice point and ω_i the context in which it was encountered. Computing (4) requires $\mathcal{D}(x)$ and $\Psi(e)$ to be known explicitly, which is not always the case in practice. Luckily, it can be shown (proof in [1]) that we can ignore $\mathcal{D}(x)$ and $\Psi(e)$ in practice, i.e. use $Pr(e|c) = \prod_{i=1}^{n} \pi_c(\omega_i, a_i)$ instead to compute $w_c(e)$, as doing so will make both nominator, $Pr(e|c)$, and denominator, $G(e)$, $\mathcal{D}(e)\Psi(e)$ times too small, and as such their ratio, $w_c(e)$, correct. In summary, to compute $\hat{o}$, the ability to compute π_c ($\forall c \in C$) and store $(\vec{a}, \vec{\omega}, f(e))$ $\forall e \in E'$, suffices in general.

6 The Proof of Concept

Thus far we have introduced a novel performance estimation technique in Sect. 3.2, discussed its potential merits in Sect. 4 and established its theoretical feasibility in Sect. 5. In what follows we wish to convince the reader that it is

---

[1] However, it is not a necessary one. It can be shown (proof in [1]) that the ability to compute only the relative likelihood $\frac{Pr(e|c)}{Pr(e|c')}$ ($\forall e \in E$ $\forall c, c' \in C$) suffices.

[2] In a deterministic setting $\Psi(e) = 1$.

in fact practical. First, in Sect. 6.1, we discuss some challenges encountered when using IS estimates in practice, in an ADP setting, and how we have addressed them in the implementation of a Proof of Concept (PoC).

While performance estimation is a key, it is also only one piece of the puzzle. The ADP is a search problem, i.e. we must search the design space for the design maximizing performance. In Sect. 6.2 we describe the high-level search strategy used in our PoC. As a whole, our PoC resembles a simple SMBO-like framework, using importance sampling estimates, in place of regression model predictions, to guide its search. To complement this description and facilitate reproduction, the source code of our PoC is made publicly available.[3]

6.1 Practical Challenges

Choice of $\mathcal{G}$: As mentioned in Sect. 3.2, we choose a mixture of C as our $\mathcal{G}$ (see (3)). Two issues arise when trying to decide what values $\lambda(c)$ should take. First, C might be extremely large, making (3) expensive to compute. Second, we want to make $\lambda(c)$ dependent on E', e.g. to focus computational efforts on unexplored/promising parts of the design space. In our PoC, we view $\mathcal{G}$, at any point, as the distribution that generated E', as opposed to the one that will be used to generate future executions. To this purpose, we keep track of C' the designs used to generate E' and use $\lambda(c) = \frac{n(c)}{|E'|}$, where $n(c)$ is the number of executions performed using c. As such, we can at any point in the search process freely select which design to evaluate next, yet have a consistent estimate for any c for which $\lim_{|E'| \to \infty} \frac{n(c)}{|E'|} > \epsilon > 0$. This roughly corresponds to what is known in IS literature as an Adaptive Defensive Mixture approach [7].

Measuring Estimate Quality: In most state-of-the-art configurators, search is not only guided by performance predictions, but also the accuracy of those predictions is taken into account. Sample size N and variance σ^2 are two classical indicators of the accuracy of a sample average. From the Central Limit Theorem (CLT) it follows that given a sufficiently large sample $F \sim \mathcal{F}$, its sample average $\bar{F} \sim \mathcal{N}\left(\mu(\mathcal{F}), \frac{\sigma(\mathcal{F})^2}{N}\right)$. As such $\frac{\sigma(\mathcal{F})}{\sqrt{N}}$, a.k.a. the Standard Error (SE), can be used as a measure of uncertainty. As $\sigma(\mathcal{F})$ is typically unknown, its sample estimate $(\overline{std})$ is used instead. Sadly, the CLT does not directly apply to IS estimates, and predicting their expected accuracy has been an active research area. In our PoC, we use SE as a measure of uncertainty nonetheless, assuming that it is a reasonable heuristic, despite the lack of theoretical guarantees. However, what is the sample size, and how to estimate the standard deviation, of an IS estimate? IS estimates are all based on the same samples E', however not all of these are relevant for all designs. In fact, a few observations $E^{c'} \subseteq E'$ might dominate $\hat{o}(c)$. In our PoC we estimate $|E^{c'}|$, a.k.a. the effective sample size, using the formula below

$$\hat{N}(c) = \frac{(\sum_{e \in E'} w_c(e))^2}{\sum_{e \in E'} w_c(e)^2}. \tag{5}$$

---

[3] github.com/Steven-Adriaensen/IS4APE.

As estimate for the standard deviation we use

$$\widehat{std}(c) = \sqrt{\frac{\sum_{e \in E'} w_c(e) * (f(e) - \hat{o}(c))^2}{\sum_{e \in E'} w_c(e)}}. \tag{6}$$

which is the standard deviation of the distribution over E', where the relative likelihood of drawing $e \in E'$ is given by $w_c(e)$. Remark that if designs are disjoint, both (5) and (6) reduce to their sample average equivalents N and $\overline{std}$.

Computational Overhead: As discussed before, the use of IS estimates as described in this paper is beneficial in case of overlap and reduces to an ordinary sample average otherwise. However, computing an IS estimate is computationally more intensive, increasing the overhead. Again, as long as this overhead is reasonable compared to the cost of an evaluation, performance gains may be significant. Naively computing $\hat{o}(c)$ using (2) would take $O(|E'||C'|)$. Furthermore, we need to recompute $\hat{o}(c)$ every time E' and thus $\mathcal{G}$ changes. In our PoC, we memoize and update $\mathcal{G}(e)$, which can be done in $O(|E'|)$ for any new e, allowing us to compute $\hat{o}(c)$ in $O(|E'|)$. A hidden constant is the cost of computing $Pr(e|c)$, which may be significant in some ADPs. One solution would be to memoize them as well. A final issue is that $Pr(e|c)$ might be extremely small, yet relevant, because $Pr(e|c')$ may be even smaller for all other c'. To avoid underflows, one could use $\log(Pr(e|c))$ instead.

Algorithm 1. high-level search strategy for our PoC

1: **function** PoC($x,c_{\text{init}},\mathcal{GP},\mathcal{LP},f,Pr,Z,NProp,PSize,MaxEvals$)
2:     $c_{\text{inc}} \leftarrow c_{\text{init}}$
3:     $\hat{M} \leftarrow IS\langle f, Pr, \varnothing, \varnothing \rangle$
4:     $C_{pool} \leftarrow \varnothing$
5:     **for** $eval = 1 : MaxEvals$ **do**
6:         $\mathcal{P} \leftarrow$ equal mixture of $\mathcal{GP}$ and $\mathcal{LP}(c_{\text{inc}})$
7:         $C_{prop} \sim \mathcal{P}$ with $|C_{prop}| = NProp$
8:         $c_{\text{inc}} \leftarrow \arg\max_{c' \in C_{prop} \cup \{c_{\text{inc}}\}} \hat{M}.lb_Z(c')$
9:         $C_{pool} \leftarrow \arg\max_{c' \in C_{pool} \cup C_{prop}}^{PSize} \hat{M}.EI(c')$
10:        $c \leftarrow \arg\max_{c' \in C_{pool} \cup \{c_{\text{inc}}\}} \hat{M}.EI(c')$
11:        $e \leftarrow$ execute(c,x)
12:        $\hat{M} \leftarrow IS\langle f, Pr, E' \cup \{e\}, C' \cup \{c\} \rangle$
13:        $c_{\text{inc}} \leftarrow \arg\max_{c' \in C_{pool} \cup \{c_{\text{inc}}\}} \hat{M}.lb_Z(c')$
14:     **end for**
15:     **return** c_{inc}
16: **end function**

6.2   High-Level Search Strategy

Finally, we describe the high-level search strategy used in our PoC (see Algorithm 1). When implementing a search strategy we were faced with the following design choices: Given our observations thus far, which design should we evaluate next? On which input? What is our incumbent? Decisions made for all of these are critical in the realization of a framework competitive with the state-of-the-art. As this paper is about performance estimation, which is largely orthogonal to these other decisions, doing so was not our main objective; e.g. to keep it simple our PoC currently only supports optimization on a single input.[4] In the remainder of this section we briefly discuss the decisions made for the other two, followed by a detailed description of the high-level search strategy as a whole.

Choice of Incumbent: Clearly, we would like our incumbent (c_{inc}) to be the best design encountered thus far. However, since we only have an estimate of performance, we also need to take estimate accuracy into account, making it a multi-objective problem. In our PoC, we measure the uncertainty about the estimate as $\widehat{unc}(c) = \frac{\widehat{std}(c)}{\sqrt{\hat{N}(c)}}$ and update the incumbent each time we come across a design c with greater $\widehat{lb}_Z(c) = \hat{o}(c) - Z * \widehat{unc}(c)$, where $Z \in [0, +\infty[$ is a scalarization factor, passed as an argument to the framework (default 1.96). Here, $\widehat{lb}_Z(c)$ can be seen as an approximate lower bound of the confidence interval for $\hat{o}$ with z-score Z (default $\simeq 95\%$ confidence interval).

Choice of Design to be Evaluated Next: To decide which design to use to generate the next execution, we use the Expected Improvement (EI) criterion [11], also used in SMAC. At each point, we attempt to evaluate the design, which maximizes the Expected (positive) Improvement over our incumbent:

$$\widehat{EI}(c) = (\hat{o}(c) - \hat{o}(c_{inc})) * \Phi\left(\frac{\hat{o}(c) - \hat{o}(c_{inc})}{\widehat{unc}(c)}\right) + \widehat{unc}(c) * \phi\left(\frac{\hat{o}(c) - \hat{o}(c_{inc})}{\widehat{unc}(c)}\right). \quad (7)$$

where ϕ and Φ are the standard normal density and cumulative distribution functions. $\widehat{EI}$ will be high for designs estimated to perform well and for those with high estimated uncertainty; and as such this criterion offers an automatic balance between exploration and exploitation.

Line by Line Description: At line 2 we initialize the incumbent to be the design c_{init} passed as starting point to the solver. At line 3 we initialize the IS "model", denoted by $\hat{M}$, and defined as a 4-tuple $IS\langle f, Pr, E', C'\rangle$, where E' is the sample of executions generated thus far and C' the bag of designs used to generate them. Initially $E' = C' = \varnothing$. Remark that $\hat{M}$ captures all information required to compute $\hat{o}$ and $\widehat{unc}$ and as such $\widehat{EI}$ and $\widehat{lb}_Z$.

Each iteration, we consider designs $C_{pool} \subseteq C$ as candidates to be evaluated next. Ideally $C_{pool} = C$, however as C might be huge, this may be intractable. On

---

[4] As this input may be a "random input generator" (e.g. $\mathcal{D}$) we in essence do not lose any generality here.

the other hand, we often lack the prior knowledge to manually select $C_{pool} \subset C$. Therefore we adapt C_{pool} dynamically with $|C_{pool}| \leq$ PSize (default 10). At line 4 we initialize C_{pool} to be empty. Each iteration, at lines 6–7, we sample $NProp$ (default 100) designs (C_{prop}) from $\mathcal{P}$, a distribution over C, to be considered for inclusion in C_{pool}. Here, $\mathcal{P}$ is an equal mixture of 2 distributions passed as arguments to the framework, which can be used to inject heuristic information:

- Global Prior ($\mathcal{GP}$): A distribution over C, allowing the user to encode prior knowledge about where good designs are most likely located in C.
- Local Prior ($\mathcal{LP}$): A distribution over C, conditioned on c_{inc}, allowing the user to encode prior knowledge of how c_{inc} can most likely be improved.

Note that our search strategy only interacts with C through these distributions, making it design representation independent. In particular, it does not assume them to be configurations, let alone makes any assumptions about the type of parameters. At line 8 we update c_{inc} to be the design with the greatest $\widehat{lb}_Z$ in $C_{prop} \cup \{c_{inc}\}$. Having generated C_{prop}, we must decide whether to include them in C_{pool} or not. At line 9 we update C_{pool} as the $PSize$ designs from $C_{prop} \cup C_{pool}$ with maximal $\widehat{EI}$. At line 10 we select the design to be evaluated next as the design in $C_{pool} \cup \{c_{inc}\}$ maximizing $\widehat{EI}$, we evaluate this design at line 11 by executing it on given input x and we update E', C' (i.e. $\hat{M}$) accordingly at line 12. At line 13, we update c_{inc} w.r.t. the new $\hat{M}$. Finally, after $MaxEvals$ iterations, we return c_{inc} at line 15.

7   Experiments

In this section, we briefly evaluate the PoC, described previously, experimentally. We detail our setup in Sect. 7.1 and discuss our results in Sect. 7.2.

7.1   Experimental Setup

Benchmark: We consider micro-benchmark 1 from [2] as ADP (see Fig. 9). It consists of a simple loop of at most 20 iterations. At each iteration i we encounter a choice point and have to decide whether to continue the loop ($a_i = 1$) or not ($a_i = 0$), if we do, we receive a "reward" r_i drawn from $\mathcal{N}(1, 4)$, otherwise execution terminates. The desirability of the execution is simply the sum of these rewards, i.e. $f(e) = \sum_{i=1}^{20} r_i$. The only input is the random generator **rng** used to generate the stochastic events during the execution. Every design is represented as a configuration $\vec{c}$ (with $|\vec{c}| = 20$), where each parameter value $c_i \in [0, 1]$ indicates the likelihood of continuing the loop at iteration i. Since one of the optimizers in our baseline (ParamILS) does not support continuous parameters, we also consider a discretized variant, where the range of each parameter is $\{0, 0.1, ..., 0.9, 1\}$ resulting in a total of 11^{20} configurations.

We used benchmark 1 because it is conceptually simple and the actual performance can be computed as $o(\vec{c}) = \sum_{i=1}^{20} \prod_{j=1}^{i} c_j$. Clearly, we have chosen a benchmark in which the design space is huge, yet is rendered tractable thanks to the high similarity between designs, making it an ideal use-case for the proposed IS approach. Remark that the optimal design is $c^* : c_i = 1, \forall i$ with $o(c^*) = 20$. Also, executions are cheap, allowing us to repeat our experiments multiple times to filter out the noise in our observations. While this benchmark exhibits various other specific features, making it trivial to solve, the *only* information required/exploited by the IS approach is Pr and f, which it treats as black box functions. In order to compute these, we stored the number of iterations performed (#it), and the sum of rewards received (sum_r), for each execution. The latter equals $f(e)$, while the former suffices to compute Pr as

$$Pr(e|\vec{c}) = \begin{cases} (1 - c_{\#it+1}) \prod_{i=1}^{\#it} c_i & 0 \le \#it \le 19 \\ \prod_{i=1}^{\#it} c_i & \#it = 20 \end{cases}$$

Baseline: We compare the performance of our PoC to that of 2 state-of-the-art configurators, (Focused) ParamILS [10] and SMAC [9], chosen as representatives for the use of independent sample averages and regression models respectively. We perform 1000 independent runs of 10000 evaluations for every framework, using their default parameter settings and $c_{init} : c_i = 0.5, \forall i$. For a fair comparison we chose the global and local prior ($\mathcal{GP}, \mathcal{LP}$) to be uninformative, uniform and one-exchange distributions respectively. As a consequence, the local and global search operators used by all 3 frameworks are essentially the same.

7.2 Results and Discussion

Figure 10 shows the (actual) performance of the incumbent obtained by each framework, after x evaluations, averaged over 1000 runs. Both SMAC and our PoC were evaluated on both the continuous (full line) and discretized (dashed line) setup, ParamILS only on the latter.

In the discretized setup we find that ParamILS nor SMAC found the optimal solution (c^*) within 10000 evaluations, obtaining an average performance of 2 and 6 respectively. Our PoC required only about 10 and 50 evaluations respectively to obtain a similar performance, and the vast majority of the runs found

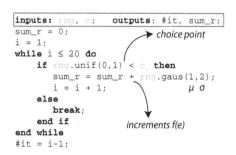

Fig. 9. Code for benchmark 1

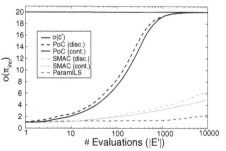

Fig. 10. Results for benchmark 1

c^* within the 2000 first evaluations (the worst needed 7573 evaluations). In the continuous setting both SMAC and our PoC performed (only) slightly worse than in the discretized setup.

Note that the results for ParamILS are worse than those reported in [2] and that our PoC performs worse than WB-PURS described therein. This because only deterministic policies, i.e. $c_i \in \{0,1\} \forall i$, were considered[5] in [2], and therefore the design space was much smaller (2^{20} vs. 11^{20}), yet included c^*.

Finally, note that the run times differed for all 3 frameworks. A run using our PoC, ParamILS and SMAC took on average about 5, 20 and 30 min respectively on our machine. However, as a single evaluation takes virtually no time, comparing optimizers on this benchmark based on actual run times would be unfair, and bias results towards those optimizers having the lowest overhead (e.g. least IO operations), which is furthermore very machine dependent. The actual overhead per evaluation was small for all frameworks (30–180 ms), and as such negligible, as long as an evaluation takes at least a few seconds, which is typically the case in more realistic ADP settings.

8   Conclusion

In this paper we proposed a novel way of estimating the performance of a family of related algorithms. Here, we viewed algorithms as distributions of executions; and algorithm performance as the expectation of some measure of the desirability of an execution, over this distribution. We showed how Importance Sampling (IS) can be used to generalize performance observations across algorithms with partially overlapping distributions, amortizing the cost of obtaining them. Finally, we described the implementation of an actual framework using this technique, which we validated experimentally.

This framework, while presenting a few interesting ideas of its own (e.g. the use of prior distributions to inject prior knowledge), is merely a Proof of Concept (PoC), and misses some features commonly found in state-of-the-art configurators, such as optimization across a set of training inputs, preliminary cutoff of poor executions and parallel execution support. Extending our PoC to support these features is a work in progress. We will also look into integrating IS into existing frameworks, e.g. to automatically exploit the conditionality of parameters. The latter should be relatively straightforward, as all we need to do is keep track of which parameters were actually used during an execution e, and determine whether a configuration c has the correct values for these ($Pr(e|c) = 1$) or not ($Pr(e|c) = 0$). Finally, we also plan to investigate the use of IS in an analysis context, where one is interested in approximating the response surface as a whole, rather than finding its highest peak.

The experimental validation presented in this paper is minimal and is by no means intended as a proof of the "superiority" of the proposed approach in real-world ADP settings. Rather, the crux of this paper was to present a novel

---

[5] As WB-PURS does not support stochastic policies it was not be included as baseline.

and interesting idea in a technically sound manner, and to show that it is at least possible to implement it in practice. That being said, we see no reason why we would not be able to improve performance in more realistic ADP settings. Clearly, the potential gain crucially depends on the design space considered, if all designs are sufficiently different, we can do no better, but overhead aside, we see no reason why we would do worse either, as all estimators presented in this paper, reduce to their sample average equivalents in this case.

References

1. Supplementary Material (2017). http://ai.vub.ac.be/node/1566
2. Adriaensen, S., Nowé, A.: Towards a white box approach to automated algorithm design. In: International Joint Conference on Artificial Intelligence (IJCAI), pp. 554–560 (2016)
3. Ansótegui, C., Sellmann, M., Tierney, K.: A gender-based genetic algorithm for the automatic configuration of algorithms. In: Gent, I.P. (ed.) CP 2009. LNCS, vol. 5732, pp. 142–157. Springer, Heidelberg (2009). doi:10.1007/978-3-642-04244-7_14
4. Bhattacharyya, A.: On a measure of divergence between two multinomial populations. Sankhyā Indian J. Stat. **7**, 401–406 (1946)
5. Breiman, L.: Random forests. Mach. Learn. **45**(1), 5–32 (2001)
6. Denny, M.: Introduction to importance sampling in rare-event simulations. Eur. J. Phys. **22**(4), 403 (2001)
7. Hesterberg, T.: Weighted average importance sampling and defensive mixture distributions. Technometrics **37**(2), 185–194 (1995)
8. Hoos, H.H.: Programming by optimization. Commun. ACM **55**(2), 70–80 (2012)
9. Hutter, F., Hoos, H.H., Leyton-Brown, K.: Sequential model-based optimization for general algorithm configuration. In: Coello, C.A.C. (ed.) LION 2011. LNCS, vol. 6683, pp. 507–523. Springer, Heidelberg (2011). doi:10.1007/978-3-642-25566-3_40
10. Hutter, F., Hoos, H.H., Leyton-Brown, K., Stützle, T.: ParamILS: an automatic algorithm configuration framework. J. Artif. Intell. Res. **36**(1), 267–306 (2009)
11. Jones, D.R., Schonlau, M., Welch, W.J.: Efficient global optimization of expensive black-box functions. J. Glob. Optim. **13**(4), 455–492 (1998)
12. López-Ibáñez, M., Dubois-Lacoste, J., Stützle, T., Birattari, M.: The irace package, iterated race for automatic algorithm configuration. Technical report, Universit Libre de Bruxelles (2011)
13. Rasmussen, C.E.: Gaussian Processes for Machine Learning. MIT Press, Cambridge (2006)
14. Rice, J.R.: The algorithm selection problem. Adv. Comput. **15**, 65–118 (1976)
15. Van Laarhoven, P.J., Aarts, E.H.: Simulated annealing. In: van Laarhoven, P.J.M., Aarts, E.H.L. (eds.) Simulated Annealing: Theory and Applications, pp. 7–15. Springer, Dordrecht (1987). doi:10.1007/978-94-015-7744-1_2

Test Problems for Parallel Algorithms of Constrained Global Optimization

Konstantin Barkalov[(✉)] and Roman Strongin

Lobachevsky State University of Nizhny Novgorod, Nizhny Novgorod, Russia
barkalov@vmk.unn.ru, strongin@unn.ru

Abstract. This work considers the problem of building a class of test problems for global optimization algorithms. The authors present an approach to building multidimensional multiextremal problems, which can clearly demonstrate the nature of the best current approximation, regardless of the problems dimensionality. As part of this approach, the objective function and constraints arise in the process of solving an auxiliary approximation problem. The proposed generator allows the problem to be simplified or complicated, which results in changes to its dimensionality and changes in the feasible domain. The generator was tested by building and solving 100 problems using a parallel global optimization index algorithm. The algorithm's results are presented using different numbers of computing cores, which clearly demonstrate its acceleration and non-redundancy.

Keywords: Global optimization · Multiextremal functions · Non-convex constraints

1 Introduction

One of the general approaches to studying and comparing multiextremal optimization algorithms is based on applying these methods to solve a set of test problems, selected at random from a certain specially constructed class. In this case, each test problem can be viewed as a random function created by a special generator. Using multiextremal optimization algorithms with large samples of such problems allows the operating characteristics of the methods to be evaluated (the likelihood of properly identifying the global optimizer within a given number of iterations), thus characterizing the efficiency of each particular algorithm.

The generator for one-dimensional problems was suggested by Hill [1]. These test functions are typical for many engineering problems; they are particularly reminiscent of reduced stress functions in problems with multiple concentrated loads (see [2] for example). Another widely known class of one-dimensional test problems is produced using a generator developed by Shekel [3].

A special GLOBALIZER software suite [4] was developed to study various one-dimensional algorithms with random samples of functions produced by the Hill and Shekel generators. A comprehensive description of this system, its

© Springer International Publishing AG 2017
R. Battiti et al. (Eds.): LION 2017, LNCS 10556, pp. 18–33, 2017.
https://doi.org/10.1007/978-3-319-69404-7_2

capabilities and example uses is provided in [5]. It should be noted that the Hill functions were successfully used in the design of a one-dimensional constrained problem generator with controlled measure of a feasible domain [6].

Another generator for random samples of two-dimensional test functions, successfully used in the studies by a number of authors, was developed and investigated in [7–10]. A generator for functions with arbitrary dimensionality was suggested in [11]. It was used to study certain multidimensional algorithms as described in [12–15]. Well-known collections of test problems for constrained global optimization algorithms were proposed in [16,17].

All of these generators produce the function to be optimized. In the case when the dimensionality is greater than two the optimization process itself cannot be clearly observed. In this regard, it is interesting to examine a different approach, initiated in [5]. In this approach, the objective function appears as a solution to a certain supporting approximation problem, which allows the nature of the best current estimate and the final result to be observed, regardless of the number of variables. Complicating the problem statement (including non-convex constraints) results in an increase of its dimensionality. In fact, the proposed generator produces an approximation problem to which the objective function is related.

2 Problem Statement

Let's consider the mathematical model of a charged particle moving through a magnetic field along the u axis in the form

$$m\ddot{u} = -eu + F \tag{1}$$

where $m > 0$ is the particles mass, $u(t)$ is the particle's current position at a moment of time $t \geq 0$, $-eu -$ is an attractive force affecting the particle, F is an external force applied to the particle along u axis (control action). It is assumed that control action F is a function of time and is represented as

$$F = m \sum_{i=1}^{n} A_i \sin(\omega_i t + \varphi_i).$$

Here $n > 0$ is the dimensionality of the vectors ω and determines the number of frequencies in the control action.

Substituting $\omega_0^2 = e/m$ the Eq. (1) is reduced to a known equation of forced oscillations

$$\ddot{u} + \omega_0^2 u = \sum_{i=1}^{n} A_i \sin(\omega_i t + \varphi_i). \tag{2}$$

The problem is to find own frequency, control action and initial conditions such that:

1. at $t \in [a, b]$ the particle would deviate from the position q_0 by no more than $\delta > 0$;
2. at $t = t_1, t_2, t_3$, the particle would deviate from positions q_1, q_2, q_3 respectively by no more than $\delta > 0$;
3. at $t = t_3$ the particle speed would be maximized.

This problem statement can be interpreted as follows: the trajectory of particle movement $u(t)$ shall pass within a "tube", then through the three "windows", with maximum slope in the last of the "windows". The illustration in Fig. 1 shows a graph of the function $u(t)$ of the solution to problem (2), which passes through the "tube" and "windows" shown in the chart by dashed lines.

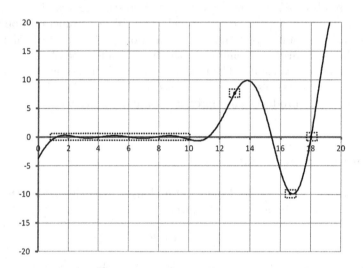

Fig. 1. Problem solution $u(t)$

Thus, for a particle of a given mass $m > 0$ it is necessary to determine its own frequency ω_0, the amplitudes A_i, frequencies ω_i and phases φ_i of the control action, as well as the initial conditions $u_0 = u(0), \dot{u}_0 = \dot{u}(0)$ for the Eq. (2), such that conditions 1–3 are true.

Using the following notation for the Eq. (2) solution

$$u(t, \omega, c) = \sum_{i=0}^{n} \left[c_{2i+1} \sin(\omega_i t) + c_{2i+2} \cos(\omega_i t) \right], \tag{3}$$

where $c = (c_1, \ldots, c_{2n+2})$, $\omega = (\omega_0, \omega_1, \ldots, \omega_n)$, we can represent the original problem as a constrained maximization problem with parameters c and ω:

$$|\dot{u}(t_3, \omega, c)| \rightarrow \max$$
$$|u(t_i, \omega, c) - q_i| \leq \delta, \quad i = 1, 2, 3, \tag{4}$$
$$|u(t, \omega, c) - q_0| \leq \delta, \quad t \in [a, b].$$

Solving the optimization problem (4) and finding the vectors w, c the solution to the original problem (2) can be written in accordance with the following relationships:

$$u_0 = \sum_{i=0}^{n} c_{2i+2}, \quad \dot{u}_0 = \sum_{i=0}^{n} c_{2i+1} w_i$$

$$A_i = \left(w_0^2 - w_i^2\right)\sqrt{c_{2i+1}^2 + c_{2i+2}^2}, \quad 1 \leq i \leq n, \tag{5}$$

$$\varphi_i = \arcsin \frac{c_{2i+2}}{\sqrt{c_{2i+1}^2 + c_{2i+2}^2}}, \quad 1 \leq i \leq n.$$

As follows from formula (3), the parameters c are included in the equation solution linearly, and the parameters w – non-linearly. Given the constraints (4) this allows the problem to be reformulated and c to be found without using a numerical optimization method.

Let's consider a set of points $(\tau_j, u_j), 0 \leq j \leq m$, with the coordinates defined as follows:

$$\tau_j = a + jh, \ u_j = q_0, \ 0 \leq j \leq m - 3, \tag{6}$$

$$\tau_{m-2} = t_1, u_{m-2} = q_1,$$

$$\tau_{m-1} = t_2, u_{m-1} = q_2,$$

$$\tau_m = t_3, u_m = q_3,$$

where $h = (b - a)/(m - 3)$, i.e. the first $m - 3$ points are located at equal distances in the center of the "tube", the other three align with the centers of the "windows".

The requirement is that the trajectory of particle $u(t)$ passes "near" the points $(\tau_j, u_j), 0 \leq j \leq m$. If the measure of deviation from the points is defined as the sum of the squared deviations

$$\Delta(c, w) = \sum_{j=0}^{m} [u_j - u(\tau_j, w, c)]^2,$$

then the parameters c can be found (given fixed values of w), by solving the least squares problem

$$c^*(w) = \arg\min \Delta(c, w). \tag{7}$$

According to the least squares method, the solution to problem (7) can be obtained by solving a system of linear algebraic equations regarding the unknown c, which can be done, e.g., by Gaussian elimination.

It should also be considered that the components of frequency vector w can be placed in ascending order, so as to avoid duplicate solutions corresponding to the vector w with similar components in different order. In addition, it is natural to assume that frequencies in the control action must not just be ordered but differ by a certain positive value, as an actual physical device can only generate control signals with a certain precision. This assumption can be represented in the form of a requirement for the following inequalities to be true:

$$\omega_{i-1}(1+\alpha) - \omega_i(1-\alpha) \leq 0, \ 1 \leq i \leq n. \tag{8}$$

Here $\alpha \in (0,1)$ is a parameter reflecting the precision of signal generation by the control device.

Then the original problem can be reformulated as follows:

$$\omega^* = \arg\max_{\omega \in \Omega} |\dot{u}(t_3, \omega, c^*(\omega))| \tag{9}$$

$$\omega_{i-1}(1+\alpha) - \omega_i(1-\alpha) \leq 0, \ 1 \leq i \leq n,$$
$$|u(t_i, \omega, c^*(\omega)) - q_i| \leq \delta, \ i = 1,2,3,$$
$$\max_{t \in [a,b]} u(t, \omega, c^*(\omega)) - \min_{t \in [a,b]} u(t, \omega, c^*(\omega)) \leq \delta,$$

where $c^*(\omega)$ is determined from (7), and the number of constraints will be dependent on the number of frequencies n in the control action.

Figure 1 shows trajectory $u(t)$, which corresponds to the solution to problem (9) with parameters $a = 1$, $b = 10$, $t_1 = 13$, $t_2 = 16.65$, $t_3 = 18$, $q_0 = q_3 = 0$, $q_1 = 7.65$, $q_2 = -9.86$. The solution (with three significant digits)

$$u(t) = 34.997 \sin(0.01t - 0.061) + 11.323 \sin(0.902t - 0.777)$$
$$- 19.489 \sin(1.023t + 0.054) + 9.147 \sin(1.139t + 0.633)$$

is determined by the optimal vectors $\omega^* = (0.01, 0.902, 1.023, 1.139)$ and $c^* = (34.93, -2.151, -8.075, -7.936, 19.461, 1.048, -7.375, 5.411)$. As follows from (5), the original problem with the solution $u(t)$ is noted as

$$\ddot{u} + 10^{-4}u = -9.213 \sin(0.902t - 0.777) - 20.383 \sin(1.023t + 0.054)$$
$$- 11.842 \sin(1.139t + 0.633),$$
$$u_0 = -3.62, \ \dot{u}_0 = 4.556.$$

3   Generating a Series of Problems

The numeric experiments described below used a generator based on the approximation problem from Sect. 2. Apparently, the variation in any parameter of the original problem (2) will change the optimization problem (9), so it is sufficient to vary just a few of them.

The centers of the first two "windows", i.e. the pairs (t_1, q_1) and (t_2, q_2), were chosen as the parameters for determining the specific problem statement. The values q_1 and q_2 were chosen independently and uniformly from the ranges

$[1, 10]$ and $[-10, -1]$, respectively. The values t_1 and t_2 were dependent: first, the value t_1 was chosen from the range $[b + 1, t_3 - 2]$, then, the value t_2 was chosen from the range $[t_1 + 1, t_3 - 1]$. All other parameters in problem (2) were fixed: $a = 1$, $b = 10$, $t_3 = 18$, $q_0 = q_3 = 0$, $\delta = 0.3$. Parameter α from (8) was chosen at 0.05. The number of points in the additional grid (6) for solving the least square problem (7) was set at 20. The problem of one-dimensional maximization and minimization from (9) were solved by a scanning over a uniform grid of 100 nodes within the interval $[a, b]$.

An important feature determining the existence of a feasible solution for the problem being considered is the number and range of frequency variation in the vector w. If the range is too small, or the number of frequencies is insufficient, the feasible domain in the problem (9) will be empty. In the experiments carried out, the number of frequencies was chosen to be $n = 3$, which corresponds to $w = (w_0, w_1, w_2, w_3)$, while the variable frequency change range was set from 0.01 to 2, i.e. $w_i \in [0.01, 2]$, $0 \le i \le 3$.

Let's note some important properties of the problems produced by the generator under consideration.

Remark 1. Problem constraints (9) are different in terms of the time required to verify them. For example, checking each of the first n constraints in (9) (let's call these constraints *geometric*)

$$w_{i-1}(1 + \alpha) - w_i(1 - \alpha) \le 0, \ 1 \le i \le n,$$

requires performing only three operations with real numbers. Testing other constraints (we will call them the *main constraints*)

$$|u(t_i, w, c^*(w)) - q_i| \le \delta, \ i = 1, 2, 3,$$

$$\max_{t \in [a,b]} u(t, w, c^*(w)) - \min_{t \in [a,b]} u(t, w, c^*(w)) \le \delta$$

is far more labor-intensive. First, this requires producing a system of linear equations to solve the problem (7) ($\sim m(2n + 2)^2$ computing of sin and cos). Second, this system needs to be solved ($\sim \frac{2}{3}(2n+2)^3$ operations on real numbers). Third, two one-dimensional optimization problems need to be solved at the last constraint in (9) (~ 100 computing of sin and cos).

Remark 2. The problems are characterized by multiextremal constraints, which form a non-convex feasible domain. For example, Fig. 2 show level lines for the functions in the right-hand sides of the main constraints for one of the problems, while Fig. 3 shows level lines of the objective function. These lines are provided within the feasible domain of the geometric constraints.

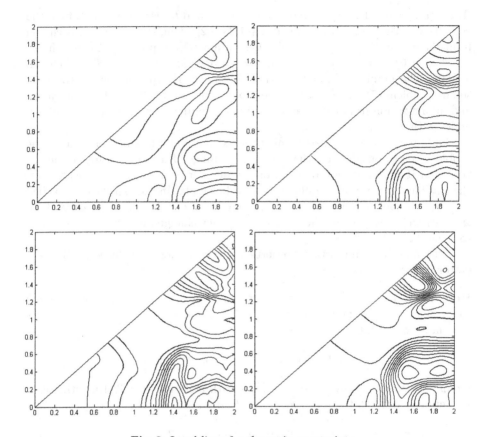

Fig. 2. Level lines for the main constraints

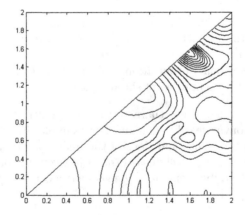

Fig. 3. Level lines for the objective function

Remark 3. Increasing the frequency change range results in new solutions appearing in an area of higher frequencies; the feasible domain of the optimization problem becomes multiply connected. For example, Fig. 4 shows two trajectories, the solid line corresponds to the vector $\omega = (0.01, 0.902, 1.023, 1.139)$ and solution

$$u(t) = 34.997 \sin(0.01t - 0.061) + 11.323 \sin(0.902t - 0.777)$$
$$- 19.489 \sin(1.023t + 0.054) + 9.147 \sin(1.139t + 0.633)$$

obtained at $\omega_i \in [0.01, 2]$, $0 \le i \le 3$, dashed line corresponds to the frequency vector $\omega = (1.749, 1.946, 2.151, 2.377)$ and solution

$$u(t) = 5.434 \sin(1.749t + 0.832) + 12.958 \sin(1.946t + 0.241)$$
$$+ 11.844 \sin(2.151t - 1.377) + 4.302 \sin(2.377t + 0.501)$$

obtained at $\omega_i \in [0.01, 4]$, $0 \le i \le 3$. Obviously, the second solution is better, as the trajectory at the last point has the largest slope.

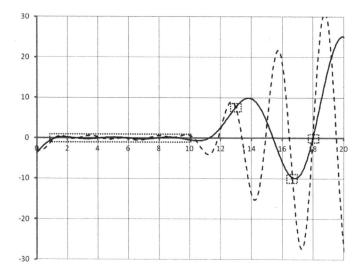

Fig. 4. Solutions with different frequencies

These optimization problem properties allow us to conclude that this generator may be applied for testing parallel global optimization algorithms. Interested readers may find the review of approaches to parallelization of optimization algorithms in [18]. In this study we will use a parallel algorithm based on information-statistical approach [5, 19].

4   Parallel Global Optimization Index Algorithm

Let's consider a multiextremal optimization problem in the form

$$\varphi(y^*) = \min\{\varphi(y) : y \in D, g_j(y) \leq 0, 1 \leq j \leq m\}, \tag{10}$$

where the search domain D is represented by a hyperparallelepiped

$$D = \{y \in R^N : a_i \leq y_i \leq b_i, 1 \leq i \leq N\}. \tag{11}$$

Suppose, that the objective function $\varphi(y)$ (henceforth denoted by $g_{m+1}(y)$) and the left-hand sides $g_j(y)$, $1 \leq j \leq m$, of the constraints satisfy Lipschitz condition

$$|g_j(x_1) - g_j(x_2)| \leq L_j \|y_1 - y_2\|, 1 \leq j \leq m+1, y_1, y_2 \in D.$$

with respective constants L_j, $1 \leq j \leq m+1$, and may be multiextremal. Then, it is suggested that even a single computing of a problem function value may be a time-consuming operation since it is related to the necessity of numerical modeling in the applied problems (see, for example, [20,21]).

Using a continuous single-valued mapping $y(x)$ (Peano-type space-filling curve) of the interval $[0,1]$ onto D from (11), a multidimensional problem (10) can be reduced to a one-dimensional problem

$$\varphi(y(x^*)) = \min\{\varphi(y(x)) : x \in [0,1], g_j(y(x)) \leq 0, 1 \leq j \leq m\}, \tag{12}$$

The reduction of dimensionality matches the multidimensional problem with a Lipschitzian objective function and Lipschitzian constraints with a one-dimensional problem where the respective functions satisfy the uniform Hölder condition (see [5]), i.e.

$$|g_j(y(x_1)) - g_j(y(x_2))| \leq K_j |x_1 - x_2|^{1/N}, \quad x_1, x_2 \in [0,1], \quad 1 \leq j \leq m+1.$$

Here N is the dimensionality of the original multidimensional problem, and the coefficients K_j are related to Lipschitz constants L_j with formulae

$$K_j \leq 2L_j\sqrt{N+3}, 1 \leq j \leq m+1.$$

The issues around the numeric construction of a Peano-type curve and the corresponding theory are considered in detail in [5,19]. Here we can just state that the numerically computed curve (*evolvent*) is an approximation of the theoretical Peano curve with a precision at least 2^{-m} for each coordinate (the parameter m is called the *evolvent density*).

Let's introduce the classification of points x from the search domain $[0,1]$ using the *index* $\nu = \nu(x)$. This index ν is determined by the following conditions:

$$g_j(y(x)) \leq 0, \quad 1 \leq j \leq \nu - 1, \quad g_\nu(y(x)) > 0,$$

where the last inequality is negligible if $\nu = m + 1$, and meets the inequalities $1 \leq \nu = \nu(x) \leq m + 1$. This classification produces a function

$$f(y(x)) = g_\nu(y(x)), \ \nu = \nu(x),$$

which is determined and computed along the interval $[0, 1]$. Its value in a point x is either the value of the left part of the constraint violated at this point (in the case, when $\nu \leq m$), or the value of the objective function (in the case, when $\nu = m + 1$). Therefore, determining the value $f(y(x))$ is reduced to a sequential computation of the values

$$g_j(y(x)), \ 1 \leq j \leq \nu = \nu(x),$$

i.e. the subsequent value $g_{j+1}(y(x))$ is only computed if $g_j(y(x)) \leq 0$. The computation process is completed either when the inequality $g_j(y(x)) > 0$ becomes true, or when the value of $\nu(x) = m + 1$ is reached.

The procedure called *trial* at point x automatically results in determining the index ν for this point. The pair of values

$$z = g_\nu(y(x)), \ \nu = \nu(x), \tag{13}$$

produced by the trial in point $x \in [0, 1]$, is called the *trial result*.

A serial index algorithm for solving one-dimensional conditional optimization problems (12) is described in detail in [6]. This algorithm belongs to a class of characteristical algorithms (see [7]). It can be parallelized using the approach described in [7] for solving unconstrained global optimization problems. Let's briefly describe the rules of the resulting *parallel index algorithm* (PIA).

Suppose we have $p \geq 1$ computational elements (e.g., CPU cores), which can be used to run p trials simultaneously. In the first iteration of the method, p trials are run in parallel at various random points $x^i \in (0, 1)$, $1 \leq i \leq p$. Suppose $n \geq 1$ iterations of the method have been completed, and as a result of which, trials were carried out in $k = k(n)$ points $x^i, 1 \leq i \leq k$. Then the points $x^{k+1}, \ldots, x^{k+p}$ of the search trials in the next $(n+1)$-th iteration will be determined according to the rules below.

1. Renumber the points $x^1, \ldots, x^k$ from previous iterations with lower indices, lowest to highest coordinate values, i.e.

$$0 = x_0 < x_1 < \ldots < x_i < \ldots < x_k < x_{k+1} = 1, \tag{14}$$

and match them with the values $z_i = g_\nu(y(x_i))$, $\nu = \nu(x_i)$, $1 \leq i \leq k$, from (13), calculated at these points; points $x_0 = 0$ $x_{k+1} = 1$ are introduced additionally; the values z_0 z_{k+1} are indeterminate.

2. Classify the numbers $i, 1 \leq i \leq k$, of the trial points from (14) by the number of problem constraints fulfilled at these points, by building the sets

$$I_\nu = \{i : 1 \leq i \leq k, \ \nu = \nu(x_i)\}, \ 1 \leq \nu \leq m + 1, \tag{15}$$

containing the numbers of all points $x_i, 1 \leq i \leq k$, with the same values of ν. The end points $x_0 = 0$ and $x_{k+1} = 1$ are interpreted as those with zero indices, and they are matched to an additional set $I_0 = 0, k+1$.

Identify the maximum current value of the index

$$M = \max \{\nu = \nu(x_i), \ 1 \leq i \leq k\}. \tag{16}$$

3. For all values of ν, $1 \leq \nu \leq m+1$, calculate the values

$$\mu_\nu = \max \left\{ \frac{|z_i - z_j|}{(x_i - x_j)^{1/N}} : i, j \in I_\nu, j < i \right\}. \tag{17}$$

If the set I_ν contains less than two elements or μ_ν from (17) equals zero, then assume $\mu_\nu = 1$.

4. For all non-empty sets I_ν, $1 \leq \nu \leq m+1$, determine the values

$$z_\nu^* = \begin{cases} -\epsilon_\nu, \nu < M, \\ \min\{g_\nu(x_i) : i \in I_\nu\}, \nu = M, \end{cases} \tag{18}$$

where M is the maximum current value of the index, and the vector

$$\epsilon_R = (\epsilon_1, \ldots, \epsilon_m), \tag{19}$$

with positive coordinates is called the *reserve vector* and is used as a parameter in the algorithm.

5. For each interval (x_{i-1}, x_i), $1 \leq i \leq k+1$, calculate the *characteristic* $R(i)$:

$$R(i) = \Delta_i + \frac{(z_i - z_{i-1})^2}{(r_\nu \mu_\nu)^2 \Delta_i} - 2\frac{z_i + z_{i-1} - 2z_\nu^*}{r_\nu \mu_\nu}, \quad \nu = \nu(x_{i-1}) = \nu(x_i),$$

$$R(i) = 2\Delta_i - 4\frac{z_i - z_\nu^*}{r_\nu \mu_\nu}, \quad \nu(x_{i-1}) < \nu(x_i) = \nu,$$

$$R(i) = 2\Delta_i - 4\frac{z_{i-1} - z_\nu^*}{r_\nu \mu_\nu}, \quad \nu = \nu(x_{i-1}) > \nu(x_i).$$

where $\Delta_i = (x_i - x_{i-1})^{1/N}$, and the values $r_\nu > 1, 1 \leq \nu \leq m+1$, are used as parameters in the algorithm.

6. Reorder the characteristics $R(i)$, $1 \leq i \leq k+1$, from highest to lowest

$$R(t_1) \geq R(t_2) \geq \ldots \geq R(t_k) \geq R(t_{k+1}) \tag{20}$$

and choose p largest characteristics with interval numbers $t_j, 1 \leq j \leq p$.

7. Carry out p new trials in parallel at the points $x^{k+j}, 1 \leq j \leq p$, calculated by the formulae

$$x^{k+j} = \frac{x_{t_j} + x_{t_j-1}}{2}, \quad \nu(x_{t_j-1}) \neq \nu(x_{t_j}),$$

$$x^{k+j} = \frac{x_{t_j} + x_{t_j-1}}{2} - \frac{\operatorname{sign}(z_{t_j} - z_{t_j-1})}{2r_\nu}\left[\frac{|z_{t_j} - z_{t_j-1}|}{\mu_\nu}\right]^N, \quad \nu(x_{t_j-1}) = \nu(x_{t_j}) = \nu.$$

The algorithm stops if the condition $\Delta_{t_j} \leq \epsilon$ becomes true for at least one number $t_j, 1 \leq j \leq p$; here $\epsilon > 0$ has an order of magnitude of the desired coordinate accuracy.

Let's formulate the conditions for algorithm convergence. For this, in addition to the exact solution y^* of the problem (10), we will also consider the ϵ-*reserved solution*, determined by the conditions

$$\varphi(y_\epsilon) = \min\{\varphi(y) : y \in D, \ g_j(y) \leq -\epsilon_j, \ 1 \leq j \leq m\},$$

where $\epsilon_1, \dots, \epsilon_m$ are positive numbers ("reserves" for each constraint). Let's also introduce the set

$$Y_\epsilon = \{y \in D : g_j(y) \leq 0, \ \varphi(y) \leq \varphi(y_\epsilon)\} \tag{21}$$

of all feasible points for the problem (10), which are no worse (in terms of the objective function's value) than ϵ-reserved solution.

Using this notation, the convergence conditions can be formulated as the theorem below.

Theorem. Suppose the following conditions are true:

1. The problem (10) has ϵ-reserved solution y_ϵ.
2. Each function $g_j(y)$, $1 \leq j \leq m+1$, satisfies Lipschitz condition with the respective constant L_j.
3. The parameters ϵ_j, $1 \leq j \leq m$, from (19) have the values of the respective coordinates from the reserve vector ϵ_R.
4. For the values μ_ν from (17) starting from a certain iteration, the following inequalities are true:

$$r_\nu \mu_\nu > 2^{3-1/N} L_\nu \sqrt{N+3}, \ 1 \leq \nu \leq m+1.$$

5. Each iteration uses a fixed number of computational elements p, $1 < p < \infty$, and each trial is completed by a single computational element within a finite time.

Then any accumulation point $\bar{y}$ of the sequence $\{y^k\}$ generated by parallel index method while solving the problem (10) is admissible and satisfies the conditions

$$\varphi(\bar{y}) = \inf\{\varphi(y^k) : g_j(y^k) \leq 0, 1 \leq j \leq m, k = 1, 2, \dots\} \leq \varphi(y_\epsilon).$$

i.e., $\bar{y}$ belongs to the set Y_ϵ from (21).

These convergence conditions proceed from the theorem of convergence of the serial index algorithm [5] and the theorem of convergence of a synchronous characteristical algorithm [7].

5   Results of Numerical Experiments

The procedure applied to evaluate the efficiency of the algorithm uses an *operating characteristics* method, originally described in [22], which consists of the following.

Suppose a problem from the series under consideration be solved by a certain algorithm. The problem is associated with the sequence of trial points $\{y(x^k)\}$ produced by the algorithm. The sequence is truncated either when a trial point falls (for the first time) into a certain ϵ-vicinity of the solution y^* or when a certain number of trials K does not produce such a point. In our experiments we used $K = 10^6$.

The results obtained by solving all problems in a series by means of the algorithm is represented by the function $P(k)$ defined as the fraction of problems in which some trial point fall within the given ϵ-neighborhood of the solution in the first k steps. This function is called the *operating characteristic* of the algorithm.

Since the specification of an ϵ-vicinity requires that y^* be known, its value was estimated in advance (for each problem) by searching through all nodes of a uniform grid (defined on the search domain), e.g. with a step 10^{-2} by coordinates.

As discussed above (see Remark 1 from Sect. 3), the constraints in the problem being addressed have a different nature. The first three constraints are geometric, and checking whether they are feasible is not hard. Checking other constraints is far more labor-intensive. Therefore, when building operating characteristics for this class of problems, we will only consider the trials that resulted in checking labor-intensive constraints. The trials that were completed while checking geometric constraints are not included in the total number of trials.

The experiments were carried out for the parallel index algorithm (PIA) described above, with the number of used cores p varying from 1 to 8. The parameters used were $r_\nu = 2.5, 1 \leq \nu \leq 8$. Components of the reserve vector from (19) were selected adaptively under the rule $\epsilon_\nu = 0.005\mu_\nu, 1 \leq \nu \leq 7$, where μ_ν is from (17) (see [5] for a justification of this component selection for the reserve vector). Computing experiments were carried out on one of the nodes of a high-performance cluster of the Nizhny Novgorod State University. The cluster node includes Intel Sandy Bridge E5-2660 2.2 GHz CPU and 64 Gb RAM. For implementation of the parallel algorithm OpenMP was used.

The algorithm's operating characteristics when using different number of computing cores p obtained with a series of 100 problems, are shown in Fig. 5. The location of the curves shows that any of the parallel algorithms solve 100% of the problems in less than $7 \cdot 10^5$ trials, with more than 80% of the problems completing within $2 \cdot 10^5$ trials. The operating characteristics also show that the algorithm is *non-redundant* – the number of trials in the parallel algorithm does not grow (compared to serial algorithm) when additional cores are employed.

Now let's evaluate the speedup achieved by using a parallel index algorithm, depending on the number p of computing cores used. Table 1 shows the average number of iterations $n(p)$ performed by the algorithm when solving a series of 100 problems, and speedup by the iterations $s(p)$ of a parallel algorithm.

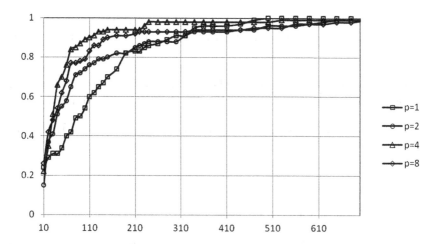

Fig. 5. Operating characteristics of PIA, which uses p cores

Table 1. Speedup of the algorithm

p	$n(p)$	$s(p)$
1	241239	–
2	94064	2.56
4	45805	5.27
8	22628	10.66

The results show that the speedup is greater than the number of cores used (hyper-acceleration). This situation is explained by the fact that the algorithm performs an adaptive evaluation of the behavior of the objective function (calculating the lower bounds for the Lipschitz constant (17) and the current minimum value (18)). For example, if the Lipschitz constant is better estimated in a parallel version, then the parallel algorithm using p cores can be accelerated by more than p times.

6   Conclusion

In summary, we must note that the method proposed in this work to generate multidimensional conditional global optimization problems allows:

- clear visualization of the best current estimate and the final solution to the problem, regardless of the number of variables;
- increased dimensionality of the optimization problem being addressed by varying the original approximation problem;
- control of the feasible domain by adding extra non-convex constraints.

The functions included in the optimization problem are computationally intensive, which also differentiates the mechanism proposed from other known mechanisms.

The proposed generator was used to build and subsequently solve 100 problems using a parallel index algorithm. The operating characteristics of the parallel algorithm have been built, clearly demonstrating its non-redundancy.

Acknowledgments. This study was supported by the Russian Science Foundation, project No. 16-11-10150.

References

1. Hill, J.D.: A search technique for multimodal surfaces. IEEE Trans. Syst. Sci. Cybern. **5**(1), 2–8 (1969)
2. Toropov, V.V.: Simulation approach to structural optimization. Struct. Optim. **1**, 37–46 (1989)
3. Shekel, J.: Test functions for multimodal search technique. In: Proceedings of the 5th Princeton Conference on Information Science Systems, pp. 354–359. Princeton University Press, Princeton (1971)
4. Strongin, R.G., Gergel, V.P., Tropichev, A.V.: Globalizer. Investigation of minimizing sequences generated by global search algorithms for univariate functions. User's guide. Nizhny Novgorod University Press, Nizhny Novgorod (1995)
5. Strongin, R.G., Sergeyev, Y.D.: Global Optimization with Non-convex Constraints: Sequential and Parallel Algorithms. Kluwer Academic Publishers, Dordrecht (2000)
6. Barkalov, K.A., Strongin, R.G.: A global optimization technique with an adaptive order of checking for constraints. Comput. Math. Math. Phys. **42**(9), 1289–1300 (2002)
7. Grishagin, V.A., Sergeyev, Y.D., Strongin, R.G.: Parallel characteristical algorithms for solving problems of global optimization. J. Glob. Optim. **10**(2), 185–206 (1997)
8. Sergeyev, Y.D., Grishagin, V.A.: Sequential and parallel algorithms for global optimization. Optim. Methods Softw. **3**, 111–124 (1994)
9. Sergeyev, Y.D., Grishagin, V.A.: Parallel asynchronous global search and the nested optimization scheme. J. Comput. Anal. Appl. **3**(2), 123–145 (2001)
10. Gergel, V., Grishagin, V., Gergel, A.: Adaptive nested optimization scheme for multidimensional global search. J. Glob. Optim. **66**(1), 35–51 (2016)
11. Gaviano, M., Kvasov, D.E., Lera, D., Sergeyev, Y.D.: Software for generation of classes of test functions with known local and global minima for global optimization. ACM Trans. Math. Softw. **29**(4), 469–480 (2003)
12. Sergeyev, Y.D., Kvasov, D.E.: Global search based on efficient diagonal partitions and a set of Lipschitz constants. SIAM J. Optim. **16**(3), 910–937 (2006)
13. Lera, D., Sergeyev, Y.D.: Lipschitz and Holder global optimization using space-filling curves. Appl. Numer. Math. **60**(1–2), 115–129 (2010)
14. Paulavicius, R., Sergeyev, Y., Kvasov, D., Zilinskas, J.: Globally-biased DISIMPL algorithm for expensive global optimization. J. Glob. Optim. **59**(2–3), 545–567 (2014)
15. Sergeyev, Y.D., Kvasov, D.E.: A deterministic global optimization using smooth diagonal auxiliary functions. Commun. Nonlinear Sci. Numer. Simul. **21**(1–3), 99–111 (2015)

16. Floudas, C.A., Pardalos, P.M.: A Collection of Test Problems for Constrained Global Optimization Algorithms. LNCS, vol. 455. Springer, Heidelberg (1990). doi:10.1007/3-540-53032-0
17. Floudas, C.A., et al.: Handbook of Test Problems in Local and Global Optimization. Springer, New York (1999). doi:10.1007/978-1-4757-3040-1
18. Pardalos, P.M., Phillips, A.T., Rosen, J.B.: Topics in Parallel Computing in Mathematical Programming. Science Press, New York (1992)
19. Sergeyev, Y.D., Strongin, R.G., Lera, D.: Introduction to Global Optimization Exploiting Space-Filling Curves. Springer, New York (2013). doi:10.1007/978-1-4614-8042-6
20. Gergel, V.P., Kuzmin, M.I., Solovyov, N.A., Grishagin, V.A.: Recognition of surface defects of cold-rolling sheets based on method of localities. Int. Rev. Autom. Control 8(1), 51–55 (2015)
21. Modorskii, V.Y., Gaynutdinova, D.F., Gergel, V.P., Barkalov, K.A.: Optimization in design of scientific products for purposes of cavitation problems. In: Simos, T.E. (ed.) ICNAAM 2015. AIP Conference Proceedings, vol. 1738 (2016). Article No. 400013
22. Grishagin, V.A.: Operating characteristics of some global search algorithms. Probl. Stoch. Search 7, 198–206 (1978). (in Russian)

Automatic Configuration of Kernel-Based Clustering: An Optimization Approach

Antonio Candelieri[1]([✉]), Ilaria Giordani[1], and Francesco Archetti[1,2]

[1] Dipartimento di Informatica Sistemistica e Comunicazione, DISCo,
Università di Milano Bicocca, 20126 Milan, Italy
antonio.candelieri@unimib.it
[2] Consorzio Milano Ricerche, via R. Cozzi 53, 20125 Milan, Italy

Abstract. This paper generalizes a method originally developed by the authors to perform data driven localization of leakages in urban Water Distribution Networks. The method is based on clustering to perform exploratory analysis and a pool of Support Vector Machines to process on line sensors readings. The performance depends on certain hyperparameters which have been considered as decision variables in a sequential model based optimization process. The objective function is related to clustering performance, computed through an external validity index defined according to the leakage localization goal. Thus, as usual in hyperparameters tuning of machine learning algorithms, the objective function is black box. In this paper it is shown how a Bayesian framework offers not only a good performance but also the flexibility to consider in the optimization loop also the automatic configuration of the algorithm. Both Gaussian Processes and Random Forests have been considered to fit the surrogate model of the objective function, while results from a simple grid search have been considered as baseline.

Keywords: Hyperparameters optimization · Sequential model based optimization · Kernel based clustering · Leakage localization

1 Introduction

In this paper we consider a machine learning model based on clustering and Support Vector Machine (SVM) classification, and its application to leakage localization in an urban water distribution network.

Machine learning based approaches for analytically localizing leaks in a Water Distribution Network (WDN) have been proposed, mostly based on the idea that leaks can be detected by correlating actual modifications in flow and pressure within the WDN to the output of a simulation model whose parameters are set to evaluate the effect induced by a leak in a specific location and with a specific severity [1–6].

This paper is based on an analytical framework that uses: (1) extensive simulation of leaks for data generation; (2) kernel-based clustering to group leaks implying similar variations in pressure and flow; (3) classification learning (i.e. SVM) to discover the relation linking variations in pressure and flow to a limited set of probably leaky pipes.

© Springer International Publishing AG 2017
R. Battiti et al. (Eds.): LION 2017, LNCS 10556, pp. 34–49, 2017.
https://doi.org/10.1007/978-3-319-69404-7_3

The main contribution of this paper is the formulation of the algorithm configuration problem and a proposal of a global optimization strategy to solve it.

Automatic configuration of machine learning algorithms has been attracting a growing attention [7]: the use of default parameters, as pointed out in [8], can result in a poor generalization performance. Grid search, the most widely used strategy for hyperparameters optimization, is hardly feasible for more than 2 or 3 parameters. For these reasons optimization methods have been widely proposed. Classical optimization cannot be used as the performance measures are typically black-box functions, and/or multimodal, whose derivatives are not available. For these reasons, global optimization is now widely accepted as the main computational framework for hyperparameters optimization and algorithm configuration.

Global optimization [9–11] methods fall in 3 large families. The first is Partitional Methods which offer global convergence properties and guaranteed accuracy estimation of the global solutions, e.g., in the case of Lipschitz global optimization [12, 13]. A significant application of these methods to hyperparameter optimization is given, e.g. in [14] for the case of SVM regression or in [15] for the case of signal processing. The second family is Random Search [16–18] and the related metaheuristics like simulated annealing, evolutionary/genetic algorithms, multistart&clustering, largely applied in global optimization. Random Search has recently received fresh attention from the machine learning community and increasingly considered as a baseline for global optimization as in Hyperband [19, 20] which considers randomly sampled configurations and relies on a principled early stopping rule to evaluate each configuration and compares its performance to Bayesian Optimization (BO).

BO represents the third family which came to dominate the landscape of hyperparameter optimization in the machine learning community [21–23]. First proposed in [24], BO is a sequential model based approach: its key ingredients are a probabilistic model which captures our beliefs – given the evaluations already performed – and an acquisition function which computes the "utility" of each candidate point for the next evaluation of f. The Bayesian framework is particularly useful when evaluations of f are costly, no derivatives are available and/or f is nonconvex/multimodal.

Very relevant to the authors activity is that the BO can be applied to "unusual" design spaces which involve categorical or conditional variables. This capability makes BO the natural solution when not only the hyperparameters but also the specific algorithm itself to be automatically selected in the so called algorithmic configuration [25].

The rest of the paper is organized as follows: Sect. 2 describes the hydraulic simulation model, the general algorithmic structure and the formulation of the performance measure of the clustering. Section 3 is devoted to the sequential model based optimization for the hyperparameters optimization and the description of the software environment utilized. Section 4 analyzes the computational performance of different strategies while Sect. 5 contains concluding remarks.

2 Material and Methods

In this paper the reference application domain is an urban WDN. Water network design and optimization of the operations (pump scheduling) has received a lot of attention in the Operation Research literature [26].

The elements of the network are subject to failures, not uncommon given the typically old age of the infrastructure. Breakages of pipes, in particular, generate bursts and leaks which can inhibit supply service of the network (or a subnetwork) and induce substantial loss of water, with an economic loss (no-revenue water), water quality problems and unnecessary increase in energy costs.

The state of the WDN is usually monitored by a number of sensors which record flows and pressures. When a leak occurs, sensors record a variation from normal operating values. We named the vector of deviations the "signature" of the leak.

The main aim of the proposed machine learning approach is to use this "signature" to identify the location of the leak. Therefore, the basic idea is to move between the "physical space" (pipes of the WDN) and the space of leak "signatures" to infer a possible relation – both direct and inverse – between causes (leaks on pipes) and effects (variations in flow and pressure at the monitoring points).

Although data could be gathered looking at historical leakage events, they would be too sparse and of poor quality. Thus, a hydraulic simulation software is used to generate a wide set of data emulating the data from sensors according to different "leakage scenarios" in the WDN, consisting in placing, in turn, a leak on each pipe and varying its severity in a given range. Our choice of simulator is EPANET 2.0, widely used for modeling WDNs and downloadable for free from the Environmental Protection Agency web site (http://www.epa.gov/nrmrl/wswrd/dw/epanet.html).

In this paper we focus on optimizing, throughout Sequential Model Based Optimization (SMBO) [27], a set of design variables which are hyperparameters of a machine learning based system which, given a new leak signature, infers its location as a limited set of probably leaky pipes. Learning is performed on a dataset obtained as vectors of N components, where N is the overall number of sensors ($N = N_p + N_f$, with N_p = number of pressure sensors and N_f number of flow sensors), that is the Input Space. According to the Fig. 1, learning is performed in two stages: one unsupervised, aimed at grouping together similar "signatures" to reduce the number of different "effects", and one supervised, aimed at estimating the group of signatures which the signature of a real leak belongs to. This allows for retrieving only the scenarios related to the signatures in that cluster and, therefore, leaky pipes associated to those signatures.

The basic idea of the analytical framework has been presented in [28–30], where Spectral Clustering (SC) is used for the unsupervised learning phase and Support Vector Machine (SVM) classification is used for the supervised learning phase. The cluster assignment provided to each instance (i.e. signature) is used as label to train the SVM classifier. While the clustering is used to model the direct relation from leak scenario (i.e. leaky pipe and leak severity) to a group of similar signatures, the SVM inverts this relation. Thus, when a reading from sensors is acquired, the variations with respect to the faultless WDN model are computed and the resulting "signature" is given

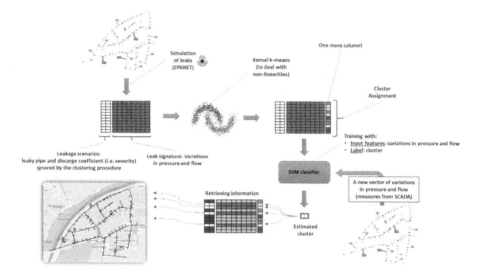

Fig. 1. Overall machine learning based leakage localization system

as input to the trained SVM which assigns the most probable cluster the signature belongs to. Finally, the pipes relative to the scenarios in that cluster are selected as "leaky pipes".

Although SC proved to be effective, it is computationally expensive and, in this study, we propose to replace SC with a Kernel k-means algorithm and apply SMBO to optimally tune its hyperparameters. This allows to reduce the computational burden induced and to infer the similarity measure – even non-linear – directly from data and implicitly through the "kernel trick", instead to define a priori a similarity measures for weighting edges of the affinity graph. It has been already reported in literature that an "equivalence" exists between SC and kernel k-means [31]. In particular, a Radial Basis Function (RBF) kernel has been chosen, characterized by its hyperparameter σ. The other hyperparameter is, naturally, the optimal number k of clusters.

Moreover, it is important to highlight that adoption of RBF kernels in global optimization algorithms has been also investigated, and compared with statistical models [32].

2.1 Notation

The WDN can be represented through a graph: $G = \langle V|E \rangle$, where:

- V is the set of "junctions", which are consumption points, reservoirs, tanks, emitters as well as simple junctions between two pipes;
- E is the set of "links", which are pipes, pumps or valves.

Let us denote with

$$\bar{E} \subset E$$

the set of pipes within the set of links. Furthermore, let us denote with

$$A = \{\alpha_1, \ldots, \alpha_l\}$$

the set of "severities" of the (simulated) leaks. Then, the set of "Leakage Scenarios" can be defined as follows:

$$S = \bar{E} \times A$$

where $s_{e,\alpha} \in S$ is related to a leak on the pipe $e \in \bar{E}$ with severity $\alpha \in A$.

Finally, the "signature" of a leak, that is the effect induced by specific leakage scenario, is defined as:

$$w_{e,\alpha} = f(s_{e,\alpha})$$

with $f(.)$ a function to compute, through EPANET, the expected variations of pressure and flow at the sensors locations.

Clustering is performed on "signatures", therefore every cluster C^k is a set of signatures (i.e. similar effects due to different leaks):

$$C^k = \{\omega_{e,\alpha} \in \mathbb{R}^N, \omega_{e,\alpha} = f(s_{e,\alpha}), e \in \bar{E}, \alpha \in A\}.$$

It is important to highlight that the clustering procedure used is "crispy", so

$$C^i \cap C^j = \emptyset \ \forall i \neq j.$$

Many clustering quality measures are given in the literature [33], divided between "internal" – basically related to inter- and intra- distances – and "external" – which need a set of labeled examples and treat the clustering as a classification problem. External measures are domain specific therefore more effective in our case but they cannot be used because as remarked above historically are sparse, unbalanced and poor quality. For these reasons we have defined ad hoc composite index to address the leakage localization objective. In particular, to obtain an effective and efficient localization of possible leaks into a WDN, clusters have to satisfy the following properties:

- The set of pipes candidate as leaky must be as limited as possible, for every cluster;
- The signatures of scenarios associated to a given pipe should be spread as less as possible among different clusters.

We refer to the first property as "compactness" and to the second one as a proxy of "accuracy".

Before to present the final index, the two following sets have to be defined:

$$\bar{E}^k = \left\{ e \in \bar{E} : \exists \alpha \in A : \omega_{e,\alpha} = f\left(s_{e,\alpha}\right) \in C^k \right\}$$

and

$$S^{k,\alpha} = \left\{ s_{e,\alpha} \in S, e \in \bar{E} : \omega_{e,\alpha} = f\left(s_{e,\alpha}\right) \in C^k \right\}.$$

The index for evaluating the fitness of clustering is the composition of two different measures:

$$I = \frac{I_C + I_P}{2}$$

where I_C measures the "compactness" of the cluster in terms of number of pipes identified as probably leaky with respect to the overall number of pipes in the WDN, while I_P measures a sort of "accuracy" that the leak is in the set of pipes identified as leaky instead to be on other pipes.

More in detail the two measures are computed as follows:

$$I_C = \frac{\sum_{k=1}^{K} I_C^k \left| \bar{E}^k \right|}{\sum_{k=1}^{K} \left| \bar{E}^k \right|}$$

where

$$I_C^k = \frac{\left| \bar{E} \right| - \left| \bar{E}^k \right|}{\left| \bar{E} \right| - 1}$$

and

$$I_P = \text{avg}_k I_P^k$$

where

$$I_P^k = \frac{\frac{1}{|A|} \sum_{\alpha \in A} \left| S^{k,\alpha} \right|}{\left| \bar{E}^k \right|}.$$

Let us suppose that C^k contains only signatures associated to all the scenarios related to only one pipe $e^* \in \bar{E}$, then:

- $\left| \bar{E}^k \right| = 1$, because $\bar{E}^k = \{e^*\}$;
- $\left| S^{k,\alpha} \right| = 1$, because $\forall \alpha \in A, S^{k,\alpha} = \left\{ s \in S : \exists! e^* \in \bar{E} : \omega = f\left(s_{e^*,\alpha}\right) \in C^k \right\}$;
- $I_P^k = \frac{\frac{1}{|A|}|A|}{|\bar{E}^k|} = 1.$

On the other side, let us suppose that C^k contains signatures associated to $|A|$ different pipes with $|A|$ different severities, then:

- $|\bar{E}^k| = |A|$, for the hypothesis;
- $|S^{k,\alpha}| = 1$, because $\forall \alpha \in A, S^{k,\alpha} = \{s \in S : \exists! e^* \in \bar{E} : \omega = f(s_{e^*,\alpha}) \in C^k\}$;
- $\sum_{\alpha \in A} |S^{k,\alpha}| = |A|$;
- $I_P^k = \frac{\frac{1}{|A|}|A|}{|\bar{E}^k|} = \frac{1}{|A|} < 1$.

In general, $|\bar{E}^k|$ can be greater than $|A|$; in the worst case $|\bar{E}^k| \gg |A|$, thus $I_P^k \ll 1$.

2.2 The Case Study and the Data Generation Process

In the following Fig. 2 the specific WDN considered in this study is reported. It is a District Metered Area (DMA), namely "Neptun", of the urban WDN in Timisoara, Romania, and was a pilot of the European Project ICeWater.

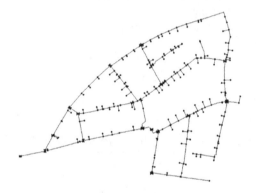

Fig. 2. The District Metered Area (DMA) "Neptun" of the urban WDN in Timisoara, Romania: the WDN case study considered in this paper

Neptune consists of 335 junctions (92 are consumption points) and 339 pipes. In the proposed approach, EPANET is used to simulate a wide set of leakage scenarios, consisting in placing, in turn, a leak on each pipe and varying its severity in a given range.

At the end of each run, EPANET provides pressures and flows at each junction and pipe, respectively, and, therefore, also at the monitoring points (i.e. sensors). Variations in pressure and flow induced by a leak (i.e. signature of the leak) are computed with respect to the simulation of the faultless WDN. Finally, a dataset is obtained having as many instances as *number-of-pipelines* times *number-of-discharge-coefficient-values*.

2.3 Kernel K-means

The kernel k-means algorithm [34] is a generalization of the standard k-means algorithm. By implicitly mapping points to a higher-dimensional space, kernel k-means can discover clusters that are non-linearly separable in input space. This provides a major advantage over standard k-means, and allows us to cluster points if we are given a positive definite matrix of similarity values.

A general weighted kernel k-means objective is mathematically equivalent to a weighted graph partitioning objective [31]; this equivalence has an important consequence: in cases where eigenvector computation is prohibitive, kernel k-means eliminates the need for any eigenvector computation required by graph partitioning.

Given a set of vectors $x_1, x_2, ..., x_n$, the kernel k-means objective can be written as a minimization of:

$$\sum_{k=1}^{K} \sum_{x \in C^k} \left\| \Phi(x_i - \bar{x}_k) \right\|^2$$

where $\Phi(x)$ is a (non-linear) function mapping vectors x_i from the Input Space to the Feature Space, and where $\bar{x}_k$ is the centroid of cluster C^k.

Expanding $\left\| \Phi(x_i - \bar{x}_k) \right\|^2$ in the objective function, one can obtain:

$$\Phi(x_i)\Phi(x_i) - \frac{2\sum_{x_j \in C^k} \Phi(x_i)\Phi(x_j)}{|C^k|} + \frac{\sum_{x_j, x_i \in C^k} \Phi(x_j)\Phi(x_i)}{|C^k|^2}.$$

Therefore, only inner products are used in the computation of the Euclidean distance between every vector and the centroid of the cluster C^k. As a conclusion, given a kernel matrix K, where $K_{ij} = \Phi(x_i)\Phi(x_j)$, these distances can be computed without knowing explicit representations of $\Phi(x)$.

3 Hyperparameter Optimization of the Unsupervised Learning Phase of the Machine Learning Pipeline

Although the overall analytical leakage localization is composed of two learning stages, the focus of this paper is on the optimization of hyperparameters of the first unsupervised learning phase (i.e. Kernel k-means clustering).

Clustering of leak signatures is aimed at grouping together similar "effects" induced by different (simulated) leaks. This allows to implement the second supervised learning phase considering a limited number of labels (i.e. the number of clusters instead of the number of pipes of the WDN). Previous Fig. 1 summarizes the overall machine learning based leakage localization approach – where Kernel k-means clustering replaces Spectral Clustering used in the preliminary papers.

3.1 Hyperparameters in the Pipeline: The Design Variables

The number and type of hyperparameters in the overall machine learning pipeline depends on the specific software used for kernel-based clustering and SVM classification.

Since the focus of this paper is to replace the original SC phase with a kernel-based k-means, just the two following hyperparameters are taken into account:

- The number k of clusters – a discrete decision variable (i.e. an integer);
- The value of the RBF kernel's parameter σ – a continuous decision variable.

The SVM hyperparameters are not part of the optimization process in this paper.

3.2 Clustering Performance: The Objective Function

The objective function, that is the clustering performance index I defined in previous Sect. 2.1, is black-box. To compute it, the execution of the kernel k-means is performed on the entire dataset, given k and σ, along with the calculation of I_C and I_P.

3.3 Sequential Model Based Optimization

The generic Sequential Model Based Optimization (SMBO) [27] consists in:

1. starting with an initial set of evaluations of the objective function;
2. fitting a regression model (i.e. surrogate of objective function) based on the overall set of evaluations performed so far;
3. querying the regression model to propose a new, promising point, usually through the optimization of an "acquisition function" (or "infill criterion");
4. evaluating objective function at the new point and adding the results to the set of evaluations;
5. if a given termination criterion (e.g. maximum number of evaluations of the objective function) is not satisfied, then return to step 2.

Several adaptations and extensions, e.g., multi-objective optimization [36], multi-point proposal [37, 38], more flexible regression models [27] or alternative ways to calculate the infill criterion [39] have been investigated recently, as reported in [40]. A specific application domain for SMBO is the hyperparameter optimization for machine learning algorithms [41–43], where the algorithm is a "black-box" and the objective function is one or multiple performance measure(s).

3.3.1 Building the Surrogate of the Objective Function: Gaussian Processes and Random Forest

Generation of a surrogate of the objective function is one the first choices in designing a SMBO process. Selection of the specific regression model to use can be suggested by the search space spanned by the decision variables. Kriging [35], based on Gaussian Processes (GP), is usually recommended, but, in the case the search space also includes categorical parameters, Random Forests (RF) are a plausible alternative [27] as they can handle such parameters directly.

In this study, two hyperparameters are considered: the number k of clusters – which is an integer categorical variable – and σ, the internal parameter of the kernel clustering – which is a continuous variable. Although RF should be preferable, due to the nature of the hyperparameter k, also GP has been investigated in this study. The initial design consists of 220 evaluations obtained through Latin Hypercube Sampling (LHS).

3.3.2 Acquisition Function: Confidence Bound

The acquisition function is the mechanism which guides the optimization process, implementing the trade-off between exploitation (choosing the new point in regions which are "promising" according to the current knowledge) and exploration (choosing the new point in less explored regions). This trade-off is usually achieved by combining, into a single formula, both the posterior mean and posterior standard deviation, as estimated through the surrogate model. A number of acquisition functions have been proposed in literature, such as Probability of Improvement (PoI), Expected Improvement (EI) and Confidence Bound (CB). In this study the Upper Confidence Bound (UCB) is used as the goal is to maximize the clustering performance index I. The guiding principle behind CB is to be optimistic in the face of uncertainty.

Finding the next promising point requires solving an optimization problem where the objective function is the acquisition function. Anyway, it can be considered inexpensive with respect to the original optimization problem. The so called "focus search" algorithm, proposed in [44], is a generic approach able to deal with numeric parameter spaces, categorical parameter spaces, as well as mixed and hierarchical spaces. This is the procedure adopted in this study.

3.3.3 Termination Criterion

Several termination criteria are possible: the maximum number of function evaluations is a very common choice. Other possibilities are a limit over the wall clock time, no improvements in the "best seen" (i.e. the best value of objective function seen so far) after a given number of consecutive SMBO iterations, or the difference from the optimum – when known – is lower than a given threshold. Other more sophisticated criteria have been also proposed, e.g. in [44] a Resource-Aware Model-Based Optimization framework that leads to efficient utilization of parallel computer architectures through resource-aware scheduling strategies, while in [45] the Lipschitz continuity assumption, quite realistic for many practical black-box problems, is considered for obtaining a global optimum estimates after performing a limited number of functions evaluations.

The termination criteria defined in this study is the maximum number of functions evaluations: this choice allows for an easy comparison between GP and RF based SMBO as well as between SMBO and a simple grid search.

3.3.4 Software Environment

The implementation of the hyperparameters optimization proposed in this study has been based on the R software environment. The recently proposed toolbox "mlrMBO" [40] has been used to implement the SMBO process, along with the kernel k-means algorithm, namely "kkmeans", provided by the R package "kernlab".

mlrMBO is a flexible and comprehensive R toolbox for model-based optimization (MBO). It can deal with both single- and multi-objective optimization, as well as continuous, categorical, mixed and conditional parameters. Additional features include multi-point batch proposal, parallelization, visualization, logging and error-handling. Very important, mlrMBO is implemented in a modular fashion, such that single components can be easily replaced or adapted by the user for specific use cases.

4   Results and Discussion

This section summarizes the results obtained. The experimental setting consists in:

- Grid search vs SMBO (using both GP and RF to generate the surrogate of the objective function): with $k = 3, ..., 13$ and σ in $[0.00001, 0.1]$, with 70 values for σ, equally distributed in the range, are used for the grid search;
- Maximization of the index I used to measure the performance of clustering output with respect to leakage localization properties (as defined in Sect. 2.1):
  - Termination criteria, based on a limit on the number of function evaluations: 770 function evaluations (equals to the number of configurations into the grid), where 220 are used as initial design.

The following Table 1 summarizes the results obtained:

Table 1. Results: best performance, hyperparameter configuration, time and iterations among the three approaches (Grid Search, GP- and RF- based SMBO)

	Clustering performance I (best seen)	k^*	σ^*	Time (sec)	Last iteration with improvement
Grid search	0,516	3	0,0000100	6165,03	NA
GP	0,505	3	0,0055322	9704,88	388
RF	0,556	3	0,0000112	12317,92	687

- The "best seen" of the performance index I, over the 770 function evaluations;
- Values of the hyperparameters, K^* and σ^*, associated to the best seen;
- Overall execution time (sec), computed as total on all the 770 evaluations;
- Last iteration with improvement of the clustering performance index I.

SMBO using RF proved to be the most effective strategy. In particular, it was able to identify a hyperparameters configuration of the kernel k-means "outside the grid" and associated to the highest value of the clustering performance index I.

On the contrary, SMBO using GP was not able to converge to a better hyperparameters configuration than the one identified by the grid search. This was probably due to the nature of k, indeed RF is usually preferred to GP in the case of categorical variables.

The following Fig. 3 compares the convergence of the different approaches. SMBO with GP converges very fast – after 388 evaluations of the objective function no more improvements are obtained, even if the best seen value of I is lower than the one obtained through the grid search.

It is important to highlight that SMBO with RF provides the following benefits:

- it is able to find a hyperparameters configuration outperforming grid search – as well as SMBO with GP – in terms of effectiveness (clustering performance index, I);

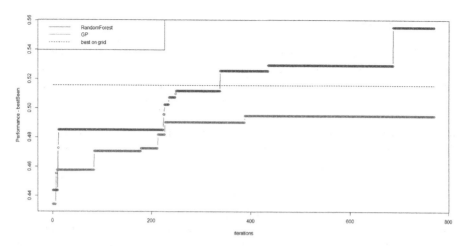

Fig. 3. Best seen over the iterations for SMBO with RF, SMB with GP and the best value obtained over the grid search (independent on the iterations)

- a hyperparameters configuration outperforming the grid search can be obtained after only 338 function evaluations, well lower the limit on the number of evaluations.

The second point is really important: although the best seen after 338 evaluations is lower than the final one, this means that the SMBO with RF outperforms grid search also in terms of efficiency, just using about half of the maximum number of evaluations available. Indeed, even with a more tightening termination criteria (e.g. 385 evaluations, that is 770/2), without any modification to the grid search, SMBO with RF should was able to outperform grid search, with a consequent drastic reduction of the wall clock time, too (approximately 3370 s).

Finally, to further highlight the benefits provided by SMBO for hyperparameters configuration, we decided "to give more chances" to the grid search, increasing the number of configurations to 2310 (770 × 3) by choosing 210 values for σ, equally distributed, in the range 0.00001 to 0.1. As result, the grid search was not able to improve in terms of clustering performance index I even if the overall wall clock time went from 6165,63 s to 17555,37 s.

This further result confirms the main drawback of the grid search for hyperparameters optimization: it performs many function evaluations in non-promising areas of the search space, basically wasting computational resources.

With respect to the SVM, it has been trained according to the cluster assignment provided by the kernel k-means – instead of SC. The same SVM configuration used in the previous study was adopted. Classification performances were similar, slightly better in the case of the kernel k-means, confirming the equivalence between SC and kernel k-means while enabling a significant reduction in terms of computational complexity in the clustering procedure.

5   Conclusions

Any performance comparison between Bayesian Optimization and other global optimization strategies can be only platform and problem dependent and thus difficult to generalize: in [19] it is stated that *"random search offers a simple, parallelizable and theoretically sound launching point"* while Bayesian Optimization may offer *"improved empirical accuracy"* but *"its selection models are intrinsically sequential and thus difficult to parallelize"*. The main problem is that Bayesian Optimization scales poorly with the number of dimensions: in [46] it is stated that *"the approach is restricted to problems of moderate dimensions up to 10 dimensions"* and a random embedding is proposed to identify a work space of a much smaller number of dimensions. Other proposal to scale Bayesian Optimization to higher dimensions are in [19, 47].

The computational results reported in this paper substantiate the known fact that the Bayesian framework is suitable to objective functions which are costly to evaluate and black box. Moreover, they can be applied to "unusual" design spaces which involve categorical or conditional inputs and are therefore able to deal with such diverse domains as A/B testing, recommender systems, reinforcement learning, environmental monitoring, sensor networks, preference learning and interactive interfaces.

The next activities will leverage the capability RF based SMBO to handle conditional parameters as well; we will also consider the optimization of the whole machine learning pipeline, moving towards an automatic algorithm configuration setting.

References

1. Xia, L., Xiao-dong, W., Xin-hua, Z., Guo-jin, L.: Bayesian theorem based on-line leakage detection and localization of municipal water supply network. Water Wastewater Eng. **12** (2006)
2. Sivapragasam, C., Maheswaran, R., Venkatesh, V.: ANN-based model for aiding leak detection in water distribution networks. Asian J. Water Environ. Pollut. **5**(3), 111–114 (2007)
3. Xia, L., Guo-jin, L.: Leak detection of municipal water supply network based on the cluster-analysis and fuzzy pattern recognition. In: 2010 International Conference on E-Product E-Service and E-Entertainment (ICEEE), vol. 1(5), pp. 7–9 (2010)
4. Lijuan, W., Hongwei, Z., Hui, J.: A leak detection method based on EPANET and genetic algorithm in water distribution systems. In: Wu, Y. (ed.) Software Engineering and Knowledge Engineering: Theory and Practice. AISC, vol. 114, pp. 459–465. Springer, Heidelberg (2012). doi:10.1007/978-3-642-03718-4_57
5. Nasir, A., Soong, B.H., Ramachandran, S.: Framework of WSN based human centric cyber physical in-pipe water monitoring system. In: 11th International Conference on Control, Automation, Robotics and Vision, pp. 1257–1261 (2010)
6. Soldevila, A., Fernandez-Canti, R.M., Blesa, J., Tornil-Sin, S., Puig, V.: Leak localization in water distribution networks using Bayesian classifiers. J. Process Control **55**, 1–9 (2017)
7. Franzin, A., Cáceres, L.P., Stützle, T.: Effect of Transformations of Numerical Parameters in Automatic Algorithm Configuration, IRIDIA Technical Report 2017-006 (2017)
8. Bagnall, A., Cawley, G.C.: On the Use of Default Parameter Settings in the Empirical Evaluation of Classification Algorithms. arXiv:1703.06777v1 [cs.LG] (2017)

9. Strongin, R.G., Sergeyev, Y.D.: Global Optimization with Non-Convex Constraints - Sequential and Parallel Algorithms. Springer, US (2000)
10. Zhigljavsky, A., Žilinskas, A.: Stochastic Global Optimization. Springer, US (2008)
11. Locatelli, M., Schoen, F.: Global Optimization - Theory, Algorithms and Applications. MOS-SIAM Series on Optimization. Society for Industrial & Applied Mathematics, Philadelphia (2013)
12. Sergeyev, Y.D., Kvasov, D.E.: Global search based on efficient diagonal partitions and a set of Lipschitz constants. SIAM J. Optim. **16**(3), 910–937 (2006)
13. Sergeyev, Y.D., Kvasov, D.E.: A deterministic global optimization using smooth diagonal auxiliary functions. Commun. Nonlinear Sci. Numer. Simul. **21**(1), 99–111 (2015)
14. Barkalov, K., Polovinkin, A., Meyerov, I., Sidorov, S., Zolotykh, N.: SVM regression parameters optimization using parallel global search algorithm. In: Malyshkin, V. (ed.) PaCT 2013. LNCS, vol. 7979, pp. 154–166. Springer, Heidelberg (2013). doi:10.1007/978-3-642-39958-9_14
15. Gillard, J.W., Kvasov, D.E.: Lipschitz optimization methods for fitting a sum of damped sinusoids to a series of observations. Stat. Interface **10**, 59–70 (2017)
16. Zabinsky, Z.B.: Stochastic Adaptive Search for Global Optimization, vol. 72. Springer, New York (2013)
17. Steponavičė, I., Shirazi-Manesh, M., Hyndman, R.J., Smith-Miles, K., Villanova, L.: On sampling methods for costly multiobjective black-box optimization. In: Pardalos, P.M., Zhigljavsky, A., Zilinskas, J. (eds.) Advances in Stochastic and Deterministic Global Optimization. SOIA, vol. 107, pp. 273–296. Springer, Cham (2016). doi:10.1007/978-3-319-29975-4_15
18. Csendes, T., Pál, L., Sendin, J.O.H., Banga, J.R.: The GLOBAL optimization method revisited. Optim. Lett. **2**(4), 445–454 (2008)
19. Li, L., Jamieson, K., DeSalvo, G., Rostamizadeh, A., Talwalkar, A.: Hyperband: A Novel Bandit-Based Approach to Hyperparameter Optimization. arXiv preprint arXiv:1603.06560 (2016)
20. Feurer, M., Klein, A., Eggensperger, K., Springenberg, J., Blum, M., Hutter, F.: Efficient and robust automated machine learning. In: Advances in Neural Information Processing Systems, pp. 2962–2970 (2015)
21. Shahriari, B., Swersky, K., Wang, Z., Adams, R.P., de Freitas, N.: Taking the human out of the loop: A review of bayesian optimization. Proc. IEEE **104**(1), 148–175 (2016)
22. Snoek, J., Larochelle, H., Adams, R.P.: Practical bayesian optimization of machine learning algorithms. In: Advances in Neural Information Processing Systems, pp. 2951–2959 (2012)
23. Brochu, E., Cora, V.M., De Freitas, N.: A tutorial on Bayesian optimization of expensive cost functions, with application to active user modeling and hierarchical reinforcement learning. arXiv preprint arXiv:1012.2599 (2010)
24. Mockus, J.: On Bayesian methods of optimization. In: Dixon, L.C.W., Szegö, G.P. (eds.) Towards Global Optimization. North-Holland, Amsterdam (1975)
25. Stützle, T.: Automated algorithm configuration: advances and prospects. In: Camacho, D., Braubach, L., Venticinque, S., Badica, C. (eds.) Intelligent Distributed Computing VIII. SCI, vol. 570, p. 5. Springer, Cham (2015). doi:10.1007/978-3-319-10422-5_2
26. Mala-Jetmarova, H., Sultanova, N., Savic, D.: Lost in optimization of water distribution systems? a literature review of system operations. Environ. Modell. Softw. **93**, 209–254 (2017)
27. Hutter, F., Hoos, H.H., Leyton-Brown, K.: Sequential model-based optimization for general algorithm configuration. In: Coello, C.A.C. (ed.) LION 2011. LNCS, vol. 6683, pp. 507–523. Springer, Heidelberg (2011). doi:10.1007/978-3-642-25566-3_40

28. Candelieri, A., Soldi, D., Archetti, F.: Cost-effective sensors placement and leak localization - The Neptun pilot of the ICeWater project. J. Water Supply: Res. Technol. AQUA **64**(5), 567–582 (2015)
29. Candelieri, A., Soldi, D., Conti, D., Archetti, F.: Analytical leakages localization in water distribution networks through spectral clustering and support vector machines. The icewater approach. Procedia Eng. **89**, 1080–1088 (2014)
30. Candelieri, A., Archetti, F., Messina, E.: Improving leakage management in urban water distribution networks through data analytics and hydraulic simulation. WIT Trans. Ecol. Environ. **171**, 107–117 (2013)
31. Dhillon, I.S., Guan, Y., Kulis, B.: Kernel k-means: spectral clustering and normalized cuts. In: Proceedings of the Tenth ACM SIGKDD International Conference on Knowledge Discovery and Data Mining, pp. 551–556 (2004)
32. Žilinskas, A.: On similarities between two models of global optimization: statistical models and radial basis functions. J. Global Optim. **48**(1), 173–182 (2010)
33. Arbelaitz, O., Gurrutxaga, I., Muguerza, J., Pérez, J.M., Perona, I.: An extensive comparative study of cluster validity indices. Pattern Recogn. **46**(1), 243–256 (2013)
34. Scholkopf, B., Smola, A., Muller, K.R.: Nonlinear component analysis as a kernel eigenvalue problem. Neural Comput. **10**, 1299–1319 (1998)
35. Jones, D.R., Schonlau, M., Welch, W.J.: Efficient global optimization of expensive black-box functions. J. Global Optim. **13**(4), 455–492 (1998)
36. Horn, D., Wagner, T., Biermann, D., Weihs, C., Bischl, B.: Model-based multi-objective optimization: taxonomy, multi-point proposal, toolbox and benchmark. In: Gaspar-Cunha, A., Henggeler Antunes, C., Coello, C.C. (eds.) EMO 2015. LNCS, vol. 9018, pp. 64–78. Springer, Cham (2015). doi:10.1007/978-3-319-15934-8_5
37. Ginsbourger, D., Le Riche, R., Carraro, L.: Kriging is well-suited to parallelize optimization. In: Tenne, Y., Goh, C.K. (eds.) Computational Intelligence in Expensive Optimization Problems. ALO, vol. 2, pp. 131–162. Springer, Heidelberg (2010). doi:10.1007/978-3-642-10701-6_6
38. Bischl, B., Wessing, S., Bauer, N., Friedrichs, K., Weihs, C.: MOI-MBO: multiobjective infill for parallel model-based optimization. In: Pardalos, Panos M., Resende, M.G.C., Vogiatzis, C., Walteros, J.L. (eds.) LION 2014. LNCS, vol. 8426, pp. 173–186. Springer, Cham (2014). doi:10.1007/978-3-319-09584-4_17
39. Bergstra, J.S., Bardenet, R., Bengio, Y., Kégl, B.: Algorithms for hyperparameter optimization. In: Advances in Neural Information Processing Systems, pp. 2546–2554 (2011)
40. Bischl, B., Richter, J., Bossek, J., Horn, D., Thomas, J., Lang, M.: mlrMBO: A Modular Framework for Model-Based Optimization of Expensive Black-Box Functions. arXiv preprint arXiv:1703.03373 (2017)
41. Thornton, C., Hutter, F., Hoos, H.H., Leyton-Brown, K.: Auto-WEKA: Combined selection and hyperparameter optimization of classification algorithms. In: Proceedings of ACM SIGKDD, pp. 847–855 (2013)
42. Lang, M., Kotthaus, H., Marwedel, P., Weihs, C., Rahnenführer, J., Bischl, B.: Automatic model selection for high-dimensional survival analysis. J. Stat. Comput. Simul. **85**(1), 62–76 (2015)
43. Horn, D., Bischl, B.: Multi-objective parameter configuration of machine learning algorithms using model-based optimization. In: 2016 IEEE Symposium Series on Computational Intelligence (SSCI), pp. 1–8 (2016)

44. Richter, J., Kotthaus, H., Bischl, B., Marwedel, P., Rahnenführer, J., Lang, M.: Faster model-based optimization through resource-aware scheduling strategies. In: Festa, P., Sellmann, M., Vanschoren, J. (eds.) LION 2016. LNCS, vol. 10079, pp. 267–273. Springer, Cham (2016). doi:10.1007/978-3-319-50349-3_22
45. Kvasov, D.E., Sergeyev, Y.D.: Deterministic approaches for solving practical black-box global optimization problems. Adv. Eng. Softw. **80**, 58–66 (2015)
46. Wang, Z., Zoghi, M., Hutter, F., Matheson, D., De Freitas, N.: Bayesian optimization in high dimensions via random embeddings. In: AAAI Press/International Joint Conferences on Artificial Intelligence (2013)
47. Klein, A., Falkner, S., Bartels, S., Hennig, P., Hutter, F.: Fast Bayesian Optimization of Machine Learning Hyperparameters on Large Datasets. arXiv:1605.07079 (2017)

Solution of the Convergecast Scheduling Problem on a Square Unit Grid When the Transmission Range is 2

Adil Erzin[⊠]

Sobolev Institute of Mathematics, Novosibirsk State University, Novosibirsk, Russia
adilerzin@math.nsc.ru

Abstract. In this paper a conflict-free data aggregation problem, known as a Convergecast Scheduling Problem, is considered. It is NP-hard in the arbitrary wireless network. The paper deals with a special case of the problem when the communication graph is a square grid with unit cells and when the transmission range is 2 (in L_1 metric). Earlier for the case under consideration we proposed a polynomial time algorithm with a guaranteed accuracy bound. In this paper we have shown that the proposed algorithm constructs an *optimal* solution to the problem.

Keywords: Wireless networks · Data aggregation · Conflict-free scheduling

1 Introduction

In the wireless sensor networks (WSN) the data collected by the sensors should be delivered to the analytical center. The process of transferring the packets of information from the sensors in such a center - a *base station* (BS) is called an *aggregation* of the data. Aggregation time (or latency), i.e. a period during which the data from all sensors fall in the BS, is the most important criterion in the quick response networks. The shorter aggregation time the more effectively WSN can react to the possible events.

A synthesis of the network through which the data is transmitted, as a rule, carried out object to the criterion of minimum communication power consumption [1–3]. As a result, a constructed communication graph (CG) is highly sparse. Consequently, not all vertices (sensors) can transmit the collected data directly to the BS. Packets from the most vertices are going through the other vertices, and the path from some vertex to the BS may consist of a big number of edges. In the formulations of the aggregation problem the volume of the transmitted data, as usually, does not taken into account. So the packets are considered to be equal in length for all vertices of the CG, and each packet is transmitted along any edge during one time round (slot).

In the majority of the wireless networks, an element (vertex or node) cannot transmit and receive packets at the same time (half duplex mode), and a vertex

© Springer International Publishing AG 2017
R. Battiti et al. (Eds.): LION 2017, LNCS 10556, pp. 50–63, 2017.
https://doi.org/10.1007/978-3-319-69404-7_4

cannot receive more than one packet simultaneously. Moreover, due to the need of energy saving, each sensor sends the packet once during the aggregation period (session). This means that the packets are transmitted along the edges of a desired *aggregation tree* (AT) rooted in BS, and an arbitrary vertex in the AT must first receive the packets from all its children, and only then can send the aggregated packet to its parent node. So, we assume that the arcs of the AT are oriented towards the root (BS), and if the AT is known, then the partial order on the set of arcs of the AT is known too.

In the most WSNs, the sensors use a common radio frequency to transmit the messages. So, if in the sensor's reception area are working more than one transmitter, then (due to the radio wave interference phenomenon), the receiver cannot get any correct data packet. Such a situation is called a *conflict* or a *collision*.

In the conflict-free data aggregation problem it is necessary to find an aggregation tree and a conflict-free schedule of minimal length [4–6]. This problem is known as a Convergecast Scheduling Problem (CSP), and it is NP-hard even in the case when the AT is given [7].

If the AT is known, it is possible to construct a graph of conflicts (GC) as follows. Each node in the GC is associated with the arc in the AT. Two vertices in the GC are linked by an edge, if the simultaneous transmission of the respective arcs in the AT implies a conflict. There is an arc going from the node i to the node j in the GC, if the end of the arc i coincides with the beginning of the arc j in the AT.

It is obvious that the CSP in the case of a given AT reduces to the problem of mixed coloring of the GC [8], which is also NP-hard, and stated as follows. Let the mixed graph $G = (V, A \cup E)$ with vertex set V, the set of arcs A and the set of edges E is given. Graph G is k-colorable if exists a function $f : V \rightarrow \{1, \ldots, k\}$. In the problem of mixed graph coloring it is required to find a minimal k, for which exists such k-coloring, that if two vertices i and j are joined by an edge $(i, j) \in E$, then $f(i) \neq f(j)$, and if there is an arc $(i, j) \in A$, then $f(i) < f(j)$.

The problem of conflict-free data aggregation has been intensively studied by both theoreticians and practitioners [1–3,6–8,10,11]. A number of heuristic algorithms was proposed to construct an approximate solution [4–6,9–12]. For some of them the guaranteed accuracy bounds in terms of degree and radius of the CG were found [13]. To assess the quality of the other heuristics the numerical experiments were carried out [4,9].

The literature also addresses the special cases of the problem. For example, when the conflicts occur only between the children of common parent in the AT [9]. Such a situation occurs when the sensors use different radio frequencies for data transmission [12]. In this case, if AT is given, the problem can be solved in a polynomial time, but when AT is a desired tree the problem remains NP-hard [14].

In [15] a special case of CG in the form of a unit square grid, in which in each node a sensor is located, and the transmission range of each sensor is 1, is considered. A simple polynomial algorithm for constructing an *optimal* solution to this problem was proposed. In [7] a similar problem in the case when the transmission distance d is an arbitrary integer not less than 1 was considered. On the one hand the increase of d can reduce the length of the schedule, but on the other hand the number of collisions increases too. The methods of constructing a conflict-free schedule of the data transmission with the guaranteed accuracy bounds depending on d are proposed. In particular, for the case when $d = 2$ was proposed an algorithm which on the $(n + 1) \times (m + 1)$ grid, where n and m are even, and the BS is in the origin $(0, 0)$, building a schedule of length $(n + m)/2 + 3$.

In this paper, we show that the lower bound for the length of the schedule when the communication graph is a unit square grid and the transmission range of sensors is 2 (in L_1 metric), is $(n + m)/2 + 3$, i.e. the considered special case of the problem is polynomially solvable, and the algorithm proposed in [7] builds an *optimal* solution to the problem.

2   General Problem Formulation

Let a communication graph (CG), in which the vertices are the images of the sensors, and two vertices are connected by an edge if a communication between the sensors can be carried out in both directions, is given. Among the vertices we distinguish a base station (BS) into which the data from all vertices should be delivered. We assume that:

- a time is discrete;
- a transmission time of any packet along any edge of the CG is one time round;
- a sensor cannot receive and transmit at the same time round;
- a sensor cannot receive more than one packet during one time round;
- each sensor transmits packet only once during the aggregation session;
- a subset of vertices may transmit simultaneously if that does not entail a conflict;

The CSP is to find a feasible data transmission schedule of minimum length.

To illustrate the problem, let us refer to the Fig. 1. Figure 1a shows an example of a CG, in which BS is marked in yellow and the AT – in the bold lines. If in this CG transfers, for example, red vertex, then it is heard by 5 other vertices. Figure 1b shows a feasible schedule, where the number inside the vertex is the time round of transmission of this vertex. The length of the schedule in Fig. 1b is 6.

According to the CG a graph of conflicts (GC) can be constructed (see. Fig. 2). Some edges (dashed in Fig. 2b) can be deleted, since the corresponding conflicts are taken into account by the transmission order. For example, the transmission of vertex 8 precedes the transmission of node 3, then the edge $(8, 3)$ can be eliminated. Figure 2c shows a feasible mixed coloring of the vertices of the GC, in which the color number corresponds to the transfer round when the corresponding vertex sends a packet (in Fig. 2c the number of each node is the number of the color).

As noted above, the CSP in general and in many special cases is NP-hard. However, in some special cases, the problem is polynomially solvable. This is so,

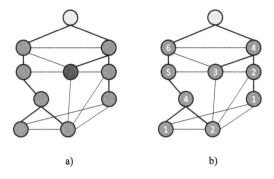

Fig. 1. (a) CG and AT (bold lines); (b) Feasible schedule of length 6. (Color figure online)

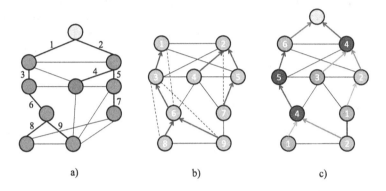

Fig. 2. (a) CG and AT; (b) Graph of Conflicts (GC); (c) Feasible coloring. (Color figure online)

for example, when the CG is the unit square grid, and a transmission distance is 1 [15]. In this paper, we are interested in the problem when a CG is also a unit square grid, but the transmission distance is 2 (in L_1 metric). In [7] for the $(n + 1) \times (m + 1)$ grid with the BS at the origin $(0, 0)$ and when transmission distance equals 2, an algorithm for constructing a schedule, the length of which (when n and m are even) equals $(n + m)/2 + 3$, is proposed.

3   CSP in the Unit Square Grid When the Transmission Distance is 2

Let us consider a grid graph (Fig. 3a), in which each node (x, y), $x = 0, 1, \ldots, n$, $y = 0, 1, \ldots, m$ contains a sensor. Transmission distance of each vertex is 2 (in L_1 metric). Each sensor must send the collected data to the base station (BS), which is located at the origin (for simplicity). The time of arrival of the last packet in the BS is the aggregation time or the *schedule length*. We denote the minimum length of the schedule as $L(n, m)$.

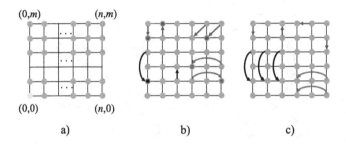

Fig. 3. (a) Grid graph; (b) Conflict (infeasible) transmissions; (c) Conflict-free (feasible) transmissions. (Color figure online)

The restrictions from the previous section naturally have to be met also. For example, in Fig. 3b the unacceptable transmissions are displayed (the transmissions shown by arrows of the same color cannot be performed simultaneously). And in Fig. 3c the conflict-free transmissions are indicated (the arrows of the same color) which can be performed simultaneously.

3.1   The Exact Lower Bound for the Schedule Length

The further results are valid (with minor adjustments) for the arbitrary n and m, but in this paper for the sake of simplicity and brevity, we set n and m to be even.

Definition 1. *The distance from the vertex to the BS is the minimum number of transmissions from this vertex.*

Then the vertex (x, y), $x = 0, 1, \ldots, n$, $y = 0, 1, \ldots, m$, is at a distance $](x + y)/2[$, where $]a[$ is the smallest integer not less than a.

Obviously, the length of the schedule cannot be less than $D = (n + m)/2$, because the most remote vertex (n, m) is located at the distance D, and the packet form it could be delivered to the BS at least during D time rounds.

Fairly obvious the next.

Property 1. If at least two vertices in the arbitrary graph are at the distance R from the BS, then the aggregation time cannot be less than $R + 1$.

Since in the considered grid three vertices are at the distance D from the BS, then follows the obvious.

Corollary 1. *The aggregation time in the unit square grid $(n + 1) \times (m + 1)$ with the BS at the origin $(0, 0)$, when the transmission distance is 2, is at least $D + 1$.*

Based on the Property 1 it is also easy to prove.

Corollary 2. *The aggregation time in the unit square grid $(n+1) \times (m+1)$ with the BS at the origin $(0,0)$, when the transmission distance is 2, is at least $D+2$.*

The main purpose of this paper is to prove the following

Lemma 1. *The aggregation time in the unit square grid $(n+1) \times (m+1)$ with the BS at the origin $(0,0)$, when the transmission distance is 2, is at least $D+3$.*

Proof. We first consider the different (up to symmetry) transmissions from the vertex (n, m) at the moment (time round) 1 (Fig. 4). At that, let us encode each possible action (transmission) as a $(t = a; b)$, where a is a time round, and b is the number/code of the allowable transmission. For example, when $(t = 1; 0)$ in Fig. 4 the vertex (n, m) is silent (don't send a packet), and if $(t = 1; 1)$ it transfers the packet to another blue vertex which is also located at the distance D from the BS (the origin).

In the figures the node with the red circle cannot transmit a packet during the moments $1, 2, \ldots, t$, but vertex with the green circle can transmit at the time round t. For example, in the case $(t = 1; 1)$ the receiver may hear 6 extra vertices besides the sender. Thus, these vertices (5 yellow and one blue) must be silent and in Fig. 4 they have the red circles.

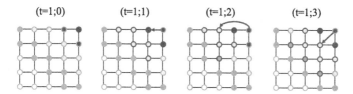

Fig. 4. Possible transmissions from the vertex (n, m) at the moment 1. (Color figure online)

When we proved the statement, we have considered all possible cases (more than 100 000), but they, of course, cannot be presented all in a single paper, so here we will describe only a few of them as the examples. So, we present not a complete proof, but only illustrate the proof. The analysis of all cases is planned to place in Internet in the foreseeable future.

Case $(t = 1; 0)$. Let us consider the possible actions of the vertex $(n-1, m)$ at the time round 1 (Fig. 5), and then consider the case $(t = 1; 0.0.0)$ in detail (when all blue nodes (at the distance D) are silent (Fig. 6) at the moment 1).

The possible actions of the vertex (n, m) at the time round 2 are shown in Fig. 7.

Consider the possible actions of the vertex (n, m) at the time 2 (Fig. 7), and the detailed case $(t = 2; 0)$ when the node (n, m) is silent. Consider the behavior

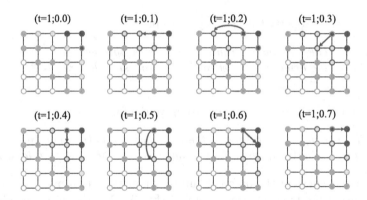

Fig. 5. Possible transmissions from the node $(n-1, m)$ at the moment 1, when vertex (n, m) is silent.

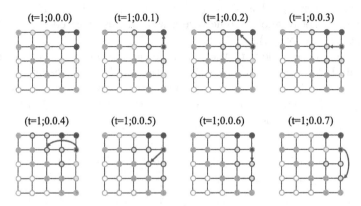

Fig. 6. Possible transmissions from the node $(n, m-1)$ at the moment 1, when vertices (n, m) and $(n-1, m)$ are silent. (Color figure online)

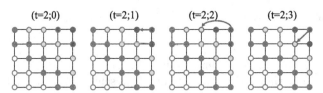

Fig. 7. Possible transmissions from the node (n, m) at the moment 2, when all blue nodes were silent at moment 1. (Color figure online)

of the vertex $(n-1, m)$ at the time round 2. If it is silent, then after two time rounds remain at least two vertices at the distance D, which have not started transmitting. Therefore, by Property 1, the length of the schedule cannot be less than $2 + D + 1 = D + 3$ (Fig. 8).

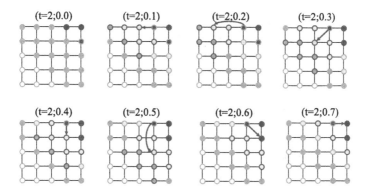

Fig. 8. Possible transmissions from the node $(n-1, m)$ at the moment 2, when all blue nodes were silent at moment 1, and vertex (n, m) is silent at moment 2. (Color figure online)

Let us consider the case $(t = 2; 0.1)$. Since the vertex (n, m) is silent, then, according to Property 1, both vertices $(n - 1, m)$ and $(n, m - 1)$ must transmit. The transmission cases are shown in Fig. 9.

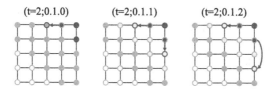

Fig. 9. Possible transmissions from the node $(n, m - 1)$ at the moment 2 in case $(t = 2; 0.1)$.

Further, as an example, we consider in detail the case $(t = 2; 0.1.1)$. At the moment 3 the vertex (n, m) must transmit a packet, else the length of the schedule will be at least $D + 3$. The possible transmissions are shown in Fig. 10.

In the case $(t = 3; 2)$ after the time round 3 remain at least 3 yellow vertices (at the distance $D - 1$) which has not transmitted, so the schedule length is at least $3 + D - 1 + 1 = D + 3$. The case $(t = 3; 1)$ we will examine in detail. If after the 3rd time round remain at least 2 yellow vertices (at the distance $D - 1$), then the length of the schedule cannot be less than $D + 3$. To avoid this the circled vertices (Fig. 11) in the case $(t = 1; 0.0.0)$ must send the packets and this can be done at the time round 1 in the manner shown in Fig. 11 by blue arrows. But then the other yellow vertices, which are outlined in the case $(t = 3; 1)$ will not be able transmitting before the 3rd time round. However, at the moment 3 they all cannot transmit the packets. As a result, after the 3rd time round at least two nodes at the distance $D - 1$ will not start transmission and therefore the schedule length is not less than $D + 3$.

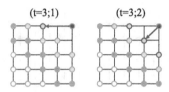

Fig. 10. Possible transmissions from the node (n, m) at the moment 3 in case $(t = 2; 0.1.1)$.

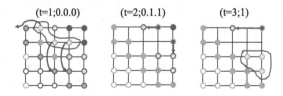

Fig. 11. Transmissions in the cases $(t = 1; 0.0.0)$, $(t = 2; 0.1.1)$, $(t = 3; 1)$. (Color figure online)

Let us consider one more case.

Case $(t = 1; 3)$. In this case, at the first time round the vertices $(n - 1, m)$ and $(n, m - 1)$ cannot transmit. The possible transmissions from the node $(n - 1, m)$ at the moment 2 are shown in Fig. 12.

Next, consider, for example, the case $(t = 2; 1.1)$ in Fig. 13. The possible transmissions from the node $(n - 1, m - 1)$ at the time round 3 are shown in Fig. 14.

The cases $(t = 3; 1)$ – $(t = 3; 7)$ are simple, since after the 3rd time round remain at least two vertices at the distance $D - 1$, which have not yet broadcast and Property 1 implies that the length of the schedule is not less than $D + 3$.

The case $(t = 3; 0)$, when the node $(n - 1, m - 1)$ is silent, is difficult. Let us consider it. That after the 3rd moment remains not more than one silent yellow vertex (at a distance $D - 1$), it is necessary to send the packets from all other yellow vertices (at the distance $D - 1$), except the vertex $(n - 1, m - 1)$, not later than at the time round 3. From the vertices $(n - 2, m)$, $(n - 2, m - 1)$, $(n - 1, m - 2)$ and $(n, m - 2)$ the packets can be sent only at the moment 3, and the only way to do so is shown in Fig. 15 $(t = 3; 0)$ with the blue arrows. Then from the vertices $(n - 3, m)$ and $(n, m - 3)$ the packets can be sent only at the moment 1, for example, as it is shown in Fig. 15 $(t = 1; 3)$ with the blue arrows. Let us consider the moment 4 and the variants of sending a packet from the vertex $(n - 1, m - 1)$. If this node will not send the packet at the moment 4, the schedule length cannot be less than $D + 3$. The possible transmissions are shown in Fig. 16. As a result, after the 4th time round at least two green nodes (at the distance $D - 2$) remain silent. So (by Property 1), we have that the length of the schedule is at least $4 + D - 2 + 1 = D + 3$.

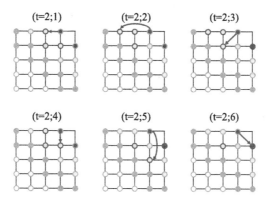

Fig. 12. Possible transmissions from the node $(n-1, m)$ at the moment 2.

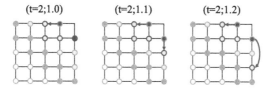

Fig. 13. Possible transmissions from the node $(n, m-1)$ in the case $(t = 2; 1)$.

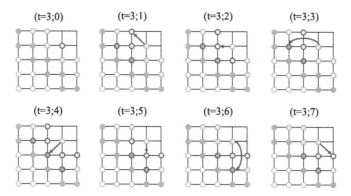

Fig. 14. Possible transmissions from the node $(n-1, m-1)$ at the moment 3 in the case $(t = 2; 1.1)$.

The author have considered all possible transmission from the vertices at a distance of at least $D - 2$ from the BS (the origin), and it is shown that the length of the schedule cannot be less than $D + 3$, which proves the lemma.

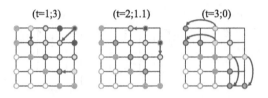

Fig. 15. Cases $(t = 1; 3)$, $(t = 2; 1.1)$ and $(t = 3; 0)$. (Color figure online)

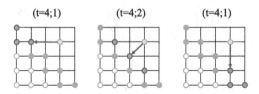

Fig. 16. Possible transmissions from the node $(n - 1, m - 1)$ at the moment 4. (Color figure online)

3.2 Algorithm *A*

In this section, we present a version of the pseudo-code of the algorithm *A* which is proposed in [7]. The set of vertices with identical ordinates for convenience we call a *layer*.

Algorithm *A*.
Step 1. Set $t = 1$.
Send from all even vertices $(0, m - 1)$, $(2, m - 1)$, ..., $(n, m - 1)$ at the layer $m - 1$ the packets up to a distance 1 to the corresponding vertices at the layer m.

Send from all vertices at the layer $m - 3$ the packets down to a distance of 2, i.e., to the corresponding nodes at the layer $m - 5$.

Send from all even vertices $(0, 1)$, $(2, 1)$, ..., $(n, 1)$ at the layer 1 the packets up to a distance 1 to the corresponding vertices at the layer 2.

Step 2. Set $t = 2$.
Send from all odd vertices $(1, m - 1)$, $(3, m - 1)$, ..., $(n - 1, m - 1)$ at the layer $m - 1$ the packets up to a distance of 1 to the corresponding vertices at the layer m.

Send from all vertices at the layer $m - 5$ the packets down to a distance of 2, i.e. to the corresponding vertices at the layer $m - 7$.

Send from all odd vertices $(1, 1)$, $(3, 1)$, ..., $(n - 1, 1)$ at the layer 1 the packets up to a distance 1 to the corresponding nodes at the layer 2.

Step 3. Set $t = t + 1$ and $k = m - 2(t - 3)$.
Send from all vertices at the layer k the packets down to a distance 2 to the corresponding vertices at the layer $k - 2$.

Send from all vertices at the layer $k - 2t - 1$ the packets down to a distance of 2 to the corresponding nodes at the layer $k - 2t - 3$.

If $k - 2t - 1 > 3$, then go to Step 3.

Step 4. Set $t = t + 1$ and $k = m - 2(t - 3)$.

Send from all vertices at the layer k the packets down at distance 2 to the corresponding vertices at the layer $k - 2$.

Send from all even vertices $(0, 3)$, $(2, 3)$, ..., $(n, 3)$ at the layer 3 the packets down to a distance 1 to the corresponding vertices at the layer 2.

Step 5. Set $t = t + 1$ and $k = m - 2(t - 3)$.

Send from all vertices at the layer k the packets down at distance 2 to the corresponding vertices at the layer $k - 2$.

Send from all odd vertices $(1, 3)$, $(3, 3)$, ..., $(n-1, 3)$ at the layer 3 the packets down to a distance 2 to the corresponding nodes at the layer 2.

Step 6. Set $t = t + 1$ and $k = m - 2(t - 3)$.

Send from all vertices at the layer k the packets down at distance 2 to the corresponding vertices at the layer $k - 2$.

If $k > 2$, then go to Step 6.

After $t = m/2 + 2$ time rounds we have a linear graph with $n + 1$ vertices (vertex 0 is a BS), which consists of the set of vertices with the coordinates $(x, 0)$, $x = 0, 1, \ldots, n$, and wherein the distance between the adjacent vertices is 1. We enumerate the vertices of this graph, starting from the BS, by the natural numbers $0, 1, \ldots, n$.

Step 7. Set $t = t + 1$.

Send the packets from the odd vertices $n - 1, n - 7, \ldots, n - 6k - 1, \ldots, 1$, $k \leq (n - 2)/6$, at the distance 1 to the right, and the packets from the odd vertices $n - 3, n - 9, \ldots, n - 6k - 4, \ldots, 3$, $k \leq (n - 7)/6$, send left at a distance of 1.

Step 8. Set $t = t + 1$.

Send the packets from the odd vertices that were silent during the previous steps, to the left at a distance of 1.

Send a packet from the vertex n to the vertex $n - 2$.

Set $k = n$.

Step 9. Set $t = t + 1$ and $k = k - 2$.

Send a packet from the vertex k to the vertex $k - 2$.

If $k > 2$, then go to Step 9.

Stop.

Note that the vertical aggregation is carried out with time complexity $O(m)$, and the horizontal aggregation is carried out with time complexity $O(n)$. Therefore, the complexity of algorithm A is $O(n + m)$.

Algorithm A returns a schedule which length is $D + 3$. From Lemma 1 we know that the aggregation time cannot be less than $D + 3$. Hence, the following theorem holds.

Theorem 1. *Algorithm A builds an optimal conflict-free schedule of data aggregation in the unit square grid $(n + 1) \times (m + 1)$, with even n and m, with the BS at the origin $(0, 0)$ and a transmission distance equals 2, the length of which is $(n + m)/2 + 3$.*

4   Conclusion

In this paper we found an exact lower bound for the length of the conflict-free schedule of data aggregation in the unit square grid (unit disk graph in L_1 metric) when the transmission range of each vertex is 2. This lower bound coincides with the length of the schedule constructed by algorithm A in a polynomial time. Consequently, polynomial time algorithm A constructs an optimal schedule, the length of which is $L(n, m) = (n + m)/2 + 3$.

Acknowledgments. This work was supported by RFBR (grant 16-07-00552) and the Ministry of Education and Science of the Republic of Kazakhstan (project No. 0115PK00550).

References

1. Erzin, A., Plotnikov, R.: Using VNS for the optimal synthesis of the communication tree in wireless sensor networks. Electro. Notes Discrete Math. **47**, 21–28 (2015)
2. Erzin, A., Plotnikov, R., Mladenovic, N.: Variable neighborhood search variants for min-power symmetric connectivity problem. Comput. Oper. Res. **78**, 557–563 (2017)
3. Plotnikov, R., Erzin, A., Mladenovic, N.: Variable neighborhood search-based heuristics for min-power symmetric connectivity problem in wireless networks. In: Kochetov, Y., Khachay, M., Beresnev, V., Nurminski, E., Pardalos, P. (eds.) DOOR 2016. LNCS, vol. 9869, pp. 220–232. Springer, Cham (2016). doi:10.1007/978-3-319-44914-2_18
4. De Souza, E., Nikolaidis, I.: An exploration of aggregation convergecast scheduling. Ad Hoc Netw. **11**, 2391–2407 (2013)
5. Malhotra, B., Nikolaidis, I., Nascimento, M.A.: Aggregation convergecast scheduling in wireless sensor networks. Wirel. Netw. **17**, 319–335 (2011)
6. Cheng, C.-T., Tse, C.K., Lau, F.C.M.: A delay-aware data collection network structure for wireless sensor networks. IEEE Sens. J. **11**(3), 699–710 (2011)
7. Erzin, A., Pyatkin, A.: Convergecast scheduling problem in case of given aggregation tree. The complexity status and some special cases. In: 10th International Symposium on Communication Systems, Networks and Digital Signal Processing, Article 16, 6 p. IEEE-Xplore, Prague (2016)
8. Hansen, P., Kuplinsky, J., De Werra, D.: Mixed graph colorings. Math. Methods Oper. Res. **45**, 145–160 (1997)
9. Incel, O.D., Ghosh, A., Krishnamachari, B., Chintalapudi, K.: Fast data collection in tree-based wireless sensor networks. IEEE Trans. Mob. Comput. **11**(1), 86–99 (2012)
10. Wang, P., He, Y., Huang, L.: Near optimal scheduling of data aggregation in wireless sensor networks. Ad Hoc Netw. **11**, 1287–1296 (2013)
11. Li, H., Hua, Q.-S., Wu, C., Lau, F.C.M.: Minimum-Latency Aggregation Scheduling in Wireless Sensor Networks under Physical Interference Model. HKU CS Technical report TR-2010-07, Source: DBLP (2010)
12. Ghods, F., Yousefi, H., Mohammad, A., Hemmatyar, A., Movaghar, A.: MC-MLAS: multi-channel minimum latency aggregation scheduling in wireless sensor networks. Comput. Netw. **57**, 3812–3825 (2013)

13. Xu, X., Li, X.-Y., Mao, X., Tang, S., Wang, S.: A delay-efficient algorithm for data aggregation in multihop wireless sensor networks. IEEE Trans. Parallel Distrib. Syst. **22**, 163–175 (2011)
14. Slater, P.J., Cockayne, E.J., Hedetniemi, S.T.: Information dissemination in trees. SIAM J. Comput. **10**(4), 692–701 (1981)
15. Gagnon, J., Narayanan, L.: Minimum latency aggregation scheduling in wireless sensor networks. In: Gao, J., Efrat, A., Fekete, S.P., Zhang, Y. (eds.) ALGOSENSORS 2014. LNCS, vol. 8847, pp. 152–168. Springer, Heidelberg (2015). doi:10.1007/978-3-662-46018-4_10

A GRASP for the Minimum Cost SAT Problem

Giovanni Felici[1], Daniele Ferone[2], Paola Festa[2(✉)], Antonio Napoletano[2],
and Tommaso Pastore[2]

[1] Institute for Systems Analysis and Computer Science,
IASI-CNR, 00185 Rome, Italy
giovanni.felici@iasi.cnr.it
[2] Department of Mathematics and Applications, University of Napoli Federico II,
Compl. MSA, Via Cintia, 80126 Napoli, Italy
{daniele.ferone,paola.festa,antonio.napoletano2,tommaso.pastore}@unina.it

Abstract. A substantial connection exists between supervised learning from data represented in logic form and the solution of the Minimum Cost Satisfiability Problem (MinCostSAT). Methods based on such connection have been developed and successfully applied in many contexts. The deployment of such methods to large-scale learning problem is often hindered by the computational challenge of solving MinCostSAT, a problem well known to be NP-complete. In this paper, we propose a GRASP-based metaheuristic designed for such problem, that proves successful in leveraging the very distinctive structure of the MinCostSAT problems arising in supervised learning. The algorithm is equipped with an original stopping criterion based on probabilistic assumptions which results very effective for deciding when the search space has been explored enough. Although the proposed solver may approach MinCostSAT of general form, in this paper we limit our analysis to some instances that have been created from artificial supervised learning problems, and show that our method outperforms more general purpose well established solvers.

Keywords: Hard combinatorial optimization · SAT problems · GRASP · Local search · Probabilistic stopping criterion · Approximate solutions

1 Introduction

Propositional Satisfiability (SAT) and its derivations are well known problems in logic and optimization, and belong to the special class of NP-complete problems [11]. Beside playing a special role in the theory of complexity, they often arise in applications, where they are used to model complex problems whose solution is of particular interest.

One such case surfaces in logic supervised learning. Here, we have a dataset of samples, each represented by a finite number of logic variables, and a particular extension of the classic SAT problem - the Minimum Cost Satisfiability Problem

© Springer International Publishing AG 2017
R. Battiti et al. (Eds.): LION 2017, LNCS 10556, pp. 64–78, 2017.
https://doi.org/10.1007/978-3-319-69404-7_5

(MinCostSAT) - can be used to iteratively identify the different clauses of a compact formula in Disjunctive Normal Form (DNF) that possesses the desirable property of assuming the value *True* on one specific subset of the dataset and the value *False* on the rest.

The use of MinCostSAT for learning propositional formula from data is described in [6,25], and its description is beyond the scope of this paper. Suffice it to say, however, that there are several reasons that motivate the validity of such an approach to supervised learning, and that it proved to be very effective in several applications, particularly on those derived from biological and medical data analysis [1,3,4,27–29].

One of the main drawbacks of the approach described in [6] lies in the difficulty of solving MinCostSAT exactly or with an appropriate quality level. Such drawback is becoming more and more evident as, in the era of Big Data, the size of the datasets that one is to analyze steadily increases. Feature selection techniques may be used to reduce the space in which the samples are represented (one such method specifically designed for data in logic form is described in [2]).

While the literature proposes both exact approaches ([10,20,25,26]) and heuristics ([23]), still the need for efficient MinCostSAT solvers remains, and in particular for solvers that may take advantage of the specific structure of those MinCostSAT representing supervised learning problems.

In this paper, we try to fill this gap and propose a GRASP-based metaheuristic designed to solve MinCostSAT problems that arise in supervised learning. In doing so, we developed a new probabilistic stopping criterion that proves to be very effective in limiting the exploration of the solution space - whose explosion is a frequent problem in metaheuristic approaches. The method has been tested on several instances derived from artificial supervised problems in logic form, and successfully compared with four established solvers in the literature (**Z3** from Microsoft Research [19], **bsolo** [12], **MiniSat+** [5], and **PWBO** [15–17]).

The paper is organized as follows. Section 2 presents a very simple formulation of the problem; Sect. 3 describes the architecture of the GRASP algorithm; Sect. 4 is devoted to the explanation of the probabilistic stopping criterion, while experimental results and conclusions are covered in Sects. 5 and 6, respectively.

2   Mathematical Formulation of the Problem

The Minimum Cost SAT Problem (MinCostSAT) - also known as Binate Covering Problem - is a special case of the well known Boolean Satisfiability Problem. Given a set of n boolean variables $X = \{x_1, \ldots, x_n\}$, a non-negative cost function $c : X \mapsto \mathbb{R}^+$ such that $c(x_i) = c_i \geq 0$, $i = 1, \ldots, n$, and a boolean formula $\varphi(X)$ expressed in CNF, the MinCostSAT problem consists in finding a truth assignment for the variables in X such that the total cost is minimized while

$\varphi(X)$ is satisfied. Accordingly, the mathematical formulation of the problem is given as follows:

$$(\text{MinCostSAT}) \quad z = \min \sum_{i=1}^{n} c_i x_i$$

subject to:

$$\varphi(X) = 1,$$
$$x_i \in \{0,1\}, \qquad \forall i = 1, \dots, n.$$

It is easy to see that a general SAT problem can be reduced to a MinCostSAT problem whose costs c_i are all equal to 0. Furthermore, the decision version of the MinCostSAT problem is NP-complete [10]. While the boolean satisfiability problem is an evergreen in the landscape of scientific literature, MinCostSAT has received less attention.

3   A GRASP for MinCostSAT

GRASP is a well established iterative multistart metaheuristic method for difficult combinatorial optimization problems [7]. The reader can refer to [8,9] for a study of a generic GRASP metaheuristic framework and its applications.

Such method is characterized by the repeated execution of two main phases: a construction and a local search phase. The construction phase iteratively adds one component at a time to the current solution under construction. At each iteration, an element is randomly selected from a *restricted candidate list* (RCL), composed by the best candidates, according to some greedy function that measures the myopic benefit of selecting each element.

Once a complete solution is obtained, the local search procedure attempts to improve it by producing a locally optimal solution with respect to some suitably defined neighborhood structure. Construction and local search phases are repeatedly applied. The best locally optimal solution found is returned as final result. Figure 1 depicts the pseudo-code of a generic GRASP for a minimization problem.

In order to allow a better and easier implementation of our GRASP, we treat the MinCostSAT as particular covering problem with incompatibility constraints. Indeed, we consider each literal $(x, \neg x)$ as a separate element, and a clause is covered if at least one literal in the clause is contained in the solution. The algorithm tries to add literals to the solution in order to cover all the clauses and, once the literal x is added to the solution, then the literal $\neg x$ cannot be inserted (and vice versa). Therefore, if the literal x is in solution, the variable x is assigned to true and all clauses covered by x are satisfied. Similarly, if the literal $\neg x$ is in solution, the variable x is assigned to false, and clauses containing $\neg x$ are satisfied.

The construction phase adds a literal at a time, until all clauses are covered or no more literals can be assigned. At each iteration of the construction, if a

```
 1  Algorithm GRASP(β)
 2  │   x* ← Nil ;
 3  │   z(x*) ← +∞ ;
 4  │   while a stopping criterion is not satisfied do
 5  │   │   Build a greedy randomized solution x ;
 6  │   │   x ← LocalSearch(x) ;
 7  │   │   if z(x) < z(x*) then
 8  │   │   │   x* ← x ;
 9  │   │   │   z(x*) ← z(x) ;
10  │   return x*
```

Fig. 1. A generic GRASP for a minimization problem.

clause can be covered only by a single literal x – due to the choices made in previous iterations – then x is selected to cover the clause. Otherwise, if there are not clauses covered by only a single literal, the addition of literals to the solution takes place according to a penalty function $penalty(\cdot)$, which greedily sorts all the candidates literals, as described below.

Let $cr(x)$ be the number of clauses yet to be covered that contain x. We then compute:

$$penalty(x) = \frac{c(x) + cr(\neg x)}{cr(x)}. \tag{1}$$

This penalty function evaluates both the benefits and disadvantages that can result from the choice of a literal rather than another. The benefits are proportional to the number of uncovered clauses that the chosen literal could cover, while the disadvantages are related to both the cost of the literal and the number of uncovered clauses that could be covered by $\neg x$. The smaller the penalty function $penalty(x)$, the more favorable is the literal x. According to the GRASP scheme, the selection of the literal to add is not purely greedy, but a Restricted Candidate List (RCL) is created with the most promising elements, and an element is randomly selected among them. Concerning the tuning of the parameter β, whose task is to adjust the greediness of the construction phase, we performed an extensive analysis over a set of ten different random seeds. Such testing showed how a nearly totally greedy setup ($\beta = 0.1$) allowed the algorithm to attain better quality solutions in smallest running times.

Let $|\mathcal{C}| = m$ be the number of clauses. Since $|X| = 2n$, in the worst case scenario the loop while (Fig. 2, line 3) in the construct-solution function pseudo-coded in Fig. 2 runs m times and in each run the most expensive operation consists in the construction of the RCL. Therefore, the total computational complexity is $\mathcal{O}(m \cdot n)$.

In the local search phase, the algorithm uses a 1-exchange (flip) neighborhood function, where two solutions are neighbors if and only if they differ in at most one component. Therefore, if there exists a better solution $\bar{x}$ that differs only for one literal from the current solution x, the current solution s is set to $\bar{s}$ and the procedure restarts. If such a solution does not exists, the procedure ends and returns the

```
1  Function construct-solution(C, X, β)
       /* C is the set of uncovered clauses                            */
       /* X is the set of candidate literals                           */
2      s ← ∅ ;
3      while C ≠ ∅ do
4          if c ∈ C can be covered only by x ∈ X then
5              s ← s ∪ {x};
6              X ← X \ {x, ¬x};
7              C ← C \ {c̄ | x ∈ c̄};
8          else
9              compute penalty(x) ∀ x ∈ X;
10             th ← min{penalty(x)} + β(max{penalty(x)} − min{penalty(x)}) ;
                   x∈X                    x∈X              x∈X
11             RCL ← { x ∈ X : penalty(x) ≤ th } ;
12             x̂ ← rand(RCL) ;
13             s ← s ∪ {x̂};
14             X ← X \ {x̂, ¬x̂};
15             C ← C \ {c̄ | x̂ ∈ c̄};
16     return s
```

Fig. 2. Pseudo-code of the GRASP construction phase.

current solution s. The local search procedure would also re-establish feasibility if the current solution is not covering all clauses of $\varphi(X)$. During our experimentation we tested the one-flip local search using two different neighborhood exploration strategies: first improvement and best improvement. With the former strategy, the current solution is replaced by the first improving solution found in its neighborhood; such improving solution is then used as a starting point for the next local exploration. On the other hand, with the best improvement strategy, the current solution x is replaced with the solution $\bar{x} \in \mathcal{N}(x)$ corresponding to the greatest improvement in terms of objective function value; $\bar{x}$ is then used as a starting point for the next local exploration. Our results showed how the first improvement strategy is slightly faster, as expected, while attaining solution of the same quality of those given by the best improvement strategy. Based on this rationale, we selected first improvement as exploration strategy in our testing phase.

4   Probabilistic Stopping Rule

Although being very fast and effective, most metaheuristics present a shortcoming in the effectiveness of their stopping rule. Usually, the stopping criterion is based on a bound on the maximum number of iterations, a limit on total execution time, or a given maximum number of consecutive iterations without improvement. In this paper, we propose a probabilistic stopping criterion, inspired by [21].

The stopping criterion is composed of two phases, described in the next subsections. It can be sketched as follows. First, let $\mathcal{X}$ be a random variable representing the value of a solution obtained at the end of a generic GRASP iteration. In the first phase – the `fitting-data` procedure – the probability distribution $f_{\mathcal{X}}(\cdot)$ of $\mathcal{X}$ is estimated, while during the second phase – `improve-probability` procedure – the probability of obtaining an improvement of the current solution value is computed. Then, accordingly to a threshold, the algorithm either stops or continues its execution.

4.1 Fitting Data Procedure

The first step to be performed in order to properly represent the random variable $\mathcal{X}$ with a theoretical distribution consists in an empirical observation of the algorithm. Examining the objective function values obtained at the end of each iteration, and counting up the respective frequencies, it is possible to select a promising parametric family of distributions. Afterwards, by means of a Maximum Likelihood Estimation (MLE), see for example [22], a choice is made regarding the parameters characterizing the best fitting distribution of the chosen family.

In order to carry on the empirical analysis of the objective function value obtained in a generic iteration of GRASP, which will result in a first guess concerning the parametric family of distributions, we represent the data obtained in the following way.

Let $\mathcal{I}$ be a fixed instance and $\mathcal{F}$ the set of solutions obtained by the algorithm up to the current iteration, and let $\mathcal{Z}$ be the multiset of the objective function values associated to $\mathcal{F}$. Since we are dealing with a minimum optimization problem, it is harder to find good quality solutions, whose cost is small in term of objective function, rather than expensive ones. This means that during the analysis of the values in $\mathcal{Z}$ we expect to find an higher concentration of elements between the mean value μ and the $\max(\mathcal{Z})$. In order to represent the values in $\mathcal{Z}$ with a positive distribution function, that presents higher frequencies in a right neighborhood of zero and a single tail which decays for growing values of the random variable, we perform a reflection on the data in $\mathcal{Z}$ by means the following transformation:

$$\bar{z} = \max(\mathcal{Z}) - z, \ \forall \, z \in \mathcal{Z}. \tag{2}$$

The behaviour of the distribution of $\bar{z}$ in our instances has then a very recognizable behaviour. A representative of such distribution is given in Fig. 3 where the histogram of absolute and relative frequencies of $\bar{z}$ are plotted. It is easy to observe how the gamma distribution family represents a reasonable educated guess for our random variable.

Once we have chosen the gamma distribution family, we estimate its parameters performing a MLE. In order to accomplish the estimation, we collect an initial sample of solution values and on-line execute a function, developed in R (whose pseudo-code is reported in Fig. 4), which carries out the MLE and returns the characteristic shape and scale parameters, k and θ, which pinpoint the specific distribution of the gamma family that best suits the data.

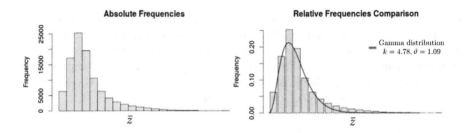

Fig. 3. Empirical analysis of frequencies of the solutions.

```
1  Function fitting-data(Z̄)
      /* Z̄ is the initial sample of the objective function values     */
2     foreach z ∈ Z̄ do
3       |  z = max(Z̄) − z;
4     {k, θ} ← MLE(Z, "gamma");
5     return {k, θ}
```

Fig. 4. Fitting data procedure.

4.2   Improve Probability Procedure

The second phase of the probabilistic stop takes place once that the probability distribution function of the random variable $\mathcal{X}$, $f_{\mathcal{X}}(\cdot)$ has been estimated.

Let $\hat{z}$ be the best solution value found so far. It is possible to compute an approximation of the probability of improving the incumbent solution by

$$p = 1 - \int_0^{\max(\mathcal{Z}) - \hat{z}} f_{\mathcal{X}}(t) \, \mathrm{d}t. \tag{3}$$

The result of the procedures `fitting-data` and `improve-probability` consists in an estimate of the probability of incurring in an improving solution in the next iterations. Such probability is compared with a user-defined threshold, α, and if $p < \alpha$ the algorithm stops. More specifically, in our implementation the stopping criterion works as follows:

(a) let q be an user-defined positive integer, and let $\bar{\mathcal{Z}}$ be the sample of initial solution values obtained by the GRASP in the first q iterations;
(b) call the `fitting-data` procedure, whose input is $\bar{\mathcal{Z}}$ is called one-off to estimate shape and scale parameters, k and θ, of the best fitting gamma distribution;
(c) every time that an incumbent is improved, `improve-probability` procedure (pseudo-code in Fig. 5) is performed and the probability p of further improvements is computed. If p is less than or equal to α the stopping criterion is satisfied. For the purpose of determining p, we have used the function `pgamma` of R package `stats`.

```
1  Function improve-probability(k, θ, z*)
      /* z* is the value of the incumbent                    */
2      p ← pgamma(z*, shape = k, scale = θ);
3      return p
```

Fig. 5. Improve probability procedure.

5 Results

Our GRASP has been implemented in C++ and compiled with `gcc 5.4.0` with the flag `-std=c++14`. All tests were run on a cluster of nodes, connected by 10 Gigabit Infiniband technology, each of them with two processors Intel Xeon E5-4610v2@2.30 GHz.

We performed two different kinds of experimental tests. In the first one, we compared the algorithm with different solvers proposed in literature, without use of probabilistic stop. In particular, we used: **Z3** solver freely available from Microsoft Research [19], **bsolo** solver kindly provided by its authors [12], the **MiniSat+** [5] available at web page http://minisat.se/, and **PWBO** available at web page http://sat.inesc-id.pt/pwbo/index.html. The aim of this first set of computational experiment is the evaluation of the quality of the solutions obtained by our algorithm within a certain time limit. More specifically, the stopping criterion for GRASP, **bsolo** and **PWBO** is a time limit of 3 h, for **Z3** and **MiniSat+** is the reaching of an optimal solution.

Z3 is a satisfiability modulo theories (SMT) solver from Microsoft Research that generalizes boolean satisfiability by adding equality reasoning, arithmetic, fixed-size bit-vectors, arrays, quantifiers, and other useful first-order theories. **Z3** integrates modern backtracking-based search algorithm for solving the CNF-SAT problem, namely DPLL-algorithm; in addition it provides a standard search pruning methods, such as two-watching literals, lemma learning using conflict clauses, phase caching for guiding case splits, and performs non-chronological backtracking.

bsolo [12,13] is an algorithmic scheme resulting from the integration of several features from SAT-algorithms in a branch-and-bound procedure to solve the binate covering problem. It incorporates the most important characteristics of a branch-and-bound and SAT algorithm, bounding and reduction techniques for the former, and search pruning techniques for the latter. In particular, it incorporates the search pruning techniques of the Generic seaRch Algorithm-SAT proposed in [14].

MiniSat+ [5,24] is a minimalistic implementation of a Chaff-like SAT solver based on the two-literal watch scheme for fast boolean constraint propagation [18], and conflict clauses driven learning [14]. In fact the **MiniSat** solver provides a mechanism which allows to minimize the clauses conflicts.

PWBO [15–17] is a Parallel Weighted Boolean Optimization Solver. The algorithm uses two threads in order to simultaneously estimate a lower and an upper bound, by means of an unsatisfiability-based procedure and a linear search, respectively. Moreover, learned clauses are shared between threads during the search.

For testing, we have initially considered the datasets used to test feature selection methods in [2], where an extensive description of the generation procedure can be found. Such testbed is composed of 4 types of problems (A,B,C,D), for each of which 10 random repetitions have been generated. Problems of type A and B are of moderate size (100 positive examples, 100 negative examples, 100 logic features), but differ in the form of the formula used to classify the samples into the positive and negative classes (the formula being more complex for B than for A). Problems of type C and D are much larger (200 positive examples, 200 negative examples, 2500 logic features), and D has a more complex generating logic formula than C.

Table 1 reports both the value of the solutions and the time needed to achieve them (in the case of GRASP, it is average over ten runs).[1] For problems of moderate size (A and B), the results show that GRASP finds an optimal solution whenever one of the exact solvers converges. Moreover, GRASP is very fast in finding the optimal solution, although here it runs the full allotted time before stopping the search. For larger instances (C and D), GRASP always provides a solution within the bounds, while two of the other tested solvers fail in doing so and the two that are successful (**bsolo, PWBO**) always obtain values of inferior quality.

The second set of experimental tests was performed for the purpose of evaluating the impact of the probabilistic stopping rule. In order to do so, we have chosen five different values for threshold α, two distinct sizes for the set $\bar{Z}$ of initial solution, and executed GRASP using ten different random seeds imposing a maximum number of iterations as stopping criterion. This experimental setup yielded for each instance, and for each threshold value, 20 executions of the algorithm. About such runs, the data collected were: the number of executions in which the probabilistic stopping rule was verified ("stops"), the mean value of the objective function of the best solution found (μ_z), and the average computational time needed (μ_t). To carry out the evaluation of the stopping rule, we executed the algorithm only using the maximum number of iterations as stopping criterion for each instance and for each random seed. About this second setup, the data collected are, as for the first one, the objective function of the best solution found ($\mu_{\hat{z}}$) and the average computational time needed ($\mu_{\hat{t}}$). For the sake of comparison, we considered the percentage gaps between the results collected with and without the probabilistic stopping rule. The second set of experimental tests is summarized in Table 2 and in Fig. 7. For each pair of columns (3,4), (6,7), (9,10), (12, 13), the table reports the percentage of loss in terms of objective function value and the percentage of gain in terms of computation times using the probabilistic stopping criterion, respectively. The analysis of the gaps shows how the probabilistic stop yields little or no changes in the objective function value while bringing dramatic improvements in the total computational time.

---

[1] For missing values, the algorithm was not able to find the optimal solution in 24 h.

Table 1. Comparison between GRASP and other solvers.

Inst.	GRASP Time	GRASP Value	Z3 Time	Z3 Value	bsolo Time	bsolo Value	MiniSat+ Time	MiniSat+ Value	pwbo-2T Time	pwbo-2T Value
A1	6.56	78.0	10767.75	78.0	0.09	78.0	0.19	78.0	0.03	78.0
A2	1.71	71.0	611.29	71.0	109.59	71.0	75.46	71.0	121.58	71.0
A3	0.64	65.0	49.75	65.0	598.71	65.0	10.22	65.0	5.14	65.0
A4	0.18	58.0	4.00	58.0	205.77	58.0	137.82	58.0	56.64	58.0
A5	0.29	66.0	69.31	66.0	331.51	66.0	9.03	66.0	30.64	66.0
A6	21.97	77.0	5500.17	77.0	328.93	77.0	32.82	77.0	359.97	77.0
A7	0.21	63.0	30.57	63.0	134.20	63.0	19.34	63.0	24.12	63.0
A8	0.25	62.0	6.57	62.0	307.69	62.0	16.84	62.0	11.81	62.0
A9	12.79	72.0	1088.83	72.0	3118.32	72.0	288.76	72.0	208.63	72.0
A10	0.33	66.0	42.23	66.0	62.03	66.0	37.75	66.0	1.81	66.0
B1	6.17	78.0	8600.60	78.0	304.36	78.0	121.25	78.0	20.01	78.0
B2	493.56	80.0	18789.20	80.0	4107.41	80.0	48.21	80.0	823.66	80.0
B3	205.37	77.0	7037.00	77.0	515.25	77.0	132.74	77.0	1.69	77.0
B4	38.26	77.0	7762.03	77.0	376.00	77.0	119.49	77.0	1462.18	77.0
B5	19.89	79.0	15785.35	79.0	3025.26	79.0	214.52	79.0	45.05	79.0
B6	28.45	76.0	4087.14	76.0	394.45	76.0	162.31	76.0	83.72	76.0
B7	129.76	78.0	10114.84	78.0	490.30	78.0	266.25	78.0	455.92	81.0*
B8	44.42	76.0	5186.45	76.0	5821.19	76.0	1319.21	76.0	259.07	76.0
B9	152.77	80.0	14802.00	80.0	5216.95	82.0	36.28	80.0	557.02	80.0
B10	7.55	73.0	1632.87	73.0	760.28	79.0	370.30	73.0	72.09	73.0
C1	366.24	132.0	86400	–	8616.25	178.0*	86400	–	343.38	178.0*
C2	543.11	131.0	86400	–	323.90	150.0*	86400	–	1742.68	174.0*
C3	5883.6	174.1	86400	–	6166.06	177.0*	86400	–	421.64	177.0*
C4	4507.63	176.3	86400	–	6209.69	178.0*	86400	–	2443.20	177.0*
C5	5707.51	171.2	86400	–	314.18	179.0*	86400	–	67.73	178.0*
C6	6269.91	172.1	86400	–	1547.90	177.0*	86400	–	2188.82	177.0*
C7	6193.15	165.9	86400	–	794.90	177.0*	86400	–	730.36	178.0*
C8	596.58	137.0	86400	–	306.27	169.0*	86400	–	837.71	178.0*
C9	466.3	136.0	86400	–	433.32	179.0*	86400	–	3455.92	178.0*
C10	938.54	136.0	86400	–	3703.94	180.0*	86400	–	4617.24	179.0*
D1	3801.61	145.3	86400	–	307.25	175.0*	86400	–	127.69	180.0*
D2	2040.64	139.0	86400	–	7704.92	177.0*	86400	–	2327.23	177.0*
D3	1742.78	143.0	86400	–	309.10	145.0*	86400	–	345.97	178.0*
D4	1741.95	135.0	86400	–	6457.79	177.0*	86400	–	295.76	178.0*
D5	1506.22	134.0	86400	–	6283.27	178.0*	86400	–	238.81	173.0*
D6	1960.87	144.5	86400	–	309.11	173.0*	86400	–	2413.42	178.0*
D7	1544.42	143.0	86400	–	4378.73	179.0*	86400	–	1250.07	178.0*
D8	1756.15	144.0	86400	–	1214.97	179.0*	86400	–	248.85	179.0*
D9	2779.38	137.0	86400	–	303.11	146.0*	86400	–	4.73	179.0*
D10	5896.86	149.0	86400	–	319.45	170.0*	86400	–	1239.93	176.0*
Y	16.05	0.0	0.73	0.0	9411.06	974*	1.96	0	0.23	0.0

*sub-optimal solution
– no optimal solution found in 24 h

Table 2. Probabilistic stop on instances A, B, C and D.

threshold α	inst	%-gap z	%-gap $t(s)$	inst	%-gap z	%-gap $t(s)$	inst	%-gap z	%-gap $t(s)$	inst	%gap z	%gap $t(s)$
$5 \cdot 10^{-2}$	A1	-0.0	83.1	B1	-2.1	87.1	C1	-6.6	76.0	D1	-5.0	79.3
$1 \cdot 10^{-2}$	A1	-0.0	83.1	B1	-2.1	87.1	C1	-6.6	76.1	D1	-5.0	79.3
$5 \cdot 10^{-3}$	A1	-0.0	83.0	B1	-2.1	87.1	C1	-5.0	74.8	D1	-4.9	78.7
$1 \cdot 10^{-3}$	A1	-0.0	2.5	B1	-2.1	87.1	C1	-3.8	70.7	D1	-1.7	58.9
$5 \cdot 10^{-4}$	A1	-0.0	-15.3	B1	-2.1	87.2	C1	-2.6	70.2	D1	-1.2	49.0
$1 \cdot 10^{-4}$	A1	-0.0	-11.8	B1	-0.5	86.1	C1	-1.3	52.5	D1	-0.2	31.6
$5 \cdot 10^{-2}$	A2	-0.0	84.0	B2	-0.7	87.0	C2	-3.5	76.0	D2	-0.1	79.1
$1 \cdot 10^{-2}$	A2	-0.0	84.1	B2	-0.7	87.0	C2	-3.5	76.2	D2	-0.1	79.1
$5 \cdot 10^{-3}$	A2	-0.0	83.6	B2	-0.7	86.9	C2	-3.5	76.7	D2	-0.1	79.1
$1 \cdot 10^{-3}$	A2	-0.0	84.0	B2	-0.7	87.0	C2	-1.9	76.4	D2	-0.1	79.1
$5 \cdot 10^{-4}$	A2	-0.0	84.9	B2	-0.7	87.0	C2	-1.9	76.1	D2	-0.1	75.7
$1 \cdot 10^{-4}$	A2	-0.0	57.9	B2	-0.1	71.3	C2	-1.9	65.2	D2	-0.1	53.5
$5 \cdot 10^{-2}$	A3	-0.0	83.4	B3	-2.7	87.0	C3	-2.7	76.3	D3	-1.8	75.2
$1 \cdot 10^{-2}$	A3	-0.0	83.8	B3	-2.7	87.0	C3	-2.1	73.0	D3	-1.8	75.2
$5 \cdot 10^{-3}$	A3	-0.0	82.9	B3	-2.7	87.0	C3	-1.7	68.0	D3	-1.7	74.8
$1 \cdot 10^{-3}$	A3	-0.0	8.3	B3	-2.6	86.6	C3	-0.6	40.9	D3	-0.8	38.5
$5 \cdot 10^{-4}$	A3	-0.0	-1.6	B3	-2.0	84.1	C3	-0.0	28.3	D3	-0.5	19.1
$1 \cdot 10^{-4}$	A3	-0.0	-6.8	B3	-0.7	58.4	C3	-0.0	9.9	D3	-0.3	14.5
$5 \cdot 10^{-2}$	A4	-0.0	86.4	B4	-2.3	86.9	C4	-4.3	78.8	D4	-2.2	75.0
$1 \cdot 10^{-2}$	A4	-0.0	6.4	B4	-2.3	86.9	C4	-3.3	68.0	D4	-2.2	70.9
$5 \cdot 10^{-3}$	A4	-0.0	3.5	B4	-2.3	86.9	C4	-2.2	63.9	D4	-2.2	66.8
$1 \cdot 10^{-3}$	A4	-0.0	1.4	B4	-2.3	87.0	C4	-1.0	51.2	D4	-2.0	41.0
$5 \cdot 10^{-4}$	A4	-0.0	5.6	B4	-2.3	86.9	C4	-0.8	48.6	D4	-1.2	29.1
$1 \cdot 10^{-4}$	A4	-0.0	6.4	B4	-0.6	74.8	C4	-0.3	38.1	D4	-1.2	18.9
$5 \cdot 10^{-2}$	A5	-0.0	87.6	B5	-0.7	86.6	C5	-2.6	79.7	D5	-5.6	75.2
$1 \cdot 10^{-2}$	A5	-0.0	12.2	B5	-0.7	86.6	C5	-1.5	71.5	D5	-4.9	75.1
$5 \cdot 10^{-3}$	A5	-0.0	12.5	B5	-0.7	86.6	C5	-0.4	68.1	D5	-4.9	75.2
$1 \cdot 10^{-3}$	A5	-0.0	12.4	B5	-0.7	86.6	C5	-0.2	53.2	D5	-4.7	67.6
$5 \cdot 10^{-4}$	A5	-0.0	12.3	B5	-0.6	86.3	C5	-0.0	46.8	D5	-3.8	60.0
$1 \cdot 10^{-4}$	A5	-0.0	12.5	B5	-0.1	19.0	C5	-0.0	33.2	D5	-3.3	49.8
$5 \cdot 10^{-2}$	A6	-0.9	87.2	B6	-0.8	86.6	C6	-3.3	79.9	D6	-7.9	76.0
$1 \cdot 10^{-2}$	A6	-0.9	87.2	B6	-0.8	86.6	C6	-2.0	70.5	D6	-5.9	74.8
$5 \cdot 10^{-3}$	A6	-0.9	87.2	B6	-0.8	86.6	C6	-1.3	65.4	D6	-5.0	74.0
$1 \cdot 10^{-3}$	A6	-0.8	87.1	B6	-0.7	86.3	C6	-0.2	49.6	D6	-2.5	71.1
$5 \cdot 10^{-4}$	A6	-0.5	86.8	B6	-0.1	72.1	C6	-0.2	39.9	D6	-2.5	71.2
$1 \cdot 10^{-4}$	A6	-0.0	66.1	B6	-0.0	7.6	C6	-0.0	36.6	D6	-2.5	67.3
$5 \cdot 10^{-2}$	A7	-0.0	87.5	B7	-3.1	86.2	C7	-3.8	74.4	D7	-6.5	75.5
$1 \cdot 10^{-2}$	A7	-0.0	11.7	B7	-3.1	86.2	C7	-2.4	65.7	D7	-5.3	72.1
$5 \cdot 10^{-3}$	A7	-0.0	11.7	B7	-3.1	86.2	C7	-1.9	60.7	D7	-4.0	68.0
$1 \cdot 10^{-3}$	A7	-0.0	11.3	B7	-3.1	86.2	C7	-0.8	43.0	D7	-2.8	61.2
$5 \cdot 10^{-4}$	A7	-0.0	11.5	B7	-3.0	86.0	C7	-0.0	36.4	D7	-2.2	60.6
$1 \cdot 10^{-4}$	A7	-0.0	11.4	B7	-0.8	75.8	C7	-0.0	14.0	D7	-2.2	57.4
$5 \cdot 10^{-2}$	A8	-0.0	88.1	B8	-1.5	86.7	C8	-3.6	73.9	D8	-11.5	76.2
$1 \cdot 10^{-2}$	A8	-0.0	88.1	B8	-1.5	86.7	C8	-3.3	74.7	D8	-6.7	73.4
$5 \cdot 10^{-3}$	A8	-0.0	88.1	B8	-1.5	86.7	C8	-3.3	74.4	D8	-6.7	73.4
$1 \cdot 10^{-3}$	A8	-0.0	16.4	B8	-1.2	86.4	C8	-3.3	73.7	D8	-4.4	68.2
$5 \cdot 10^{-4}$	A8	-0.0	16.6	B8	-0.8	74.5	C8	-3.2	65.6	D8	-3.4	67.9
$1 \cdot 10^{-4}$	A8	-0.0	16.5	B8	-0.0	7.8	C8	-2.2	60.5	D8	-2.4	64.9
$5 \cdot 10^{-2}$	A9	-0.0	88.0	B9	-1.9	85.9	C9	-4.1	75.3	D9	-2.1	75.2
$1 \cdot 10^{-2}$	A9	-0.0	88.0	B9	-1.9	85.9	C9	-2.7	74.8	D9	-2.1	75.2
$5 \cdot 10^{-3}$	A9	-0.0	88.0	B9	-1.9	85.9	C9	-1.1	74.4	D9	-2.1	75.2
$1 \cdot 10^{-3}$	A9	-0.0	16.0	B9	-1.9	85.9	C9	-1.1	66.6	D9	-2.1	75.2
$5 \cdot 10^{-4}$	A9	-0.0	16.0	B9	-1.7	84.9	C9	-0.2	56.5	D9	-2.1	67.7
$1 \cdot 10^{-4}$	A9	-0.0	15.9	B9	-0.5	45.2	C9	-0.2	55.7	D9	-1.9	60.4
$5 \cdot 10^{-2}$	A10	-0.0	83.3	B10	-0.3	87.7	C10	-0.4	76.3	D10	-7.1	73.7
$1 \cdot 10^{-2}$	A10	-0.0	75.4	B10	-0.3	87.6	C10	-0.4	76.2	D10	-6.9	73.8
$5 \cdot 10^{-3}$	A10	-0.0	0.5	B10	-0.3	87.7	C10	-0.3	67.9	D10	-6.4	73.1
$1 \cdot 10^{-3}$	A10	-0.0	-5.4	B10	-0.3	87.6	C10	-0.3	48.0	D10	-4.5	62.0
$5 \cdot 10^{-4}$	A10	-0.0	-4.8	B10	-0.0	87.4	C10	-0.3	48.0	D10	-4.3	57.3
$1 \cdot 10^{-4}$	A10	-0.0	-4.7	B10	-0.0	35.7	C10	-0.2	27.0	D10	-3.1	38.6

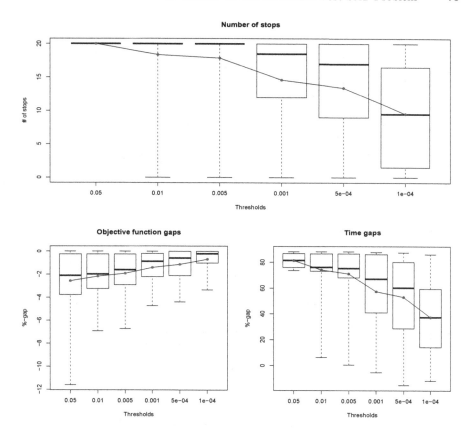

Fig. 6. Experimental evaluation of the probabilistic stopping rule. In each boxplot, the boxes represent the first and the second quartile; solid line represent median while dotted vertical line is the full variation range. Plots vary for each threshold α. The dots connected by a line represent the mean values.

The experimental evaluation of the probabilistic stop is summarized in the three distinct boxplots of Fig. 6. Each boxplot reports a sensible information related to the impact of the probabilistic stop, namely: the number of times the probabilistic criterion has been satisfied, the gaps in the objective function values, and the gaps in the computation times obtained comparing the solutions obtained with and without the use of the probabilistic stopping rule. Such information are collected, for each instance, as averages of the data obtained over 20 trials in the experimental setup described above. The first boxplot depicts the number of total stops recorded for different values of threshold α. Larger values of α, indeed, yield a less coercive stopping rule, thus recording an higher number of stops. Anyhow, even for the smallest, most conservative α, the average number of stops recorded is close to 50% of the tests performed. In the second boxplot, the objective function gap is reported. Such gap quantifies the qualitative worsening in quality of the solutions obtained with the probabilistic stopping rule.

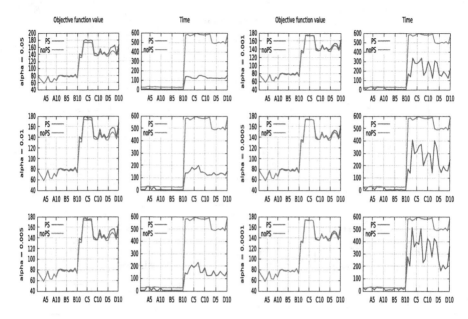

Fig. 7. Comparison of objective function values and computation times obtained with and without probabilistic stopping rule for different threshold values.

The gaps yielded show how even with the highest α, the difference in solution quality is extremely small, with a single minimum of -11.5% for the instance D8, and a very promising average gap, slightly below -2%. As expected, decreasing the α values the solutions obtained with and without the probabilistic stopping rule will align with each other, and the negative gaps will accordingly grow up to approximately -1%. The third boxplot shows the gaps obtained in the computation times. The analysis of such gaps is the key to realistically appraise the actual benefit provided by the use of the probabilistic stopping rule. Observing the results reported, it is possible to note how even in the case of the smallest threshold, i.e., using the most strict probabilistic stopping criterion, the stops recorded are such that an average time discount close to the 40% is encountered. A more direct display of this time gaps can be obtained straightly considering the total time discount in seconds: with the smallest α we have experienced a time discount of 4847.6 s over the 11595.9 total seconds needed for the execution without the probabilistic stop. Analyzing in the same fashion the values obtained under the largest threshold, we observed an excellent average discount just over 80%, which quantified in seconds amounts to an astonishing total discount of 8919.64 s over the 11595.9 total seconds registered for the execution without the probabilistic stop.

6   Conclusions

In this paper, we have investigated a strategy for a GRASP heuristic that solves large sized MinCostSAT. The method adopts a straight-forward implementation of the main ingredients of the heuristic, but proposes a new probabilistic stopping rule. Experimental results show that, for instances belonging to particular class of MinCostSAT problems, the method performs very well and the new stopping rule provides a very effective way to reduce the number of iterations of the algorithm without observing any significant decay in the value of the objective function.

The work presented has to be considered preliminary, but it clearly indicates several research directions that we intend to pursue: the refinement of the dynamic estimate of the probability distribution of the solutions found by the algorithm, the comparative testing of instances of larger size, and the extension to other classes of problems. Last, but not least, attention will be directed toward the incorporation of the proposed heuristic into methods that are specifically designed to extract logic formulas from data and to the test of the performances of the proposed algorithm in this setting.

Acknowledgements. This work has been realized thanks to the use of the S.Co.P.E. computing infrastructure at the University of Napoli FEDERICO II.

References

1. Arisi, I., D'Onofrio, M., Brandi, R., Felsani, A., Capsoni, S., Drovandi, G., Felici, G., Weitschek, E., Bertolazzi, P., Cattaneo, A.: Gene expression biomarkers in the brain of a mouse model for alzheimer's disease: Mining of microarray data by logic classification and feature selection. J. Alzheimer's Dis. **24**(4), 721–738 (2011)
2. Bertolazzi, P., Felici, G., Festa, P., Fiscon, G., Weitschek, E.: Integer programming models for feature selection: new extensions and a randomized solution algorithm. Eur. J. Oper. Res. **250**(2), 389–399 (2016)
3. Bertolazzi, P., Felici, G., Weitschek, E.: Learning to classify species with barcodes. BMC Bioinform. **10**(14), S7 (2009)
4. Cestarelli, V., Fiscon, G., Felici, G., Bertolazzi, P., Weitschek, E.: CAMUR: knowledge extraction from RNA-seq cancer data through equivalent classification rules. Bioinformatics **32**(5), 697–704 (2016)
5. Eén, N., Sörensson, N.: Translating pseudo-boolean constraints into SAT. J. Satisf. Boolean Model. Comput. **2**, 1–26 (2006)
6. Felici, G., Truemper, K.: A minsat approach for learning in logic domains. INFORMS J. Comput. **14**(1), 20–36 (2002)
7. Feo, T.A., Resende, M.G.C.: Greedy randomized adaptive search procedures. J. Glob. Optim. **6**(2), 109–133 (1995)
8. Festa, P., Resende, M.G.C.: An annotated bibliography of GRASP - part I: algorithms. Int. Trans. Oper. Res. **16**(1), 1–24 (2009)
9. Festa, P., Resende, M.G.C.: An annotated bibliography of GRASP - part II: applications. Int. Trans. Oper. Res. **16**(2), 131–172 (2009)

10. Fu, Z., Malik, S.: Solving the minimum-cost satisfiability problem using SAT based branch-and-bound search. In: 2006 IEEE/ACM International Conference on Computer Aided Design, pp. 852–859, November 2006
11. Garey, M.R., Johnson, D.S.: Computers and Intractability, vol. 29. W.H. Freeman, New York (2002)
12. Manquinho, V.M., Marques-Silva, J.P.: Search pruning techniques in SAT-based branch-and-bound algorithms for the binate covering problem. IEEE Trans. Comput. Aided Des. Integr. Circ. Syst. **21**(5), 505–516 (2002)
13. Manquinho, V.M., Flores, P.F., Silva, J.P.M., Oliveira, A.L.: Prime implicant computation using satisfiability algorithms. In: Ninth IEEE International Conference on Tools with Artificial Intelligence, 1997 Proceedings, pp. 232–239. IEEE (1997)
14. Marques-Silva, J.P., Sakallah, K.A.: GRASP: a search algorithm for propositional satisfiability. IEEE Trans. Comput. **48**(5), 506–521 (1999)
15. Martins, R., Manquinho, V., Lynce, I.: Clause sharing in deterministic parallel maximum satisfiability. In: RCRA International Workshop on Experimental Evaluation of Algorithms for Solving Problems with Combinatorial Explosion (2012)
16. Martins, R., Manquinho, V.M., Lynce, I.: Clause sharing in parallel MaxSAT. In: Hamadi, Y., Schoenauer, M. (eds.) LION 2012. LNCS, pp. 455–460. Springer, Heidelberg (2012). doi:10.1007/978-3-642-34413-8_44
17. Martins, R., Manquinho, V.M., Lynce, I.: Parallel search for maximum satisfiability. AI Commun. **25**(2), 75–95 (2012)
18. Moskewicz, M.W., Madigan, C.F., Zhao, Y., Zhang, L., Malik, S.: Chaff: engineering an efficient SAT solver. In: Proceedings of the 38th Annual Design Automation Conference, DAC 2001, pp. 530–535. ACM, New York (2001)
19. de Moura, L., Bjørner, N.: Z3: an efficient SMT solver. In: Ramakrishnan, C.R., Rehof, J. (eds.) TACAS 2008. LNCS, vol. 4963, pp. 337–340. Springer, Heidelberg (2008). doi:10.1007/978-3-540-78800-3_24
20. Pipponzi, M., Somenzi, F.: An iterative algorithm for the binate covering problem. In: Proceedings of the European Design Automation Conference, EDAC 1990, pp. 208–211, March 1990
21. Ribeiro, C.C., Rosseti, I., Souza, R.C.: Probabilistic stopping rules for GRASP heuristics and extensions. Int. Trans. Oper. Res. **20**(3), 301–323 (2013)
22. Scholz, F.W.: Maximum likelihood estimation (2004)
23. Servit, M., Zamazal, J.: Heuristic approach to binate covering problem. In: Proceedings The European Conference on Design Automation, pp. 123–129, March 1992
24. Sorensson, N., Een, N.: Minisat v1.13 - a sat solver with conflict-clause minimization. Technical report (2005(53))
25. Truemper, K.: Design of Logic-Based Intelligent Systems. Wiley-Interscience Publication, Wiley (2004)
26. Villa, T., Kam, T., Brayton, R.K., Sangiovanni-Vincenteili, A.L.: Explicit and implicit algorithms for binate covering problems. IEEE Trans. Comput. Aided Des. Integr. Circ. Syst. **16**(7), 677–691 (1997)
27. Weitschek, E., Felici, G., Bertolazzi, P.: MALA: a microarray clustering and classification software. In: 2012 23rd International Workshop on Database and Expert Systems Applications, pp. 201–205, September 2012
28. Weitschek, E., Fiscon, G., Felici, G.: Supervised DNA barcodes species classification: analysis, comparisons and results. BioData Min. **7**(1), 4 (2014)
29. Weitschek, E., Lo Presti, A., Drovandi, G., Felici, G., Ciccozzi, M., Ciotti, M., Bertolazzi, P.: Human polyomaviruses identification by logic mining techniques. Virol. J. **9**(1), 58 (2012)

A New Local Search for the p-Center Problem Based on the Critical Vertex Concept

Daniele Ferone[1], Paola Festa[1(✉)], Antonio Napoletano[1],
and Mauricio G.C. Resende[2]

[1] Department of Mathematics and Applications, University of Napoli Federico II,
Compl. MSA, Via Cintia, 80126 Naples, Italy
{daniele.ferone,paola.festa,antonio.napoletano2}@unina.it
[2] Mathematical Optimization and Planning, Amazon.com, Seattle, USA
resendem@amazon.com

Abstract. For the p-center problem, we propose a new smart local search based on the critical vertex concept and embed it in a GRASP framework. Experimental results attest the robustness of the proposed search procedure and confirm that for benchmark instances it converges to optimal or near/optimal solutions faster than the best known state-of-the-art local search.

1 Introduction

The p-center problem is one of the best-known discrete location problems first introduced in the literature in 1964 by Hakimi [13]. It consists of locating p facilities and assigning clients to them in order to minimize the maximum distance between a client and the facility to which the client is assigned (i.e., the closest facility). Useless to say that this problem arises in many different real-world contexts, whenever one designs a system for public facilities, such as schools or emergency services.

Formally, we are given a complete undirected edge-weighted bipartite graph $G = (V \cup U, E, c)$, where

- $V = \{1, 2, \ldots, n\}$ is a set of n potential locations for facilities;
- $U = \{1, 2, \ldots, m\}$ is a set of m clients or demand points;
- $E = \{(i, j) \mid i \in V, j \in U\}$ is a set of $n \times m$ edges;
- $c : E \mapsto \mathbb{R}^+ \cup \{0\}$ is a function that assigns a nonnegative distance c_{ij} to each edge $(i, j) \in E$.

The p-center problem is to find a subset $P \subseteq V$ of size p such that its weight, defined as

$$\mathbb{C}(P) = \max_{i \in U} \min_{j \in P} c_{ij} \qquad (1)$$

is minimized. The minimum value is called the *radius*. Although it is not a restrictive hypothesis, in this paper we consider the special case where $V = U$

© Springer International Publishing AG 2017
R. Battiti et al. (Eds.): LION 2017, LNCS 10556, pp. 79–92, 2017.
https://doi.org/10.1007/978-3-319-69404-7_6

is the vertex set of a complete graph $G = (V, E)$, each distance c_{ij} represents the length of a shortest path between vertices i and j ($c_{ii} = 0$), and hence the triangle inequality is satisfied.

In 1979, Kariv and Hakimi [16] proved that the problem is NP-hard, even in the case where the input instance has a simple structure (e.g., a planar graph of maximum vertex degree 3). In 1970, Minieka [20] designed the first exact method for the p-center problem viewed as a series of set covering problems. His algorithm iteratively chooses a threshold r for the radius and checks whether all clients can be covered within distance r using no more than p facilities. If so, the threshold r is decreased; otherwise, it is increased. Inspired by Minieka's idea, in 1995 Daskin [3] proposed a recursive bisection algorithm that systematically reduces the gap between upper and lower bounds on the radius. More recently, in 2010 Salhi and Al-Khedhairi [26] proposed a faster exact approach based on Daskin's algorithm that obtains tighter upper and lower bounds by incorporating information from a three-level heuristic that uses a variable neighborhood strategy in the first two levels and at the third level a perturbation mechanism for diversification purposes.

Recently, several facility location problems similar to the p-center have been used to model scenarios arising in financial markets. The main steps to use such techniques are the following: first, to describe the considered financial market via a correlation matrix of stock prices; second, to model the matrix as a graph, stocks and correlation coefficients between them are represented by nodes and edges, respectively. With this idea, Goldengorin et al. [11] used the p-median problem to analyze stock markets. Another interesting area where these problems arise is the manufacturing system with the aim of lowering production costs [12].

Due to the computational complexity of the p-center problem, several approximation and heuristic algorithms have been proposed for solving it. By exploiting the relationship between the p-center problem and the dominating set problem [15,18], nice approximation results were proved. With respect to inapproximability results, Hochbaum and Shmoys [15] proposed a 2-approximation algorithm for the problem with triangle inequality, showing that for any $\delta < 2$ the existence of a δ-approximation algorithm would imply that $P = NP$.

Although interesting in theory, approximation algorithms are often outperformed in practice by more straightforward heuristics with no particular performance guarantees. Local search is the main ingredient for most of the heuristic algorithms that have appeared in the literature. In conjunction with various techniques for escaping local optima, these heuristics provide solutions which exceed the theoretical upper bound of approximating the problem and derive from ideas used to solve the p-median problem, a similar NP-hard problem [17]. Given a set F of m potential facilities, a set U of n users (or customers), a distance function $d : U \times F \mapsto \mathbb{R}$, and a constant $p \leq m$, the p-median problem is to determine a subset of p facilities to open so as to minimize the sum of the distances from each user to its closest open facility. For the p-median problem, in 2004 Resende and Werneck [25] proposed a multistart heuristic that hybridizes GRASP with Path-Relinking as both, intensification and post-optimization phases. In 1997,

Hansen and Mladenović [14] proposed three heuristics: `Greedy`, `Alternate`, and `Interchange` (vertex substitution). To select the first facility, `Greedy` solves a 1-center problem. The remaining $p-1$ facilities are then iteratively added, one at a time, and at each iteration the location which most reduces the maximum cost is selected. In [5], Dyer and Frieze suggested a variant, where the first center is chosen at random. In the first iteration of `Alternate`, facilities are located at p vertices chosen in V, clients are assigned to the closest facility, and the 1-center problem is solved for each facility's set of clients. During the subsequent iterations, the process is repeated with the new locations of the facilities until no more changes in assignments occur. As for the `Interchange` procedure, a certain pattern of p facilities is initially given. Then, facilities are moved iteratively, one by one, to vacant sites with the objective of reducing the total (or maximum) cost. This local search process stops when no movement of a single facility decreases the value of the objective function. A multistart version of `Interchange` was also proposed, where the process is repeated a given number of times and the best solution is kept. The combination of `Greedy` and `Interchange` has been most often used for solving the p-median problem. In 2003, Mladenović et al. [21] adapted it to the p-center problem and proposed a Tabu Search (TS) and a Variable Neighborhood Search (VNS), i.e., a heuristic that uses the history of the search in order to construct a new solution and a competitor that is not history sensitive, respectively. The TS is designed by extending `Interchange` to the chain-interchange move, while in the VNS, a perturbed solution is obtained from the incumbent by a k-interchange operation and `Interchange` is used to improve it. If a better solution than the incumbent is found, the search is recentered around it. In 2011, Davidović et al. [4] proposed a Bee Colony algorithm, a random search population-based technique, where an artificial system composed of a number of precisely defined agents, also called individuals or artificial bees.

To the best of our knowledge, most of the research effort devoted towards the development of metaheuristics for this problem has been put into the design of efficient local search procedures. The purpose of this article is propose a new local search and to highlight how its performances are better than best-known local search proposed in literature (Mladenović et al.'s [21] local search based on VNS strategy), both in terms of solutions quality and convergence speed.

The remainder of the paper is organized as follows. In Sect. 2, a `GRASP` construction procedure is described. In Sect. 3, we introduce the new concept of *critical vertex* with relative definitions and describe a new local search algorithm. Computational results presented in Sect. 4 empirically demonstrate that our local search is capable of obtaining better results than the best known local search, and they are validated by a statistical significance test. Concluding remarks are made in Sect. 5.

2   GRASP Construction Phase

GRASP is a randomized multistart iterative method proposed in Feo and Resende [6, 7] and having two phases: a greedy randomized construction phase and a local search phase. For a comprehensive study of GRASP strategies and their variants, the reader is referred to the survey papers by Festa and Resende [9,10], as well as to their annotated bibliography [8].

Starting from a partial solution made of $1 \leq randElem \leq p$ facilities randomly selected from V, our GRASP construction procedure iteratively selects the remaining $p-randElem$ facilities in a greedy randomized fashion. The greedy function takes into account the contribution to the objective function achieved by selecting a particular candidate element. In more detail, given a partial solution P, $|P| < p$, for each $i \in V \setminus P$, we compute $w(i) = \mathbb{C}(P \cup \{i\})$. The pure greedy choice would consist in selecting the vertex with the smallest greedy function value. This procedure instead computes the smallest and the largest greedy function values:

$$z_{\min} = \min_{i \in V \setminus P} w(i); \quad z_{\max} = \max_{i \in V \setminus P} w(i).$$

Then, denoting by $\mu = z_{\min} + \beta(z_{\max} - z_{\min})$ the cut-off value, where β is a parameter such that $\beta \in [0,1]$, a *restricted candidate list* (RCL) is made up of all vertices whose greedy value is less than or equal to μ. The new facility to be added to P is finally randomly selected from the RCL.

The pseudo-code is shown in Fig. 1, where $\alpha \in [0,1]$.

Function greedy-randomized-build($G = \langle V, E, \mathbb{C} \rangle, p, \alpha, \beta$)

```
1  P ← ∅ ;
2  randElem := ⌊α · p⌋ ;
3  for k = 1, ..., randElem do                        // random component
4  │   f ← SelectRandom(V \ P);
5  │   P ← P ∪ {f} ;
6  while |P| < p do
7  │   z_min ← +∞ ;
8  │   z_max ← -∞ ;
9  │   for i ∈ V \ P do
10 │   │   if z_min > C(P ∪ {i}) then
11 │   │   │   z_min ← C(P ∪ {i}) ;
12 │   │   if z_max < C(P ∪ {i}) then
13 │   │   │   z_max ← C(P ∪ {i}) ;
14 │   μ ← z_min + β(z_max - z_min) ;
15 │   RCL ← {i ∈ V \ P | C(P ∪ {i}) ≤ μ} ;
16 │   f ← SelectRandom(RCL) ;
17 │   P ← P ∪ {f};
18 return P;
```

Fig. 1. Pseudo-code of the greedy randomized construction.

3   Plateau Surfer: A New Local Search Based on the Critical Vertex Concept

Given a feasible solution P, the `Interchange` local search proposed by Hansen and Mladenović [14] consists in swapping a facility $f \in P$ with a facility $\bar{f} \notin P$ which results in a decrease of the current cost function. Especially in the case of instances with many vertices, we have noticed that usually a single swap does not strictly improve the current solution, because there are several facilities whose distance is equal to the radius of the solution. In other words, the objective function is characterized by large plateaus and the `Interchange` local search cannot escape from such regions. To face this type of difficulties, inspired by Variable Formulation Search [22,23], we have decided to use a refined way for comparing between valid solutions by introducing the concept of *critical vertex*. Given a solution $P \subseteq V$, let $\delta_P : V \mapsto \mathbb{R}^+ \cup \{0\}$ be a function that assigns to each vertex $i \in V$ the distance between i and its closest facility according to solution P. Clearly, the cost of a solution P can be equivalently written as $\mathbb{C}(P) = \max\{\delta_P(i) : i \in V\}$. We also give the following definition:

Definition 1 (Critical vertex). *Let $P \subseteq V$ be a solution whose cost is $\mathbb{C}(P)$. For each vertex $i \in V$, i is said to be a* critical vertex *for P, if and only if $\delta_P(i) = \mathbb{C}(P)$.*

In the following, we will denote with $\max_{\delta_P} = |\{i \in V : \delta_P(i) = \mathbb{C}(P)\}|$ the number of vertices whose distance from their closest facility results in the objective function value corresponding to solution P. We define also the comparison operator $<_{cv}$, and we will say that $P <_{cv} P'$ if and only if $\max_{\delta_P} < \max_{\delta_{P'}}$.

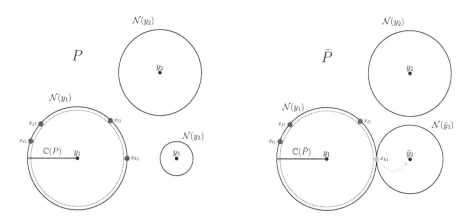

Fig. 2. An example of how the local search works. In this case, the algorithm switches from solution P to solution $\bar{P}$. In $\bar{P}$, a new facility $\bar{y}_3$ is selected in place of y_3 in P, $\bar{y}_3$ attracts one of the "critical vertices" from the neighborhood of the facility y_1. Although the cost of the two solutions is the same, the algorithm selects the new solution $\bar{P}$ because $\max_{\delta_{\bar{P}}} < \max_{\delta_P}$.

The main idea of our *plateau surfer* local search is to use the concept of
critical vertex to escape from plateaus, moving to solutions that have either
a better cost than the current solution or equal cost but less critical vertices.
Figure 2 shows a simple application of the algorithm, while in Figs. 3 and 4, for
four benchmark instances, both Mladenović's local search and our local search
are applied once taken as input the same starting feasible solution. It is evident
that both the procedures make the same first moves. However, as soon as a
plateau is met, Mladenović's local search ends, while our local search is able to
escape from the plateau moving to other solutions with the same cost value,

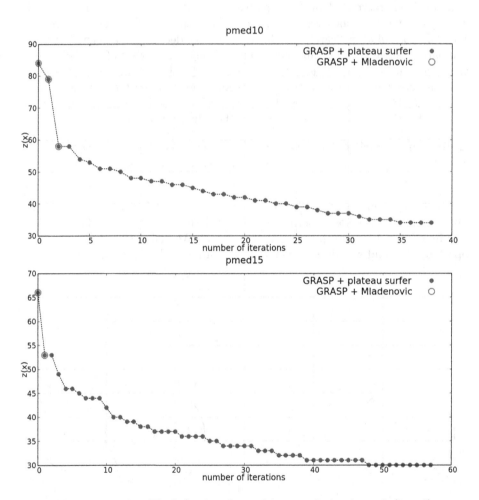

Fig. 3. Plateau escaping. The behavior of our *plateau surfer* local search (in red) compared with the Mladenović's one (in blue). Both algorithms work on the same instances
taking as input the same starting solution. Filled red dots and empty blue circles indicate the solutions found by the two algorithms. Mladenović local search stops as soon
as the first plateau is met. (Color figure online)

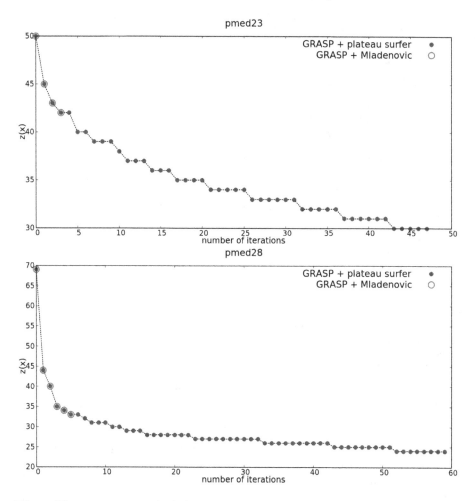

Fig. 4. Plateau escaping. The behavior of our *plateau surfer* local search (in red) compared with the Mladenović's one (in blue) on other two different instances. (Color figure online)

and restarting the procedure from a new solution that can lead to a strict cost function improvement.

Let us analyze in more detail the behavior of our local search, whose pseudo-code is reported in Fig. 5. The main part of the algorithm consists in the portion of the pseudo-code that goes from line 7 to line 14. Starting from an initial solution P, the algorithm tries to improve the solution replacing a vertex $j \notin P$ with a facility $i \in P$. Clearly, this swap is stored as an improving move if the new solution $\bar{P} = P \setminus \{i\} \cup \{j\}$ is strictly better than P according to the cost function $\mathbb{C}$. If $\mathbb{C}(\bar{P})$ is better than the current cost $\mathbb{C}(P)$, then $\bar{P}$ is compared also with the incumbent

---

Function plateau-surfer-local-search($G = \langle V, A, \mathbb{C} \rangle, P, p$)

---

1 **repeat**
2 | modified := **false**;
3 | **forall** $i \in P$ **do**
4 | | best_flip := best_cv_flip := NIL;
5 | | bestNewSolValue := $\mathbb{C}(P)$;
6 | | best_cv := $\max_\delta(\bar{P})$;
7 | | **forall** $j \in V \setminus P$ **do**
8 | | | $\bar{P} := P \setminus \{i\} \cup \{j\}$;
9 | | | **if** $\mathbb{C}(\bar{P}) <$ bestNewSolValue **then**
10 | | | | bestNewSolValue := $\mathbb{C}(\bar{P})$;
11 | | | | best_flip := j;
12 | | | **else if** best_flip = NIL **and** $\max_\delta(\bar{P}) <$ best_cv **then**
13 | | | | best_cv := $\max_\delta(\bar{P})$;
14 | | | | best_cv_flip := j;
15 | | **if** best_flip $\neq$ NIL **then**
16 | | | $P := P \setminus \{i\} \cup \{$best_flip$\}$;
17 | | | modified := **true**;
18 | | **else if** best_cv_flip $\neq$ NIL **then**
19 | | | $P := P \setminus \{i\} \cup \{$best_cv_flip$\}$;
20 | | | modified := **true**;
21 **until** modified = **false**;
22 **return** P;

Fig. 5. Pseudocode of the plateau surfer local search algorithm based on the critical vertex concept.

solution and if it is the best solution found so far, the incumbent is update and the swap that led to this improvement stored (lines 9–11).

Otherwise, the algorithm checks if it is possible to reduce the number of critical vertices. If the new solution $\bar{P}$ is such that $\bar{P} <_{cv} P$, then the algorithm checks if $\bar{P}$ is the best solution found so far (line 12), the value that counts the number of critical vertices in a solution is update (line 13), and the current swap stored as an improving move (line 14).

To study the computational complexity of our local search, let be $n = |V|$ and $p = |P|$, the number of vertices in the graph and the number of facilities in a solution, respectively. The loops at lines 3 and 7 are executed p and n times, respectively. The update of the solution takes $\mathcal{O}(n)$. In conclusion, the total complexity is $\mathcal{O}(p \cdot n^2)$.

4   Experimental Results

In this section, we describe computational experience with the local search proposed in this paper. We have compared it with the local search proposed by Mladenović et al. [21], by embedding both in a GRASP framework.

The algorithms were implemented in C++, compiled with gcc 5.2.1 under Ubuntu with -std=c++14 flag. The stopping criterion is $maxTime = 0.1 \cdot n + 0.5 \cdot p$. All the tests were run on a cluster of nodes, connected by 10 Gigabit Infiniband technology, each of them with two processors Intel Xeon E5-4610v2@2.30 GHz.

Table 1 summarizes the results on a set of ORLIB instances, originally introduced in [1]. It consists of 40 graphs with number of vertices ranging from 100 to 900, each with a suggested value of p ranging from 5 to 200. Each vertex is both a user and a potential facility, and distances are given by shortest path lengths. Tables 2 and 3 report the results on the TSP set of instances. They are just sets of points on the plane. Originally proposed for the traveling salesman problem, they are available from the TSPLIB [24]. Each vertex can be either a user or an open facility. We used the *Mersenne Twister* random number generator by Matsumoto and Nishimura [19]. Each algorithm was run with 10 different seeds, and minimum (min), average (E) and variance (σ^2) values are listed in each table. The second to last column lists the %-Gap between average solutions. To deeper investigate the statistical significance of the results obtained by the two local searches, we performed the Wilcoxon test [2,27].

Generally speaking, the Wilcoxon test is a ranking method that well applies in the case of a number of paired comparisons leading to a series of differences, some of which may be positive and some negative. Its basic idea is to substitute scores $1, 2, 3, \ldots, n$ with the actual numerical data, in order to obtain a rapid approximate idea of the significance of the differences in experiments of this kind.

More formally, let A_1 and A_2 be two algorithms, $I_1, \ldots, I_l$ be l instances of the problem to solve, and let $\delta_{A_i}(I_j)$ be the value of the solution obtained by algorithm A_i ($i = 1, 2$) on instance I_j ($j = 1, \ldots, l$). For each $j = 1, \ldots, l$, the Wilcoxon test computes the differences $\Delta_j = |\delta_{A_1}(I_j) - \delta_{A_2}(I_j)|$ and sorts them in non decreasing order. Accordingly, starting with a smallest rank equal to 1, to each difference Δ_j, it assigns a non decreasing rank R_j. Ties receive a rank equal to the average of the sorted positions they span. Then, the following quantities are computed

$$W^+ = \sum_{j=1,\ldots,l:\ \Delta_j > 0} R_j,$$

$$W^- = \sum_{j=1,\ldots,l:\ \Delta_j < 0} R_j.$$

Under the null hypothesis that $\delta_{A_1}(I_j)$ and $\delta_{A_2}(I_j)$ have the same median value, it should result that $W^+ = W^-$. If the p-value associated to the experiment is less than an a priori fixed significance level α, then the null hypothesis is rejected and the difference between W^+ and W^- is considered significant.

The last column of each table lists the p-values where the %-Gap is significant, all the values are less than $\alpha = 0.01$. This outcome of the Wilcoxon test further confirms that our local search is better performing than the local search proposed by Mladenović et al.

Table 1. Results on ORLIB instances.

Instance	GRASP + mladenovic			GRASP + plateau-surfer			%-Gap	p-value
	min	E	σ^2	min	E	σ^2		
pmed01	127	127	0	127	127	0	0.00	
pmed02	98	98	0	98	98	0	0.00	
pmed03	93	93.14	0.12	93	93.54	0.25	0.43	
pmed04	74	76.21	1.33	74	74.02	0.04	−2.87	1.20E−16
pmed05	48	48.46	0.43	48	48	0	−0.95	
pmed06	84	84	0	84	84	0	0.00	
pmed07	64	64.15	0.27	64	64	0	−0.23	
pmed08	57	59.39	1.36	55	55.54	0.73	−6.48	3.37E−18
pmed09	42	46.87	2.83	37	37.01	0.01	−21.04	2.80E−18
pmed10	29	31.21	0.81	20	20.01	0.01	−35.89	9.38E−19
pmed11	59	59	0	59	59	0	0.00	
pmed12	51	51.89	0.1	51	51.41	0.24	−0.93	
pmed13	42	44.47	0.73	36	36.94	0.06	−16.93	1.04E−18
pmed14	35	38.59	3.24	26	26.85	0.13	−30.42	2.11E−18
pmed15	28	30.23	0.7	18	18	0	−40.46	1.09E−18
pmed16	47	47	0	47	47	0	0.00	
pmed17	39	40.71	0.23	39	39	0	−4.20	8.69E−20
pmed18	36	37.95	0.29	29	29.41	0.24	−22.50	6.37E−19
pmed19	27	29.32	0.42	19	19.13	0.11	−34.75	6.25E−19
pmed20	25	27.05	0.99	14	14	0	−48.24	1.46E−18
pmed21	40	40	0	40	40	0	0.00	
pmed22	39	40.06	0.24	38	38.94	0.06	−2.80	1.30E−18
pmed23	30	32.02	0.44	23	23.21	0.17	−27.51	7.16E−19
pmed24	24	25.38	0.34	16	16	0	−36.96	4.37E−19
pmed25	22	22.62	0.24	11	11.89	0.1	−47.44	2.77E−19
pmed26	38	38	0	38	38	0	0.00	
pmed27	33	33.96	0.06	32	32	0	−5.77	2.15E−22
pmed28	26	26.78	0.17	19	19	0	−29.05	2.20E−20
pmed29	23	23.43	0.31	13	13.68	0.22	−41.61	8.00E−19
pmed30	20	21.18	0.47	10	10	0	−52.79	6.50E−19
pmed31	30	30	0	30	30	0	0.00	
pmed32	30	30.37	0.23	29	29.62	0.24	−2.47	
pmed33	23	23.76	0.2	16	16.28	0.2	−31.48	4.31E−19
pmed34	21	22.42	0.66	11	11.56	0.25	−48.44	1.59E−18
pmed35	30	30.01	0.01	30	30	0	−0.03	
pmed36	28	29.37	0.31	27	27.65	0.23	−5.86	4.52E−18
pmed37	23	24.07	0.37	16	16	0	−33.53	2.74E−19
pmed38	29	29	0	29	29	0	0.00	
pmed39	24	25.08	0.11	23	23.98	0.02	−4.39	4.68E−21
pmed40	20	21.81	0.43	14	14	0	−35.81	5.14E−19
Average							−16.78	

Table 2. Results on TSPLIB instances (1)

Instace	p	GRASP + mladenovic			GRASP + plateau-surfer			%-Gap	p-value
		min	E	σ^2	min	E	σ^2		
pcb3038	50	534.48	608.49	1068.09	355.68	374.66	51.05	−38.43	3.90E−18
	100	399.49	481.75	1285.58	259.67	270.2	17.56	−43.91	3.90E−18
	150	331.62	428.69	1741.11	206.71	215.78	23.73	−49.67	3.90E−18
	200	301.01	386.56	3161.87	177.79	190.88	10.4	−50.62	3.90E−18
	250	292.48	359.59	3323.62	155.03	163.75	19.24	−54.46	3.90E−18
	300	261.28	349.42	2902.71	143.39	151.89	10.04	−56.53	3.90E−18
	350	258.82	336.08	3755.72	123.85	136.22	22.45	−59.47	3.90E−18
	400	249.78	337.14	4033.46	119.07	122.31	1.2	−63.72	3.90E−18
	450	214.97	321.36	3373.23	115	117	0.6	−63.59	3.90E−18
	500	209.35	299.4	3378.98	102	110.38	5.78	−63.13	3.90E−18
pr1002	10	3056.55	3313.49	10132.77	2616.3	2727.45	2260.95	−17.69	3.90E−18
	20	2404.16	2668.29	8244.65	1806.93	1886.89	1516.07	−29.28	3.90E−18
	30	2124.26	2358.07	4432.11	1456.02	1505.55	910.93	−36.15	3.89E−18
	40	1960.23	2172.63	7831.77	1253.99	1302.76	751.62	−40.04	3.90E−18
	50	1755.7	1992.08	5842.66	1097.72	1156.77	815.35	−41.93	3.90E−18
	60	1697.79	1865.5	5872.47	1001.25	1042.82	257.42	−44.1	3.89E−18
	70	1569.24	1736.41	4078.39	900	954.04	307.65	−45.06	3.89E−18
	80	1486.61	1633.87	3278.4	851.47	889.5	407.29	−45.56	3.88E−18
	90	1350.93	1543.17	3922.25	764.85	809.78	382.29	−47.52	3.89E−18
	100	1312.44	1472.47	2616	743.3	767.62	77.4	−47.87	3.89E−18
pr439	10	2575.12	2931.83	38470.59	1971.83	1972.28	19.61	−32.73	3.79E−18
	20	1940.52	2577.03	23638.88	1185.59	1194.12	124.58	−53.66	3.71E−18
	30	1792.34	2510.91	23692.47	886	919.1	442.37	−63.4	3.89E−18
	40	1525.2	2413.33	53876.4	704.45	728.19	39.31	−69.83	3.88E−18
	50	1358.54	2252.46	89633.71	575	595.4	64.21	−73.57	3.82E−18
	60	1386.09	2170.85	110065.93	515.39	537.66	75.43	−75.23	3.89E−18
	70	1370.45	1898.53	116167.77	480.23	499.65	4.93	−73.68	3.73E−18
	80	1140.18	1815.1	118394.68	424.26	440.27	166.06	−75.74	3.89E−18
	90	1191.9	1699.64	91388.99	400	406.17	31.71	−76.1	3.88E−18
	100	1190.85	1679.73	94076.45	375	384.27	98.91	−77.12	3.89E−18
rat575	10	81.32	92.98	9.27	73	74.71	0.79	−19.65	3.90E−18
	20	68.07	73.86	3.7	50.54	53.04	0.63	−28.19	3.90E−18
	30	59.81	64.61	3.67	41.79	43.53	0.47	−32.63	3.90E−18
	40	54.13	58.37	3.43	36.12	37.43	0.29	−35.87	3.90E−18
	50	47.68	53.78	3.56	32.45	33.36	0.17	−37.97	3.90E−18
	60	45.62	50.03	3.21	29.15	30.17	0.19	−39.7	3.90E−18
	70	43.68	46.96	2.97	27	27.78	0.13	−40.84	3.90E−18
	80	39.81	44.2	2.75	25.02	25.99	0.11	−41.2	3.90E−18
	90	38.48	41.98	2.06	23.85	24.4	0.07	−41.88	3.90E−18
	100	37.01	39.93	1.4	22.2	23.01	0.08	−42.37	3.89E−18
rat783	10	102.22	110.93	13.17	83.49	87.82	1.65	−20.83	3.90E−18
	20	80.53	88.56	7.68	59.68	62.8	1.41	−29.09	3.90E−18
	30	69.58	76.92	7.87	49.25	51.48	0.74	−33.07	3.90E−18
	40	62.97	69.63	4.62	42.05	44.27	0.53	−36.42	3.90E−18
	50	59.41	65.26	5.59	38.29	39.6	0.42	−39.32	3.90E−18
	60	54.82	60.35	4.77	34.48	35.92	0.24	−40.48	3.90E−18
	70	49.4	56.56	7.48	32.06	33.11	0.24	−41.46	3.90E−18
	80	48.51	53.76	4.03	29.55	30.94	0.21	−42.45	3.90E−18
	90	46.07	51.82	3.53	28.18	28.85	0.11	−44.33	3.90E−18
	100	43.97	49.5	4.68	26.31	27.49	0.14	−44.46	3.90E−18
Average								−46.84	

Table 3. Results on TSPLIB instances (2)

Instance	p	GRASP + mladenovic			GRASP + plateau-surfer			%-Gap	p-value
		min	E	σ^2	min	E	σ^2		
rl1323	10	3810.84	4185.89	24655.46	3110.57	3241.79	3290.56	−22.55	3.90E−18
	20	2996.4	3348.31	23183.21	2090.87	2236.28	2798.56	−33.21	3.90E−18
	30	2689.44	2979.79	14205.75	1730.78	1808.94	1544.85	−39.29	3.90E−18
	40	2337.92	2712.93	14193.05	1479.24	1576.25	1710.4	−41.9	3.90E−18
	50	2195.91	2462.95	9835.09	1300	1363.88	950.66	−44.62	3.90E−18
	60	2021.87	2278.94	16400.27	1181.3	1244.03	657.55	−45.41	3.90E−18
	70	1900.77	2128.45	11883.58	1076.2	1127.98	475.13	−47	3.90E−18
	80	1866.8	2033.24	4501.73	988.87	1048.87	438.82	−48.41	3.89E−18
	90	1634.37	1966.13	4643.42	935.02	978.6	289.18	−50.23	3.89E−18
	100	1631.5	1909.56	8483.55	886.85	914	238.2	−52.14	3.89E−18
u1060	10	3110.65	3373.87	7541.61	2301.7	2440	599.42	−27.68	3.86E−18
	20	2652.6	2818.37	5787.51	1650.34	1749.15	2814.03	−37.94	3.90E−18
	30	2501.72	2684.87	3811.23	1302.94	1373.21	912.92	−48.85	3.90E−18
	40	2442.07	2616.15	5267.85	1118.59	1176.14	593.6	−55.04	3.90E−18
	50	2378.36	2591.96	7266.77	950.66	1021.55	418.91	−60.59	3.90E−18
	60	2301.83	2602.13	13579.82	860.49	919.97	374.54	−64.65	3.90E−18
	70	2378.36	2606.64	10944.09	790.13	828.16	441.03	−68.23	3.90E−18
	80	2351.82	2622.32	12980.39	720.94	753.64	306.94	−71.26	3.90E−18
	90	2248.61	2562.01	10260.36	667.55	708.04	107.79	−72.36	3.90E−18
	100	2060.29	2494.08	11025.91	632.11	653.15	110.65	−73.81	3.90E−18
	110	2049.18	2444.22	10385.95	570.49	613.02	148.7	−74.92	3.90E−18
	120	2122.97	2406.19	9191.4	570	579.93	96.23	−75.9	3.90E−18
	130	1839.55	2390.82	12029.95	538.82	561.62	78.78	−76.51	3.90E−18
	140	1924.48	2316.25	12982.87	500.39	527.66	172.51	−77.22	3.90E−18
	150	1942.27	2300.45	13245.06	499.65	503.26	20.49	−78.12	3.90E−18
u1817	10	592.97	646.89	325	466.96	485.44	104.33	−24.96	3.90E−18
	20	462.3	564.44	560.9	330.2	348.15	53.96	−38.32	3.90E−18
	30	418.91	530.34	1018.29	265.19	283.4	58.43	−46.56	3.90E−18
	40	407.19	526.44	956.01	232.25	245.78	43.42	−53.31	3.90E−18
	50	330.21	507.52	2889.76	204.79	217.05	26.96	−57.23	3.90E−18
	60	352.88	497.35	3539.09	184.91	197.26	21.79	−60.34	3.90E−18
	70	321.27	477.43	4139.93	170.39	181.53	13.67	−61.98	3.90E−18
	80	289.61	445.35	4866.81	154.5	166.46	22.68	−62.62	3.90E−18
	90	283.99	422.34	3828.2	148.11	153.5	13.72	−63.65	3.90E−18
	100	283.99	416.69	2660.21	136.79	146.67	7.4	−64.8	3.90E−18
Average								−54.9	

5 Concluding Remarks

In this paper, we presented a new local search heuristic for the p-center problem, whose potential applications appear in telecommunications, in transportation logistics, and whenever one must to design a system to organize some sort of public facilities, such as, for example, schools or emergency services.

The computational experiments show that the proposed local search is capable to reduce the number of local optimum solutions using the concept of *critical vertex*, and it improves the results of the best local search for the problem.

Future lines of work will be focused on a deeper investigation of the robustness of our proposal by applying it on further instances coming from financial markets and manufacturing systems.

Acknowledgements. This work has been realized thanks to the use of the S.Co.P.E. computing infrastructure at the University of Napoli FEDERICO II.

References

1. Beasley, J.: A note on solving large p-median problems. Eur. J. Oper. Res. **21**, 270–273 (1985)
2. Coffin, M., Saltzman, M.: Statistical analysis of computational tests of algorithms and heuristics. INFORMS J. Comput. **12**(1), 24–44 (2000)
3. Daskin, M.: Network and Discrete Location: Models, Algorithms, and Applications. Wiley, New York (1995)
4. Davidović, T., Ramljak, D., Šelmić, M., Teodorović, D.: Bee colony optimization for the p-center problem. Comput. Oper. Res. **38**(10), 1367–1376 (2011)
5. Dyer, M., Frieze, A.: A simple heuristic for the p-centre problem. Oper. Res. Lett. **3**(6), 285–288 (1985)
6. Feo, T., Resende, M.: A probabilistic heuristic for a computationally difficult set covering problem. Oper. Res. Lett. **8**, 67–71 (1989)
7. Feo, T., Resende, M.: Greedy randomized adaptive search procedures. J. Global Optim. **6**, 109–133 (1995)
8. Festa, P., Resende, M.: GRASP: an annotated bibliography. In: Ribeiro, C., Hansen, P. (eds.) Essays and Surveys on Metaheuristics, pp. 325–367. Kluwer Academic Publishers, London (2002)
9. Festa, P., Resende, M.: An annotated bibliography of GRASP - part I: algorithms. Int. Trans. Oper. Res. **16**(1), 1–24 (2009)
10. Festa, P., Resende, M.: An annotated bibliography of GRASP - part II: applications. Int. Trans. Oper. Res. **16**(2), 131–172 (2009)
11. Goldengorin, B., Kocheturov, A., Pardalos, P.M.: A pseudo-boolean approach to the market graph analysis by means of the p-median model. In: Aleskerov, F., Goldengorin, B., Pardalos, P.M. (eds.) Clusters, Orders, and Trees: Methods and Applications. SOIA, vol. 92, pp. 77–89. Springer, New York (2014). doi:10.1007/978-1-4939-0742-7_5
12. Goldengorin, B., Krushinsky, D., Pardalos, P.M.: Application of the PMP to cell formation in group technology. In: Goldengorin, B., Krushinsky, D., Pardalos, P.M. (eds.) Cell Formation in Industrial Engineering. SOIA, vol. 79, pp. 75–99. Springer, New York (2013). doi:10.1007/978-1-4614-8002-0_3

13. Hakimi, S.: Optimum locations of switching centers and the absolute centers and medians of a graph. Oper. Res. **12**(3), 450–459 (1964)
14. Hansen, P., Mladenović, N.: Variable neighborhood search for the p-median. Locat. Sci. **5**(4), 207–226 (1997)
15. Hochbaum, D., Shmoys, D.: A best possible heuristic for the k-Center problem. Math. Oper. Res. **10**(2), 180–184 (1985)
16. Kariv, O., Hakimi, S.: An algorithmic approach to network location problems. Part I: the p-centers. SIAM J. Appl. Math. **37**(3), 513–538 (1979)
17. Kariv, O., Hakimi, S.: An algorithmic approach to network location problems. Part II: the p-medians. SIAM J. Appl. Math. **37**(3), 539–560 (1979)
18. Martinich, J.S.: A vertex-closing approach to the p-center problem. Nav. Res. Logist. **35**(2), 185–201 (1988)
19. Matsumoto, M., Nishimura, T.: Mersenne twister: a 623-dimensionally equidistributed uniform pseudo-random number generator. ACM Trans. Model. Comput. Simul. **8**(1), 3–30 (1998)
20. Minieka, E.: The m-center problem. SIAM Rev. **12**(1), 138–139 (1970)
21. Mladenović, N., Labbé, M., Hansen, P.: Solving the p-center problem with Tabu Search and variable neighborhood search. Networks **42**(April), 48–64 (2003)
22. Mladenovic, N., Urosevic, D., Prez-Brito, D., Garca-Gonzlez, C.G.: Variable neighbourhood search for bandwidth reduction. Eur. J. Oper. Res. **200**(1), 14–27 (2010)
23. Pardo, E.G., Mladenovi, N., Pantrigo, J.J., Duarte, A.: Variable formulation search for the cutwidth minimization problem. Appl. Soft Comput. **13**(5), 2242–2252 (2013)
24. Reinelt, G.: TSPLIB—A traveling salesman problem library. ORSA J. Comput. **3**(4), 376–384 (1991)
25. Resende, M., Werneck, R.: A hybrid heuristic for the p-median problem. J. Heuristics **10**(1), 59–88 (2004)
26. Salhi, S., Al-Khedhairi, A.: Integrating heuristic information into exact methods: the case of the vertex p-centre problem. J. Oper. Res. Soc. **61**(11), 1619–1631 (2010)
27. Wilcoxon, F.: Individual comparisons by ranking methods. Biom. Bull. **1**(6), 80–83 (1945)

An Iterated Local Search Framework with Adaptive Operator Selection for Nurse Rostering

Angeliki Gretsista$^{(\boxtimes)}$ and Edmund K. Burke

School of Electronic Engineering and Computer Science,
Queen Mary University of London, London, UK
{a.gretsista,vp-se}@qmul.ac.uk

Abstract. Considerable attention has been paid to selective hyper-heuristic frameworks for addressing computationally hard scheduling problems. By using selective hyper-heuristics, we can derive benefits from the strength of low level heuristics and their components at different stages of the heuristic search. In this paper, a simple, general and effective selective hyper heuristic is presented. We introduce an iterated local search based hyper-heuristic framework that incorporates the adaptive operator selection scheme to learn through the search process. The considered iterative approach employs an action selection model to decide the perturbation strategy to apply in each step and a credit assignment module to score its performance. The designed framework allows us to employ any action selection model and credit assignment mechanism used in the literature. Empirical results and an analysis of six different action selection models against state-of-the-art approaches, across 39 problem instances, highlight the significant potential of the proposed selection hyper-heuristics. Further analysis on the adaptive behavior of the model suggests that two of the six models are able to learn the best performing perturbation strategy, resulting in significant performance gains.

Keywords: Nurse rostering · Personnel scheduling · Hyper-heuristics · Action selection models

1 Introduction

One of the main motivations for the development of hyper-heuristics was to develop search algorithms that can operate with a certain degree of generality [1,2]. Hyper-heuristics can be considered to be "high level" general approaches that are able to select or generate low-level heuristics, whilst restricting the need to use domain knowledge [3]. In particular, selection hyper-heuristics choose heuristics from a predefined set of low level heuristics within a framework, where the aim is to determine a sequence of perturbations that provide efficient solutions for a given problem. On the other hand, the idea behind generative hyper-heuristics is to develop new heuristics based on the basic components of the input low level heuristics [3].

© Springer International Publishing AG 2017
R. Battiti et al. (Eds.): LION 2017, LNCS 10556, pp. 93–108, 2017.
https://doi.org/10.1007/978-3-319-69404-7_7

Nurse rostering is a well-known and well-studied personnel scheduling problem. The main objective is to assign the appropriate level of qualified nurses to shifts to cover the demand across different medical wards in a hospital, with respect to a diverse set of hard and soft constraints [4]. Representative constraints capture work regulations, employee preferences and fairness of allocation regarding weekend/night shifts or leave and days off [5]. To address nurse rostering problems, researchers have proposed a wide variety of methodologies including hyper-heuristics. In the recent work of Asta et al. [6], a tensor based algorithm is embodied to an online learning selection hyper-heuristic framework. In [7] Lü and Hao present an algorithm that takes two neighborhood moves and adaptively switches between search strategies by considering their search history. Other recent approaches constitute evolutionary perturbative hyper-heuristics [8], harmony search-based hyper-heuristics [9] and hybrid approaches [10–12].

Several publications have appeared in recent years proposing selective hyper-heuristics to address problems across different domains. Representative works can be the sequence-based selection hyper heuristic, which is motivated by the hidden Markov model [13], the PHunter [14], the Fair-Share Iterated Local Search [15], as well as the GIHH [16], the winner of the CHeSC 2011 [17] competition. The latter, is a characteristic example of a quite effective, yet highly complex, state-of-the-art hyper-heuristic. GIHH incorporates a dynamic heuristic selection mechanism and a move acceptance strategy, which evolves based on the properties of the current search landscape. We can observe that many of the best performing selective hyper-heuristics are developed based on the Iterated Local Search (ILS) algorithm. The ILS algorithm is driven by the alternation between intensification, intensive search of the neighborhood for locating locally optimal solutions, and diversification, in order to "jump" to new unexplored neighborhoods that can lead to better local optima. Determining the appropriate amount of diversification re-actively is essential for effective search [18].

In this study, we incorporate the adaptive operator selection paradigm in the main structure of an ILS algorithm to re-actively identify the most effective perturbation strategy. The proposed approach is a generic hyper-heuristic framework, namely the Iterated Local Search based hyper-heuristic framework (HHILS). HHILS incorporates an action selection mechanism to learn and adaptively select the most efficient low-level perturbation heuristic and a credit assignment module to empirically estimate the quality of the selected perturbation according to its ability to improve the incumbent solution. We utilize six state-of-the-art action selection models that result in six HHILS variants and compare them to eight state-of-the-art hyper-heuristics, on a set of 39 challenging nurse problem instances. The experimental results clearly justify the potential of the proposed framework.

The remainder of the paper is organized as follows: The nurse rostering problem is briefly described in Sect. 2. The proposed hyper-heuristic framework is presented in Sect. 3 where the main algorithm with the credit assignment and the action selection procedures are briefly explained in Sects. 3.1 and 3.2 respectively. Section 4 presents the experimental setup in Sect. 4.1 and the results in

Sect. 4.2. Finally, Sect. 5 concludes with a summary of the experimental findings of this work and provides some pointers for future work.

2   The Nurse Rostering Problem

In this paper, we use the model proposed in [10] to address the nurse rostering problem. To capture a large variety of different constraints the model adopts only one hard constraint, i.e., each employee can be assigned only one shift per day. It incorporates the remaining constraints as soft constraints in the objective function. Thus, the objective function is a weighted sum of the soft constraints, in which the associated weights take high or low values depending on their importance. Note that, associating a very high weight to a constraint is analogous to considering it as hard, whilst associating a very low value is analogous to discarding the specific constraint. This formulation also facilitates the easy integration of additional constraints based on the needs of different hospitals.

The soft constrains can be categorized in coverage constraints, where two or more nurses are associated, and in employee working constraints, that refer to each employee separately. Mainly, the coverage constraints ensure that the required number of employees is working over each shift. Coverage constraints can be also exploited to select skilled or qualified staff for specific shifts. Likewise, the employee's working constraints refer to the preference of each employee. The aim at this point is to satisfy the individual nurse by being flexible to decide her own schedule. For instance, we aim to be able to identify specific workload, to incorporate preferred days off, to select the most convenient shift pattern, to define vacations and so on.

3   The Proposed Approach

The proposed framework incorporates the adaptive operator selection paradigm [19] within the main structure of an ILS algorithm. HHILS replaces the application of a single perturbation heuristic of the ILS algorithm with an action selection mechanism that adaptively selects among various low-level perturbation heuristics. The applied perturbation heuristic is followed by the local search and acceptance procedures of ILS and then is scored based on its ability to improve the incumbent solution. As such, the proposed framework adaptively learns which is the most effective perturbation heuristic to apply during search. A description of the framework and its main characteristics can be briefly outlined as follows.

Algorithm 1 demonstrates the main algorithmic steps of the developed framework. As a first step, HHILS initializes all the required modules and data structures. This step also comprises the input of the chosen action selection and the credit assignment module (line 1). Next, HHILS initializes a solution for the given problem instance (line 2). The initial solution can be randomly generated or constructed based on a constructive heuristic, depending on the problem

---

Algorithm 1. Pseudo code of the proposed HHILS framework.

1: Initialise the action selection, and the credit assignment module as well as create all data structures required by HHILS.

2: $s_{cur} \leftarrow$ GenerateInitialSolution() /* **Generate Initial Solution:** Initialise or construct a solution for the problem instance at hand. */

3: **while** termination criteria do not hold **do**

4:     selected$_{llh} \leftarrow$ ActionSelection(str) /* **Action Selection:** Select a low-level heuristic with the str-th action selection model (Section 3.2). */

5:     $s_{tmp} \leftarrow$ ApplyAction(s_{cur}, selected$_{llh}$) /* **Perturbation:** Perturb the current solution (s_{cur}) with the selected low-level heuristic (selected$_{llh}$). */

6:     $s_{tmp} \leftarrow$ ApplyLocalSearch(s_{tmp}) /* **Local Search:** Apply a local search procedure on the temporary solution s_{tmp}. */

7:     $s_{cur} \leftarrow$ AcceptanceCriterion(s_{cur}, s_{tmp}) /* **Acceptance:** Accept which solution, between s_{cur}, and s_{tmp}, will survive to the next iteration based on an Acceptance criterion. */

8:     CreditAssignment(selected$_{llh}$) /* **Credit Assignment:** Score the used action (selected$_{llh}$) based on feedback from the problem at hand (Section 3.1).*/

9: **end while**

10: **Return:** the best solution found so far s_{cur}.

---

at hand. After having the initial solution, each iteration of HHILS operates five main steps. There is a set S which includes the k available perturbation low-level search heuristics, $\mathcal{S} = \{llh_1, llh_2, \dots, llh_k\}$. HHILS exploits an action selection method (ActionSelection(str)) to predict and select the most fitting perturbation low-level search heuristic included in S for the next step (line 4, see Sect. 3.2 for details). Having selected the perturbation low-level heuristic ($selected_{llh}$), a new solution, s_{tmp}, is generated by applying the $selected_{llh}$ perturbation heuristic to the current solution s_{cur} (line 5). After the perturbation, the new solution (s_{tmp}) is refined by a local search method, ApplyLocalSearch(s_{tmp}) (line 6). The local search procedure utilizes a set of greedy local search heuristics which are applied in an iterative way. More specifically, given a list L of the λ available local search heuristics, $L = \{l_1, l_2, \dots, l_\lambda\}$, at each repetition, a local search is selected in a uniform random way to be applied to the current solution. If the selected local search heuristic is not able to provide a better position for the s_{tmp}, it is excluded from L, and another local search heuristic is selected. This continues until an improved solution has been produced. By the end of the iterative local search process, the result will be the incumbent solution (s_{tmp}). In the next step, HHILS will decide the best solution, between s_{cur} and s_{tmp}, to use for the next iteration through the AcceptanceCriterion(s_{cur}, s_{tmp}) procedure (line 7). Here, we have adopted a Simulating Annealing acceptance rule to allow worsening moves being accepted with a probability. The acceptance probability can be calculated as $p = e^{(f(s_{cur}) - f(s_{tmp}))/(T \cdot \mu_i)}$, where $f(s_{cur})$ and $f(s_{tmp})$ are the objective values of the incumbent (s_{cur}) and temporary (s_{tmp}) solutions, T is the temperature value with $T \in \mathcal{R}$, (here is fixed to $T = 2$) and μ_i is the mean improvement of the improving iterations [15]. The value μ essentially normalizes the objective value difference by a quantity that is not problem dependent. In the last step, HHILS assigns a score to the utilized low-level perturbation heuristics (selected$_{llh}$) involved based on its performance (incumbent improvement) through the credit assignment module CreditAssignment (selected$_{llh}$) (line 8).

Four different perturbation strategies are used here, one from the mutation and three from the ruin and recreate categories. The mutation heuristic (HM) randomly un-assigns shifts based on an intensity parameter respecting the feasibility of the solution. The three ruin and recreate heuristics (HR1–HR3) are all inspired by the one proposed in [20]. HR1 unassigns all shifts of random employees from the schedule and recreates the schedule by prioritizing the objectives related to weekdays and then to weekends. Then, greedy procedures are used to satisfy the remaining objectives. A hill climbing procedure is employed to improve the quality of the roster. HR1 destroys the solution by removing the schedule of a medium number of employees. HR2 adopts a similar procedure accepting a greater change to the solution, proportional to the number of employees in the schedule, while HR3 slightly perturbs the solution by removing the shifts from only one employee. Five different local searchers are also adopted LS1–LS5, The first three are using different neighborhood operators from the literature, i.e., *vertical*, *horizontal* and *new* swaps respectively, while the last two local searchers follow a variable depth search strategy with different neighborhood operators (LS4: vertical and new, LS5: vertical, horizontal and new) [21]. A detailed description of all the employed low-level heuristics can be found in the documentation of HyFlex [22].

3.1   Credit Assignment Module

In order to assess the quality of the last action performed each time, a credit assignment module has been employed. The most conventional way to determine the impact of each move and assign a credit to it, is to associate the search move with the solution improvement caused by its application. To this end, we can calculate the credit of an action based on the improvement of the incumbent solution weighted by the effort paid to improve it.

More precisely, HHILS rewards each low-level perturbation heuristic according to the ability to improve the incumbent solution normalized by the total time spent to achieve this improvement. Let $\mathcal{S} = \{llh_1, llh_2, \ldots, llh_k\}$ be the set of k available low-level perturbation heuristics and t_{llh_i} be the execution time consumed by action $llh_i \in \mathcal{S}$ to search the solution space. The total time consumed by action llh_i can be calculated according to $t_{llh_i} = t_{llh_i} + t_{llh_i}^{sp}$, where $t_{llh_i}^{sp}$ is the execution time consumed by action llh_i for the current iteration. The reward r_{llh_i} of action llh_i can, therefore, be calculated according to the following equation: $r_{llh_i} = \frac{1 + improvement_{llh_i}}{t_{llh_i}}$, where $improvement_{llh_i}$ simply counts the number of times action llh_i improves the incumbent solution (i.e., $f(s_{cur}) < f(s_{tmp})$, where s_{cur} is the incumbent solution and s_{tmp} is the solution produced by the search operations at the current iteration). Notice that the value "1" in the numerator is responsible for assigning non-zero rewards to actions that have not yet led to an improvement of the incumbent. It also helps define the minimal reward for each action that is proportional to the time spent on searching for a new (improved or not) solution.

In general, the estimation of the empirical quality of an action has to consider also that recent rewards influence the quality more than earlier. Thus, it is a common practice to estimate the empirical quality of an action with a more accurate and reliable way based on a simple moving average of the current and past reward values [19,23,24]. As such, the empirical quality $q_{llh_i(t)}$ of each action (llh_i) in the current time step (t) is estimated in accordance with the following relaxation mechanism:

$$q_{llh_i}(t+1) = q_{llh_i}(t) + \gamma(r_{llh_i}(t) - q_{llh_i}(t))$$
$$= (1-\gamma)q_{llh_i}(t) + \gamma r_{llh_i}(t) \tag{1}$$

where $\gamma \in (0,1]$ is the adaptation rate which can amplify the influence of the most recent rewards over their history (here γ is fixed to 0.1) [19,24].

Notice also, that the credit assignment module could employ any reward value that can be measured during the search process to score the applied search operation. Representative examples of such rewards are the fitness improvement [19], ranking successful movements [19,25], and landscape analysis measures [26].

3.2   Action Selection Methodology

HHILS utilizes an action selection model to select the most suitable perturbation low-level heuristic to apply during search. The proposed framework can adopt any available action selection model. Many action selection models have been proposed in the literature recently. They usually adopt theory and practical algorithms motivated by different scientific fields, such as statistics, artificial intelligence, and machine learning. Some recent characteristic examples include probability matching [24], adaptive pursuit [24], statistical based models [25], and various reinforcement learning approaches [19,23,27].

Here, we employ six state-of-the-art action selection models with different characteristics: the baseline model of uniform selection, proportional selection, Probability Matching [24], Adaptive Pursuit [24], Soft-Max selection [23], and the Upper Confidence Bound Multi-Armed Bandit model [19,23]. For completeness purposes, we provide a short description of the developed models. A full description of all models can be found in the original publications [19,23,24]. The first model, *uniform selection* (US) acts as a baseline model, since it selects the available actions with equal probability regardless of their empirical quality, i.e., through a uniform random distribution. It is worth noting that randomized models in either the action space or the parameter space of an algorithm can be seen as essential baseline models [28]. *Proportional selection* (PS) simply selects an action proportionally based on its empirical quality, thus the higher the empirical quality value of an action the higher the probability of being selected.

Probability Matching (PM) and *Adaptive Pursuit* (AP) [24] are two well-known and successful probabilistic schemes that update the selection probability of an action by considering its empirical quality with respect to the remaining actions. PM updates the probability of each action with respect to its empirical quality while keeping a minimal probability for all actions to provide them with the opportunity to be selected regardless of their efficiency [24]. Similarly, AP adopts a probabilistic scheme with a winner-takes-all strategy, in which only efficient actions are promoted. AP discards the minimal probability value of each action to clearly distinguish the efficient from the inefficient actions, which in PM are treated equally. Thus, instead of proportionally adapting the probabilities of all available actions, it arises the probability of the best available action, and reduces the probabilities of the remaining actions.

The final two action selection models come from the field of Reinforcement Learning techniques, which have been successfully applied as action selection models [19,23]. Specifically, we employ two well known algorithms in this study, the *Softmax* (SM) [23] and the *Upper Confidence Bound* (UCB1) *Multi-armed Bandit algorithm* (MAB) [19,27,29]. The former essentially utilizes a Gibbs distribution to transform the empirical quality of each action to a probability. The higher the empirical quality, the higher the probability of an action being selected. Softmax also utilizes a temperature parameter to amplify or condense differences between the action probabilities. High temperature values lead to probabilities which are almost equal for all available actions, whilst low temperature values encourage larger differences. The scheme becomes very greedy towards selecting the best available action as the temperature value decreases (tends to zero).

In the multi-armed bandit case the UCB1 algorithm deterministically selects an action by following the principle of optimism in the face of uncertainty. The principle acts according to an optimistic guess on the merit of the expected empirical quality of each action and deterministically selects the action with the highest guess. When the guess is not correct, the optimistic guess is being rapidly decreased and the user is compelled to switch to a different action. If the guess is correct then the user will exploit the associated action and incur limited regret. "Optimism" can be formulated by an upper confidence bound that tries to balance the trade-off between exploration and exploitation of the selection process amongst the available actions. Hence, UCB1 favors the selection of the action that potentially exhibits the best reward (optimism in the face of uncertainty), while providing the opportunity for scarcely tried actions to be applied frequently. It is also worth mentioning that UCB1 is one of the very successful multi-armed bandit techniques and achieves an optimal regret rate on the multi-armed bandit formulation [29].

4   Experimental Results

We firstly present the experimental setup of this study (Sect. 4.1), which includes details about the environment used, the considered problem instances, the proposed as well as the state-of-the-art hyper-heuristics, and the parameter configurations of all considered algorithms. We then proceed with the presentation of the experimental results and thorough statistical analysis of the algorithms on the given problem instances.

4.1 Experimental Setup

In this experimental study, we develop and evaluate eighteen different hyper-heuristic approaches. Ten of them are based on the proposed HHILS framework, while the remaining eight are state-of-the-art hyper-heuristics that have been proposed in the literature. A family of hyper-heuristics can be defined in the proposed HHILS by adopting different action selection models. In this study, we develop the following six state-of-the-art action selection models resulting in the following hyper-heuristics: **ILS-US:** The HHILS variant that employs uniform selection among the available perturbation low-level heuristics; **ILS-PS:** HHILS employing the proportional selection method (roulette wheel); **ILS-AP:** HHILS employing the Adaptive Pursuit selection method [24]; **ILS-MAB:** HHILS employing the Upper Confidence Bound Multi-Armed Bandit method [19,23,27,29]; **ILS-PM:** HHILS employing the Probability Matching selection method [24]; **ILS-SM:** HHILS employing the Soft-Max selection method [23]. The HHILS variants select among four different low-level heuristics (HM, HR1–HR3). To evaluate the effect of the adaptive operator selection procedure, we develop four different HHILS variants that instead of using the proposed adaptive procedure, they adopt only one perturbation strategy. As such, **ILS-HM** is the HHILS with the HM perturbation heuristic, **ILS-HR1** is the HHILS with the HR1 perturbation heuristic, and so on.

Additionally, we develop and compare the following eight state-of-the-art hyper-heuristics, that are considered for the diversity of their characteristics and their high performance: **HH1, HH1A:** The HH1 and HH1A (HH1adap) predetermined sequence non-worsening selection hyper-heuristic for nurse rostering problems [30]; **HH2, HH2A:** The HH2 and HH1A (HH2adap) greedy absolute largest improvement selection hyper-heuristic for nurse rostering problems [30]; **PHunter:** the Pearl Hunter hyper-heuristic [14]; **SSHH:** The sequence-based selection hyper-heuristic with Hidden Markov Model [13]; **FSILS:** The Fair-Share hyper-heuristic algorithm [15], which is a state-of-the-art methodology in the field; **GIHH:** The GIHH, or Adapt-HH, hyper-heuristic algorithm [16], which was the winner of the CHESC 2011 competition [17]. Notice that all hyper-heuristics have been implemented in JAVA using the HyFlex (v1.1) framework [17,22,31,32].

In this study, we use a set of 39 real world nurse rostering problem instances with different characteristics (Table 1 [10][1]), where optimal solutions are known for the majority of the instances. To acquire equitable comparisons across all hyper-heuristics, we retain the same common parameters in all proposed approaches. Additionally, we preserve the default parameters used in the compared algorithms, as defined in the original works [13–16,30].

---

[1] More details can be found in http://www.cs.nott.ac.uk/~tec/NRP/.

Table 1. Nurse rostering problem instances used.

BCV-1.8.1, BCV-1.8.2, BCV-1.8.3, BCV-1.8.4, BCV-2.46.1, BCV-2.46.1, BCV-3.46.2, BCV-4.13.1,
BCV-4.13.2, BCV-5.4.1, BCV-6.13.1, BCV-6.13.2, BCV-7.10.1, BCV-8.13.1, BCV-8.13.2,
BCV-A.12.1, BCV-A.12.2, ORTEC01, ORTEC02, GPost, GPost-B, QMC-1, QMC-2,
Ikeg-2Sh-DATA1, Ikeg-3Sh-DATA1, Ikeg-3Sh-DATA1.2, Valouxis-1, WHPP, LLR, Musa, Azaiez,
SINTEF, CHILD-A2, ERMGH-A, ERMGH-B, ERRVH-A, ERRVH-B, MER-A, QMC-A

To evaluate the performance of a hyper-heuristic on each problem instance, we conducted 31 independent runs. For each of these runs, we considered the best-objective value gained as its performance value, where a lower objective value denotes better performance. To facilitate comparisons among the considered hyper-heuristics we utilize the following metrics: (a) the normalized objective value $f(\cdot)$, to fair compare the results across all problem instances, (b) the regret of a hyper-heuristic reg, to compare the considered algorithm with the best performing one, and (c) the d_{avg} metric to measure the distance of the average performance of a hyper-heuristic from the best known solution for a given instance.

We normalize the objective values in a common range of values by applying a linear transformation of all the objective values gained, $S = [S_{\min}, S_{\max}]$, to a normalized range $T = [T_{\min}, T_{\max}]$, here $T = [0, 1]$, where $S_{\min}$ is the best known solution of the considered problem and $S_{\max}$ is maximum objective function value observed by all utilized hyper-heuristics. The linear transformation can be calculated according to function $f : S \rightarrow T, f(y) = (y - S_{\min})/(S_{\max} - S_{\min})$, where $y \in S \subset \mathbb{R}$ is the performance value obtained by a hyper-heuristic on a given problem instance. For a given algorithm A from a set of n algorithms to compare $A \in \mathsf{A} = \{A_j, j = 1, \dots, n\}$, and a specific execution run i, we denote the regret of A as $reg_A^p(i) = f_A^p(i) - \min_{\forall i, A_j \in \mathsf{A}}(f_A^p)$, where p is the considered problem instance, $f_A^p(i)$ is the fitness value that has been attained for A at the i-th run for the problem instance p and $\min_{\forall i, A_j \in \mathsf{A}}(f_A^p)$ is the lowest fitness value across all execution runs of the algorithms included in A. Consequently, reg denotes the regret preferring an algorithm A to the best performing algorithm for a specific problem instance. Equivalently, d_{avg} is determined as $d_{avg} = \frac{BKS - \mu}{BKS}$, where BKS is the value of the best known solution according to the literature, regardless of whether the solution is optimal or not, and μ denotes the average performance value of the utilized algorithm for all executed runs on a problem instance. Thus, μ declares the normalized distance of the average performance of an algorithm from the BKS for a specific instance.

To facilitate fair comparisons, for each hyper-heuristic, each problem instance and each execution run, we employ the same execution time budget, i.e. 10 min of CPU time as measured by the benchmarking program provided in HyFlex [17, 22]. All experiments required have been conducted on a high performance cluster that has computation nodes with Intel Xeon E5645 CPUs and 24 GB of RAM running GNU/Linux Operating System. For this hardware, the allowed time limit equals to 646 s ($t_{allowed} = 646$ s).

4.2 Experimental Results and Analysis

Table 2 exhibits summarizing performance statistics for the developed hyper-heuristics on all considered nurse rostering problem instances, in terms of the normalized objective (f), the regret (reg) and the d_{avg} metrics used. Specifically, for each algorithm and each metric, the mean (μ_X), median (m_X) and standard deviation (σ_X) values are presented, where $X \in \{f, reg, d_{avg}\}$. We divide the table in three categories, the HHILS variants using single perturbation strategies, the proposed adaptive HHILS variants and the state-of-the-art hyper-heuristics. To improve the presentation of the results, we highlight with **boldface** font the cases where either the mean or the median metric values indicate best performance (i.e., the smallest values) across each category of algorithms.

Table 2. Summarizing statistics of the normalized objective values, regret, and d_{avg} metrics of all hyper-heuristics across all considered problem instances.

Algorithm	μ_f	m_f	σ_f	μ_{reg}	m_{reg}	σ_{reg}	$\mu_{d_{avg}}$	$m_{d_{avg}}$	$\sigma_{d_{avg}}$
ILS-HM	**0.1037**	**0.0162**	0.2008	**0.0690**	**0.0040**	0.1424	0.0740	0.1670	0.0000
ILS-HR1	0.2758	0.1907	0.2986	0.2497	0.1591	0.2785	0.6253	0.4376	0.9592
ILS-HR2	0.1091	0.0203	0.1993	0.0724	0.0058	0.1433	**0.0680**	**0.1369**	0.0192
ILS-HR3	0.1110	0.0181	0.2037	0.0757	0.0066	0.1469	0.0735	0.1456	0.0221
ILS-AP	**0.1053**	**0.0168**	0.2016	**0.0692**	**0.0047**	0.1438	**0.0632**	0.1482	0.0135
ILS-MAB	0.1083	0.0185	0.2017	0.0731	0.0069	0.1441	0.0695	0.1410	0.0187
ILS-PM	0.1103	0.0190	0.2025	0.0747	0.0080	0.1451	0.0764	0.1557	0.0115
ILS-PS	0.1065	0.0185	0.2000	0.0713	0.0075	0.1414	0.0692	**0.1363**	0.0112
ILS-SM	0.1102	0.0196	0.2014	0.0739	0.0073	0.1443	0.0696	0.1458	0.0166
ILS-US	0.1104	0.0199	0.2026	0.0747	0.0084	0.1461	0.0765	0.1435	0.0157
FSILS	**0.1114**	0.0242	0.1843	**0.0776**	0.0150	0.1139	**0.0904**	**0.1451**	0.0276
GIHH	0.1286	0.0231	0.2070	0.0940	0.0114	0.1524	0.1258	0.1738	0.0545
HH1	0.1409	0.0349	0.2048	0.1087	0.0227	0.1489	0.1622	0.2071	0.0898
HH1A	0.1358	0.0322	0.2005	0.1040	0.0227	0.1409	0.1523	0.1977	0.0755
HH2	0.1334	0.0227	0.2035	0.0994	**0.0108**	0.1468	0.1321	0.1941	0.0508
HH2A	0.2866	0.1364	0.3151	0.2613	0.1250	0.2921	0.4349	0.4449	0.2222
PHunter	0.1332	**0.0219**	0.2285	0.1000	0.0122	0.1816	0.1409	0.2409	0.0726
SSHH	0.1320	0.0242	0.2139	0.0992	0.0144	0.1620	0.1467	0.2115	0.0726

Considering the HHILS variants with a single perturbation strategy, Table 2 clearly suggests that the best performing hyper-heuristics are ILS-HM and ILS-HR2, in terms of mean and median normalized objective values. Notice also that ILS-HM is the best performing algorithm comparing against all considered hyper-heuristics. The remaining two cases exhibit large differences in both mean and median normalized objective values, which indicates that there is a large difference in the effectiveness of the low-level perturbation strategies. This behavior clearly motivates the proposed approach that endeavors to learn and identify the best performing perturbation strategy from the available ones. Regarding the proposed HHILS variants that employ the adaptive operator selection mechanism, ILS-AP and ILS-PS exhibit the best performance in terms of the mean and median normalized objective values among the HHILS variants, respectively. ILS-PM, ILS-SM and ILS-MAB behave similarly with ILS-US, i.e., with the uniform selection strategy, which indicates that the adaptive schemes are not capable of identifying and learning the best performing perturbation strategy. However, it can be observed that even a random selection of the available choices is effective in this set of problem instances. In addition, it is worth mentioning that all adaptive HHILS variants outperform all state-of-the-art hyper-heuristics in terms of mean normalized objective values. Among the state-of-the-art hyper-heuristics the most promising algorithm is FSILS. Notice that FSILS follows a similar iterated local search based hyper-heuristic that does not incorporate any learning, or action selection mechanism, to identify the best perturbation strategy. HHILS uses a different local search procedure (it does not employ an active list of local searchers) than FSILS and does not incorporate any restarting mechanism [15]. The good performance of FSILS indicates that the ILS paradigm is effective in the considered problem instances. The significantly better performance of HHILS adaptive variants also denotes that the combination of different perturbation operators and the learning mechanism enhance the search process. Similar observations can be made for the remaining robustness metrics, where they suggest that selecting any of the HHILS variants will not lead to higher regret (or d_{avg}) than using the state-of-the-art hyper-heuristics.

To further understand the behavior of the adaptive HHILS variants, Fig. 1 illustrates the selection frequency of the available perturbation strategies across all problem instances, averaged over all independent runs. It is clear that ILS-AP and ILS-PS show the largest frequency variation among the available strategies. In contrast, ILS-MAB, ILS-SM and ILS-PM tend to equally select all available strategies, making them behave similarly with the ILS-US (uniform selection), which validates the observed performance values in Table 2.

The application of the Friedman rank sum test [33] clearly suggests that statistically significant differences in performance occur ($p < 0.0001$, $\chi^2 = 244.8393$ at the 5% significance level) among the considered hyper-heuristics across all problem instances. Therefore, we carry out a post-hoc analysis by performing pairwise Wilcoxon-signed rank tests between any two performance samples obtained by the considered hyper-heuristics. The tests suggest that all proposed HHILS adaptive variants exhibit significant differences in performance against all state-of-the-art hyper-heuristics. ILS-AP significantly outperforms all HHILS variants apart from ILS-PS, which behaves equally well. ILS-MAB, ILS-PM, ILS-SM behave equally well with ILS-US. Considering the HHILS variants with single perturbation strategy ILS-HM outperforms all algorithms, while ILS-HR2 behaves equally well with the majority of the HHILS with adaptive strategies,

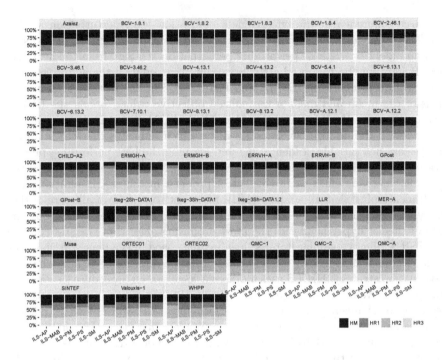

Fig. 1. Frequency graphs of the selected strategies by the adaptive HHILS variants for all problem instances averaged over all simulations

apart from ILS-AP which is superior. From the state-of-the-art hyper-heuristics GIHH, PHunter and SSHH behave similarly, while FSILS outperforms them.

To facilitate further comparisons, Table 3 exhibits descriptive statistics in terms of the best objective values attained by the considered hyper-heuristics among the majority of the tested problem instances. Notice that, we do not present results for BCV-5.4.1, BCV-8.13.1, BCV-8.13.2, LLR, Musa, Azaiez instances, since all hyper-heuristics solved these instances to the best-known solution without any deviation. The presented results align well with the afore-mentioned observations, while also suggest that there are instances where a large deviation in the observed performance among the hyper-heuristics exists (e.g. in CHILD-A2, ERRVH-A, ERRVH-B, and MER-A). This might also indicate that these problems are more challenging than others.

Table 3. Summarizing statistics of the objective values for all hyper-heuristics across all problem instances on B1 set.

Algorithm	μ_f	m_f	σ_f^2	μ_f	m_f	σ_f^2	μ_f	m_f	σ_f^2	μ_f	m_f	σ_f^2	μ_f	m_f	σ_f^2	μ_f	m_f	σ_f^2	μ_f	m_f	σ_f^2	μ_f	m_f	σ_f^2	μ_f	m_f	σ_f^2	μ_f	m_f	σ_f^2	μ_f	m_f	σ_f^2
	BCV-1.8.1			BCV-1.8.2			BCV-1.8.3			BCV-1.8.4			BCV-2.46.1			BCV-3.46.1			BCV-3.46.2			BCV-4.13.1			BCV-4.13.2			BCV-6.13.1			BCV-6.13.2		
ILS-HM	252.16	252	0.37	856.81	854	5.55	232.48	232	0.51	242.00	242	0.00	1584.26	1592	12.30	3303.61	3300	10.65	894.00	894	0.00	10.00	10	0.00	10.00	10	0.00	773.68	776	4.23	392.00	392	0.00
ILS-HR1	255.81	254	4.71	871.26	871	10.95	233.35	233	1.89	248.06	243	6.36	1609.74	1612	23.45	3349.61	3355	23.82	895.29	895	0.90	10.16	10	0.45	10.03	10	0.18	790.06	784	20.95	396.74	392	26.22
ILS-HR2	252.10	252	0.30	856.03	854	3.59	232.52	233	0.51	242.00	242	0.00	1582.35	1572	11.43	3311.94	3314	11.53	894.03	894	0.18	10.00	10	0.00	10.00	10	0.00	774.48	776	4.36	392.00	392	0.00
ILS-HR3	252.13	252	0.34	855.45	854	2.47	232.35	232	0.49	242.00	242	0.00	1575.94	1572	8.00	3319.94	3319	11.38	894.06	894	0.25	10.00	10	0.00	10.00	10	0.00	775.48	776	5.03	392.00	392	0.00
FSILS	252.52	253	0.51	856.26	855	2.49	232.55	233	0.51	242.00	242	0.00	1572.87	1572	3.60	3323.52	3323	14.39	894.16	894	0.37	10.00	10	0.00	10.00	10	0.00	775.23	776	4.78	392.00	392	0.00
GIHH	252.32	252	0.48	855.55	854	2.78	232.61	233	0.50	242.00	242	0.00	1582.58	1575	11.18	3338.81	3342	16.48	894.52	894	0.57	10.00	10	0.00	10.00	10	0.00	775.26	776	3.79	392.00	392	0.00
HH1	252.74	253	0.58	854.26	854	1.41	232.61	233	0.56	242.00	242	0.00	1585.55	1592	10.82	3347.90	3350	24.79	894.94	895	0.51	10.00	10	0.00	10.00	10	0.00	777.81	776	5.74	392.00	392	0.00
HH1A	252.81	253	0.48	854.16	854	1.27	232.81	233	0.48	242.00	242	0.00	1576.42	1572	7.82	3342.19	3344	19.94	894.68	895	0.54	10.00	10	0.00	10.00	10	0.00	775.23	776	4.88	392.00	392	0.00
HH2	252.19	252	0.40	855.55	854	4.41	232.71	233	0.46	242.00	242	0.00	1577.32	1572	8.81	3366.90	3367	12.75	894.58	895	0.50	10.00	10	0.00	10.00	10	0.00	773.48	776	4.26	392.00	392	0.00
HH2A	253.39	253	0.56	856.74	856	2.02	233.03	233	0.48	242.48	242	0.63	1581.06	1574	9.64	3387.61	3390	22.26	896.03	896	0.84	10.00	10	0.00	10.00	10	0.00	780.16	784	4.27	392.00	392	0.00
ILS-AP	252.06	252	0.25	854.61	854	2.04	232.39	232	0.50	242.00	242	0.00	1576.74	1572	8.40	3314.52	3316	11.04	894.06	894	0.25	10.00	10	0.00	10.00	10	0.00	774.87	776	3.43	392.00	392	0.00
ILS-MAB	252.26	252	0.44	854.68	854	1.97	232.26	232	0.44	242.00	242	0.00	1580.42	1572	12.39	3314.45	3315	9.45	894.06	894	0.25	10.00	10	0.00	10.00	10	0.00	775.81	776	4.50	392.00	392	0.00
ILS-PM	252.13	252	0.34	855.23	854	2.63	232.32	232	0.48	242.00	242	0.00	1583.68	1592	10.09	3315.58	3316	11.67	894.00	894	0.00	10.00	10	0.00	10.00	10	0.00	776.00	776	4.13	392.00	392	0.00
ILS-PS	252.16	252	0.37	855.00	854	1.98	232.39	232	0.50	242.00	242	0.00	1581.71	1572	12.49	3316.39	3315	13.34	894.06	894	0.25	10.00	10	0.00	10.00	10	0.00	775.29	776	4.60	392.00	392	0.00
ILS-SM	252.26	252	0.44	855.74	854	3.12	232.26	232	0.44	242.00	242	0.00	1582.32	1592	10.16	3314.19	3311	12.38	894.03	894	0.18	10.00	10	0.00	10.00	10	0.00	775.23	776	3.17	392.00	392	0.00
ILS-US	252.00	252	0.00	856.06	854	4.48	232.52	233	0.51	242.00	242	0.00	1583.71	1592	11.31	3320.19	3318	14.37	894.03	894	0.18	10.00	10	0.00	10.00	10	0.00	775.58	776	2.91	392.00	392	0.00
PHunter	252.29	252	0.46	856.10	854	3.07	232.52	233	0.51	242.00	242	0.00	1574.68	1572	6.79	3318.39	3314	16.95	894.23	894	0.43	10.00	10	0.00	10.00	10	0.00	779.35	784	5.38	392.00	392	0.00
SSHH	252.29	252	0.46	857.03	857	4.35	232.55	233	0.51	242.00	242	0.00	1584.58	1592	12.00	3315.42	3316	15.94	894.39	894	0.50	10.00	10	0.00	10.00	10	0.00	782.00	784	3.46	392.00	392	0.00
	BCV-7.10.1			BCV-A.12.1			BCV-A.12.2			ORTEC01			ORTEC02			GPost			GPost-B			QMC-1			QMC-2			Ikeg-2Sh-D1			Ikeg-3Sh-D1		
ILS-HM	383.26	381	2.53	1573.03	1550	91.26	2111.84	2093	97.63	321.61	320	12.93	344.84	345	21.07	8.61	9	0.80	5.84	5	1.29	16.26	16	0.82	30.03	30	0.48	0.26	0	0.44	15.45	15	3.15
ILS-HR1	412.10	386	40.51	2101.42	1980	493.70	2609.39	2510	359.90	901.03	460	813.73	1011.61	530	1023.55	419.55	218	503.34	242.42	217	264.11	22.87	23	2.83	31.03	31	1.20	6.39	6	2.46	28.74	29	4.44
ILS-HR2	382.29	381	2.22	1593.65	1570	94.78	2123.16	2112	97.04	332.74	335	10.63	359.42	360	17.10	8.68	9	1.11	5.84	5	1.24	17.26	17	0.93	29.84	30	0.52	0.29	0	0.53	17.77	18	3.27
ILS-HR3	382.61	381	2.38	1588.97	1564	109.56	2144.26	2134	93.32	332.58	335	10.87	357.42	355	13.90	8.65	9	1.05	6.81	7	1.42	17.58	17	0.99	29.87	30	0.50	0.29	0	0.53	18.61	19	2.53
FSILS	381.65	381	1.70	1681.29	1685	110.69	2212.32	2205	95.81	353.55	355	16.44	375.16	375	22.60	10.10	10	1.60	6.74	6	1.26	18.68	19	1.17	30.06	30	0.36	1.52	1	1.23	20.45	21	3.25
GIHH	381.48	381	1.50	1782.00	1785	158.08	2325.35	2285	176.28	345.03	345	29.04	471.94	390	340.28	9.35	9	1.43	6.97	6	1.74	17.74	18	1.79	29.68	30	0.65	1.71	2	1.49	22.32	23	4.76
HH1	381.00	381	0.00	1819.16	1770	222.05	2465.42	2415	249.75	340.48	340	16.65	390.32	395	33.51	9.32	9	1.22	6.77	6	1.84	19.74	20	1.90	29.87	30	0.62	2.68	2	2.02	24.42	23	5.18
HH1A	381.00	381	0.00	1831.48	1800	148.98	2406.81	2385	176.54	342.58	340	19.19	390.00	395	24.63	9.55	9	1.36	6.61	6	1.65	19.39	19	1.65	29.97	30	0.66	3.35	3	2.18	25.03	25	5.92
HH2	382.77	381	2.43	1802.06	1805	113.15	2402.00	2415	129.12	323.45	325	12.08	354.84	355	21.27	8.48	8	0.72	5.26	5	0.77	18.97	19	1.33	29.52	30	0.51	0.58	0	0.85	21.03	21	3.83
HH2A	381.00	381	0.00	2114.94	2125	186.35	2737.19	2755	179.95	406.45	400	24.98	485.68	490	44.14	10.61	11	1.56	10.03	10	2.50	27.03	27	1.45	31.10	31	0.65	14.06	14	2.83	50.03	51	5.39
ILS-AP	382.61	381	2.38	1585.87	1560	90.75	2136.65	2148	99.94	325.00	325	11.03	353.87	355	20.56	8.81	9	1.14	6.13	6	1.26	16.65	17	0.88	29.84	30	0.58	0.42	0	0.62	17.06	16	3.12
ILS-MAB	383.10	381	2.51	1575.03	1585	75.01	2122.77	2129	85.29	332.42	335	11.75	355.35	360	16.99	8.81	9	0.98	5.77	5	0.99	17.29	17	0.90	29.81	30	0.54	0.39	0	0.67	16.48	17	2.47
ILS-PM	383.42	381	2.54	1601.39	1560	107.19	2141.00	2150	87.26	326.94	330	10.85	353.39	355	13.99	8.61	9	1.33	5.58	5	1.18	17.16	17	0.86	29.97	30	0.55	0.42	0	0.56	17.52	17	3.11
ILS-PS	383.26	381	2.53	1574.45	1560	97.73	2141.48	2145	106.28	328.10	330	12.63	355.00	350	16.78	8.81	9	0.91	5.81	5	1.08	16.77	17	0.88	29.77	30	0.50	0.35	0	0.61	17.26	18	1.71
ILS-SM	383.42	381	2.54	1581.19	1575	84.34	2137.26	2110	88.38	331.29	330	13.48	356.13	355	18.06	8.81	9	1.33	6.13	6	1.31	17.26	17	0.77	29.84	30	0.52	0.55	0	0.85	17.58	18	2.38
ILS-US	383.42	381	2.54	1533.94	1519	85.87	2145.10	2135	107.37	331.77	330	12.62	359.03	360	18.00	9.06	9	1.59	6.35	6	1.70	17.03	17	0.84	30.00	30	0.63	0.35	0	0.61	17.03	17	2.34
PHunter	383.26	381	2.53	1623.03	1600	116.59	2219.68	2230	97.73	324.84	325	14.11	393.23	345	240.29	8.71	9	0.82	5.74	5	1.29	17.19	17	1.19	29.68	30	0.79	1.13	1	1.23	20.32	20	3.44
SSHH	383.26	381	2.53	1745.03	1750	142.48	2334.32	2355	165.31	340.97	335	32.70	406.16	360	248.94	9.52	9	1.82	13.35	7	37.27	17.77	18	1.23	29.90	30	0.70	1.35	1	1.45	19.65	19	3.67
	Ikeg-3Sh-D1.2			Valouxis			WHPP			SINTEF			CHILD-A2			ERMGH-A			ERMGH-B			ERRVH-A			ERRVH-B			MER-A			QMC-A		
ILS-HM	18.29	18	2.90	64.52	60	19.81	1873.00	2001	338.94	0.81	1	0.60	1147.45	1129	38.66	795.13	795	0.50	1358.77	1355	31.85	2218.87	2214	40.53	3195.06	3183	44.87	9614.97	9581	236.72	26.39	26	0.50
ILS-HR1	35.48	33	19.15	187.74	160	59.48	2034.84	2003	314.46	8.61	8	3.72	1189.48	1145	84.70	796.71	797	1.74	1364.58	1366	33.55	2298.00	2275	85.76	3346.03	3348	113.30	9757.55	9702	246.88	27.29	27	1.16
ILS-HR2	20.42	20	2.66	76.77	80	13.76	1485.77	1005	506.85	2.06	2	0.73	1154.35	1138	41.93	795.26	795	0.89	1367.65	1365	25.37	2229.94	2237	39.20	3224.87	3231	53.18	9644.06	9649	118.19	26.26	26	0.44
ILS-HR3	20.26	20	2.67	73.55	80	13.05	1711.42	2001	460.60	2.35	2	0.61	1153.16	1132	48.60	795.00	795	0.00	1368.87	1359	23.47	2240.16	2226	37.17	3223.00	3209	45.23	9609.10	9589	100.32	26.29	26	0.46
FSILS	24.39	24	2.62	78.71	80	16.28	1904.65	2001	299.57	2.68	3	0.98	1157.13	1136	44.80	795.00	795	0.00	1312.29	1316	10.92	2243.16	2240	38.38	3258.77	3251	57.87	9771.29	9757	154.33	26.32	26	0.48
GIHH	24.71	24	4.31	87.10	80	26.10	1872.26	2001	340.31	1.71	1	1.19	1198.94	1205	87.97	795.00	795	0.00	1353.19	1356	28.59	2268.71	2258	74.77	3275.39	3258	86.99	9656.45	9609	133.76	26.19	26	0.40
HH1	29.61	30	5.22	90.97	80	21.19	1969.61	2002	179.58	1.97	2	1.20	1239.52	1223	84.02	795.19	795	0.60	1320.48	1319	7.05	2305.81	2284	70.62	3332.23	3322	103.92	9942.03	9851	274.83	26.42	26	0.62
HH1A	28.77	29	3.98	97.42	100	23.52	1937.94	2002	249.45	2.10	2	1.11	1216.55	1214	77.52	795.13	795	0.43	1321.68	1319	7.32	2302.61	2276	101.23	3365.61	3333	107.40	10191.48	10101	394.44	26.55	27	0.51
HH2	25.65	26	3.30	63.87	60	13.08	1874.32	2002	337.51	0.87	1	0.67	1227.00	1233	68.60	795.19	795	0.60	1327.52	1326	10.20	2399.61	2368	103.40	3605.48	3570	187.45	15449.48	15508	2023.95	26.19	26	0.40
HH2A	55.81	56	5.44	154.84	160	20.64	2005.55	2005	2.17	5.16	5	0.93	1745.39	1740	144.07	795.03	795	0.18	1332.39	1330	8.96	3213.52	3194	241.26	4797.94	4810	249.14	46794.58	47194	2194.82	26.90	27	0.40
ILS-AP	19.74	19	2.25	72.26	80	15.21	1775.74	2001	423.99	1.52	2	0.89	1141.74	1128	34.65	795.00	795	0.00	1366.32	1363	26.80	2222.90	2221	31.65	3217.48	3217	53.60	9574.81	9560	110.93	26.16	26	0.37
ILS-MAB	20.35	21	2.44	76.77	80	17.20	1711.68	2001	460.26	1.94	2	0.81	1160.84	1141	43.99	795.00	795	0.00	1365.68	1366	27.91	2235.68	2251	37.48	3223.90	3233	41.19	9561.94	9574	127.39	26.35	26	0.49
ILS-PM	20.13	20	2.40	73.55	80	18.17	1807.71	2001	400.74	1.68	2	0.75	1139.81	1127	37.04	795.00	795	0.00	1363.74	1363	29.97	2243.84	2256	37.99	3206.19	3188	48.67	9608.77	9579	193.31	26.19	26	0.40
ILS-PS	21.03	21	2.75	70.97	80	17.77	1743.97	2001	443.25	1.68	2	0.60	1141.94	1130	36.53	795.00	795	0.00	1360.74	1357	30.81	2225.42	2230	36.04	3187.68	3169	46.61	9628.00	9656	123.80	26.32	26	0.48
ILS-SM	21.39	22	2.40	72.90	80	13.21	1743.71	2001	443.92	1.94	2	0.96	1149.87	1130	39.09	795.00	795	0.00	1365.94	1366	28.35	2238.29	2252	39.93	3224.06	3239	42.36	9639.87	9631	130.68	26.13	26	0.34
ILS-US	20.48	20	2.22	73.55	80	13.05	1711.68	2001	460.19	2.03	2	0.84	1149.32	1132	37.15	795.00	795	0.00	1361.71	1369	35.49	2215.61	2207	34.95	3221.00	3235	50.78	9659.94	9618	230.51	26.26	26	0.44
PHunter	23.94	24	3.53	70.32	80	17.03	1968.94	2001	179.27	1.35	1	0.80	1641.65	1491	523.42	795.00	795	0.00	1382.26	1395	18.78	2404.52	2285	225.19	3478.00	3302	457.38	10048.58	10013	290.78	26.45	26	0.62
SSHH	23.55	23	3.99	79.35	80	18.96	1969.03	2001	179.66	2.35	2	1.11	1182.68	1179	96.26	795.19	795	0.60	1370.94	1363	25.96	2264.23	2250	74.10	3249.29	3240	86.91	9632.97	9601	100.06	26.52	26	0.57

5   Conclusions

In this study, we proposed a simple and effective Iterated Local Search based selection hyper-heuristic framework that adopts the adaptive operator selection paradigm to successfully address a wide variety of nurse rostering problem instances. It employs an action selection model to select different perturbation strategies and a credit assignment module to appropriately score them. The proposed framework is able to adopt any action selection model and credit assignment mechanism available in the literature. In this study, we have tested six different action selection models resulting in new competitive hyper-heuristics. The high level nature of the framework makes it widely applicable to new or unseen problem instances/domains without requiring further modifications.

The adaptive characteristics of the proposed framework are investigated by comparing with its non-adaptive variants, while its performance is evaluated through comparisons with 8 state-of-the-art hyper-heuristics on 39 different nurse rostering problem instances. The experimental results suggest that the proposed framework operates significantly better against the state-of-the-art hyper-heuristics. The proposed adaptive mechanisms seem to be effective across the majority of the problem instances, with the Adaptive Pursuit and the simple proportional action selection model to be able to learn and identify the most promising perturbation strategies. The remaining three considered action selection models operate similarly with the uniform selection, which indicates that they are not able to identify the best performing perturbation strategy. However, even a simple random selection performs significantly better than the majority of the state-of-the-art algorithms. Therefore, further experimentation and analysis of the adaptive strategies on more nurse rostering problem instances have to be performed to draw safe conclusions about their behavior. Future work will also include comparisons with specialized state-of-the-art heuristics developed for nurse rostering problems.

References

1. Burke, E.K., Kendall, G., Newall, J., Hart, E., Ross, P., Schulenburg, S.: Hyper-heuristics: an emerging direction in modern search technology. In: Glover, F., Kochenberger, G.A. (eds.) Handbook of Metaheuristics, pp. 457–474. Springer, Boston (2003). doi:10.1007/0-306-48056-5_16
2. Burke, E.K., Gendreau, M., Hyde, M., Kendall, G., Ochoa, G., Özcan, E., Qu, R.: Hyper-heuristics: a survey of the state of the art. J. Oper. Res. Soc. 64(12), 1695–1724 (2013)
3. Burke, E.K., Hyde, M., Kendall, G., Ochoa, G., Özcan, E., Woodward, J.R.: A classification of hyper-heuristic approaches. In: Gendreau, M., Potvin, J.Y. (eds.) Handbook of Metaheuristics. International Series in Operations Research & Management Science, vol. 146, pp. 449–468. Springer, Boston (2010). doi:10.1007/978-1-4419-1665-5_15
4. Burke, E.K., Causmaecker, P.D., Berghe, G.V., Landeghem, H.V.: The state of the art of nurse rostering. J. Sched. 7(6), 441–499 (2004)

5. Ernst, A.T., Jiang, H., Krishnamoorthy, M., Sier, D.: Staff scheduling and rostering: a review of applications, methods and models. Eur. J. Oper. Res. **153**(1), 3–27 (2004)
6. Asta, S., Özcan, E., Curtois, T.: A tensor based hyper-heuristic for nurse rostering. Knowl. Based Syst. **98**, 185–199 (2016)
7. Lü, Z., Hao, J.K.: Adaptive neighborhood search for nurse rostering. Eur. J. Oper. Res. **218**(3), 865–876 (2012)
8. Rae, C., Pillay, N.: Investigation into an evolutionary algorithm hyperheuristic for the nurse rostering problem. In: Proceedings of the 10th International Conference on the Practice and Theory of Automated, PATAT 2014, pp. 527–532 (2014)
9. Anwar, K., Awadallah, M.A., Khader, A.T., Al-betar, M.A.: Hyper-heuristic approach for solving nurse rostering problem. In: 2014 IEEE Symposium on Computational Intelligence in Ensemble Learning (CIEL), pp. 1–6, December 2014
10. Burke, E.K., Curtois, T.: New approaches to nurse rostering benchmark instances. Eur. J. Oper. Res. **237**(1), 71–81 (2014)
11. Bai, R., Burke, E., Kendall, G., Li, J., McCollum, B.: A hybrid evolutionary approach to the nurse rostering problem. IEEE TEVC **14**(4), 580–590 (2010)
12. Burke, E.K., Li, J., Qu, R.: A hybrid model of integer programming and variable neighbourhood search for highly-constrained nurse rostering problems. Eur. J. Oper. Res. **203**(2), 484–493 (2010)
13. Kheiri, A., Keedwell, E.: A sequence-based selection hyper-heuristic utilising a hidden Markov model. In: Proceedings of the 2015 Annual Conference on Genetic and Evolutionary Computation, GECCO 2015, pp. 417–424. ACM, New York (2015)
14. Chan, C.Y., Xue, F., Ip, W.H., Cheung, C.F.: A hyper-heuristic inspired by pearl hunting. In: Hamadi, Y., Schoenauer, M. (eds.) LION 2012. LNCS, pp. 349–353. Springer, Heidelberg (2012). doi:10.1007/978-3-642-34413-8_26
15. Adriaensen, S., Brys, T., Nowé, A.: Fair-share ILS: a simple state-of-the-art iterated local search hyperheuristic. In: Proceedings of the 2014 Conference on Genetic and Evolutionary Computation, GECCO 2014, pp. 1303–1310. ACM (2014)
16. Mısır, M., Verbeeck, K., Causmaecker, P., Berghe, G.: An intelligent hyper-heuristic framework for CHeSC 2011. In: Hamadi, Y., Schoenauer, M. (eds.) LION 2012. LNCS, pp. 461–466. Springer, Heidelberg (2012). doi:10.1007/978-3-642-34413-8_45
17. CHeSC 2011 (2011). http://www.asap.cs.nott.ac.uk/external/chesc2011/
18. Battiti, R., Brunato, M., Mascia, F.: Reactive Search and Intelligent Optimization. Operations research/Computer Science Interfaces, vol. 45. Springer, Boston (2008). doi:10.1007/978-0-387-09624-7
19. Fialho, A.: Adaptive operator selection for optimization. Ph.D. thesis, Université Paris-Sud XI, Orsay, France, December 2010
20. Burke, E.K., Curtois, T., Post, G., Qu, R., Veltman, B.: A hybrid heuristic ordering and variable neighbourhood search for the nurse rostering problem. Eur. J. Oper. Res. **188**(2), 330–341 (2008)
21. Burke, E.K., Curtois, T., Qu, R., Vanden Berghe, G.: A time predefined variable depth search for nurse rostering. INFORMS J. Comput. **25**(3), 411–419 (2013)
22. CHeSC 2014: The second cross-domain heuristic search challenge (2014). http://www.hyflex.org/chesc2014/, http://www.hyflex.org/. Accessed 25 Mar 2015
23. Sutton, R.S., Barto, A.G.: Introduction to Reinforcement Learning, 1st edn. MIT Press, Cambridge (1998)
24. Thierens, D.: Adaptive strategies for operator allocation. In: Lobo, F., Lima, C., Michalewicz, Z. (eds.) Parameter Setting in Evolutionary Algorithms. SCI, vol. 54, pp. 77–90. Springer, UK (2007). doi:10.1007/978-3-540-69432-8_4

25. Epitropakis, M.G., Tasoulis, D.K., Pavlidis, N.G., Plagianakos, V.P., Vrahatis, M.N.: Tracking particle swarm optimizers: an adaptive approach through multinomial distribution tracking with exponential forgetting. In: 2012 IEEE Congress on Evolutionary Computation (CEC), pp. 1–8 (2012)
26. Munoz, M.A., Sun, Y., Kirley, M., Halgamuge, S.K.: Algorithm selection for black-box continuous optimization problems: a survey on methods and challenges. Inf. Sci. **317**, 224–245 (2015)
27. Fialho, A., Costa, L.D., Schoenauer, M., Sebag, M.: Analyzing bandit-based adaptive operator selection mechanisms. Ann. Math. Artif. Intell. **60**(1–2), 25–64 (2010)
28. Karafotias, G., Hoogendoorn, M., Eiben, A.E.: Why parameter control mechanisms should be benchmarked against random variation. In: 2013 IEEE Congress on Evolutionary Computation (CEC), pp. 349–355, June 2013
29. Auer, P., Cesa-Bianchi, N., Fischer, P.: Finite-time analysis of the multiarmed bandit problem. Mach. Learn. **47**(2–3), 235–256 (2002)
30. Banerjea-Brodeur, M.: Selection hyper-heuristics for healthcare scheduling. Ph.D. thesis, University of Nottingham, UK, June 2013
31. Asta, S., Özcan, E., Parkes, A.J.: Batched mode hyper-heuristics. In: Nicosia, G., Pardalos, P. (eds.) LION 2013. LNCS, vol. 7997, pp. 404–409. Springer, Heidelberg (2013). doi:10.1007/978-3-642-44973-4_43
32. Ochoa, et al.: HyFlex: a benchmark framework for cross-domain heuristic search. In: Hao, J.-K., Middendorf, M. (eds.) EvoCOP 2012. LNCS, vol. 7245, pp. 136–147. Springer, Heidelberg (2012). doi:10.1007/978-3-642-29124-1_12
33. Hollander, M., Wolfe, D.A., Chicken, E.: Nonparametric Statistical Methods, 3rd edn. Wiley, Hoboken (2013)

Learning a Reactive Restart Strategy to Improve Stochastic Search

Serdar Kadioglu[1], Meinolf Sellmann[2], and Markus Wagner[3(✉)]

[1] Department of Computer Science, Brown University, Providence, RI, USA
serdark@cs.brown.edu
[2] Cortlandt Manor, Cortlandt, NY, USA
meinolf@gmail.com
[3] Optimisation and Logistics, The University of Adelaide, Adelaide, SA, Australia
markus.wagner@adelaide.edu.au

Abstract. Building on the recent success of bet-and-run approaches for restarted local search solvers, we introduce the idea of learning online adaptive restart strategies. Universal restart strategies deploy a fixed schedule that runs with utter disregard of the characteristics that each individual run exhibits. Whether a run looks promising or abysmal, it gets run exactly until the predetermined limit is reached. Bet-and-run strategies are at least slightly less ignorant as they decide which trial to use for a long run based on the performance achieved so far. We introduce the idea of learning fully adaptive restart strategies for black-box solvers, whereby the learning is performed by a parameter tuner. Numerical results show that adaptive strategies can be learned effectively and that these significantly outperform bet-and-run strategies.

Keywords: Restart strategies · Adaptive methods · Parameter tuning

1 Introduction

Restarted search has become an integral part of combinatorial search algorithms. Even before heavy-tailed runtime distributions were found to explain the massive variance in search performance [1], in local search restarts were commonly used as a search diversification technique [2].

Fixed-schedule restart strategies were studied theoretically in [3]. For SAT and constraint programming solvers, practical studies followed. For example, one study found that there is hardly any difference between theoretically optimal schedules and simple geometrically growing limits [4]. SAT solvers used geometrically growing limits for quite some time before the community largely adapted theoretically optimal schedules (whereby the optimality guarantees are based on assumptions that actually do not hold for clause-learning solvers where consecutive restarts are not independent). Audemard and Simon [5] argued that fixed schedules are suboptimal for SAT solvers and designed adaptive restarts strategies for one SAT solver specifically.

© Springer International Publishing AG 2017
R. Battiti et al. (Eds.): LION 2017, LNCS 10556, pp. 109–123, 2017.
https://doi.org/10.1007/978-3-319-69404-7_8

In this paper, we describe a general methodology for embedding any black-box optimization solver into an adaptive stochastic restart framework. The framework monitors certain key performance metrics that are based on the evolution of the objective function values of the solutions found. Based on these observations, the method then adaptively computes scores that affect the likelihood whether we continue the current run beyond the original limit, whether we start a new run, or whether we continue the best run so far. We employ an automatic parameter tuner to learn how to adapt these probabilities dependent on the observed performance metrics.

In the following, we recap the idea of bet-and-run strategies. We continue with reviewing the recently introduced idea of hyper-parameterizing local search solvers to achieve superior online adaptive behavior. We then introduce the idea of using automatic hyper-reactive search tuning for learning adaptive restart strategies. Finally, we present experimental results that clearly show that adaptive search significantly outperforms bet-and-run strategies.

2   Restart Strategies

Nowadays, stochastic search algorithms and randomized search heuristics are frequently restarted: If a run does not conclude within a pre-determined limit, we restart the algorithm. This was shown to to help avoid heavy-tailed runtime distributions [1]. Due to the added complexity of designing an appropriate restart strategy for a given target algorithm, the two most common techniques used are to either restarts with a certain probability at the end of each iteration, or to employ a fixed schedule of restarts.

Some theoretical results exist on how to construct optimal restart strategies. For example, Luby et al. [3] showed that, for Las Vegas algorithms with known run time distribution, there is an optimal stopping time in order to minimize the expected running time. They also showed that, if the distribution is unknown, there is a universal sequence of running times which is the optimal restarting strategy up to constant factors.

Fewer results are known for the optimization case. Marti [6] and Lourenco et al. [7] present practical approaches, and a recent theoretical result is presented by Schoenauer et al. [8]. Particularly for the satisfiability problem, several studies make an empirical comparison of a number of restart policies [9,10].

Quite often, classical optimization algorithms are deterministic and thus cannot be improved by restarts. This also appears to hold for certain popular modern solvers, such as IBM ILOG CPLEX. However, characteristics can change when memory constraints or parallel computations are encountered. This was the initial idea of Lalla-Ruiz and Voß [11], who investigated different mathematical programming formulations to provide different starting points for the solver.

Many other modern optimization algorithms, while also working mostly deterministically, have some randomized component, for example by choosing a random starting point. Two very typical uses for an algorithm with time budget t are to (a) use all of time t for a single run of the algorithm (single-run strategy), or (b) to make a number of k runs of the algorithm, each with running time t/k (multi-run strategy).

Extending these two classical strategies, Fischetti et al. [12] investigated the use of the following BET-AND-RUN strategy with a total time limit t:

Phase 1 performs k runs of the algorithm for some (short) time limit t_1 with $t_1 \leq t/k$.

Phase 2 uses remaining time $t_2 = t - k \cdot t_1$ to continue *only the best run* from the first phase until timeout.

Note that the multi-run strategy of restarting from scratch k times is a special case by choosing $t_1 = t/k$ and $t_2 = 0$ and the single-run strategy corresponds to $k = 1$; thus, it suffices to consider different parameter settings of the *bet-and-run* strategy to also cover these two strategies.

Fischetti et al. [12] experimentally studied such a BET-AND-RUN strategy for mixed-integer programming. They explicitly introduce diversity in the starting conditions of the used MIP solver (IBM ILOG CPLEX) by directly accessing internal mechanisms. In their experiments, $k = 5$ performed best.

Recently, Friedrich et al. [13] investigated a comprehensive range of BET-AND-RUN strategies on the traveling salesperson problem and the minimum vertex cover problem. Their best strategy was RESTARTS$_{1\%}^{40}$, which in the first phase does 40 short runs with a time limit that is 1% of the total time budget and then uses the remaining 60% of the total time budget to continue the best run of the first phase. They investigated the use of the universal sequence of Luby et al. [3] as well, using various choices of t_1, however, it turned out inferior.

The theoretical analysis is provided by Lissovoi et al. [14], who investigated BET-AND-RUN for a family of pseudo-Boolean functions, consisting of a plateau and a slope, as an abstraction of real fitness landscapes with promising and deceptive regions. The authors showed that BET-AND-RUN with non-trivial k and t_1 are necessary to find the global optimum efficiently. Also, they showed that the choice of t_1 is linked to properties of the function, and they provided a fixed budget analysis to guide selection of the bet-and-run parameters to maximise expected fitness after $t = k \cdot t_1 + t_2$ fitness evaluations.

The goal of our present research is to address the two challenges encountered in previous works: the need to set k and t_1 in case of BET-AND-RUN, and the general issue of inflexibility in previous approaches. Our framework can decide online whether (i) the current run should be continued, (ii) the best run so far should be continued, or (iii) a completely new run should be started.

3   Learning Dynamic Parameter Updates

Our objective is to provide a generic framework for making restart strategies adaptive for any optimization solver. To this end we build on the idea to use parameter tuners for training adaptive search strategies [15,16] and a recently proposed approach for constructing a hyper-reactive dialectic search solver [17].

In [17], an existing local search meta-heuristic called dialectic search [18] was modified in such a way that the search decisions (when and how much to diversify, how strongly to intensify, when to restart, etc.) were taken with regard to the way how the optimization was observed to progress. In essence, the solver tracked features of the optimization process itself and then tied these to decisions (such as: which percentage of variables to modify to generate a new start point) via logistic regression functions. The weights of these functions, one for each meta-heuristic search decision, then became the hyper-parameters of the solver.

Key to making this work in practice is an effective method for learning the weights in the logistic regression functions. Since the only immediately meaningful observation is the overall performance of the solver, parameter tuner GGA [19] was used to "learn" which weights result in good performance.

4   A Hyper-Parameterized Restart Strategy

We now combine the two core ideas presented above. Namely, the idea of considering a batch of runs with the option to continue some of them, and the idea to automatically learn which run to continue or whether to start a new run based on the observed performance characteristics of past runs.

The first ingredient we need are features that somehow give us an idea of the big picture of what is going on when tackling the instance at hand.

4.1   Features

Whenever a restart decision has to be made, we have three options. We can either continue the current run, we can continue the best run so far, or we can start a completely new run. For each of these options we essentially track two values: The first tells us how good each run looked initially, the second what the trajectory looks like for making further progress.

For the current run and the best run so far, we record their best objective function found after the initial limit. For the new run option, we track how well any new run did after the initial limit and compute the running average.

For the trajectory of the current and the best run, we extrapolate the performance improvement achieved between the best solution found in the initial run and the best performance achieved so far. The extrapolation point is the end of the remaining time we have for the optimization.

For the new runs, to get an estimate how well we might do if (from now until the overall time limit is reached) all we did was run new runs, we consider the standard deviation in objective function performance. Then, we estimate the trajectory as the average minus the standard deviation times the square root of two times the logarithm of the number of new runs we can still afford to conduct. While not exact, this is a lower bound for the minimum of repeated stochastic experiments for which all we know are the mean and the standard deviation [20].

We thus compute six data points that we can use to decide whether to continue the current run, give the current best run more time, or start a completely new run. One complication arises. Namely, for different instances, the objective function values observed will generally operate on vastly different scales. However, to learn strategies offline, we need to compute weights, and these need to work with all kinds of instances. Consequently, rather than taking average and projected objective function values at face value, we first normalize them.

In particular, we consider the three initial values (best found in initial time interval for current and best, and running average of best found for all new runs) and normalize them between 0 and 1. That is, we shift and scale these values in such a way that their smallest will be 0, the largest will be 1, and the last will be somewhere between 0 and 1. Analogously, we normalize the trajectory values.

On top of the six features thus computed we also use the percentage of overall time that has already elapsed, the percentage of overall time afforded in the beginning where all we do is restart a new run every time, and the time a new run will be given as percentage of total time left. In total we thus arrive at nine features.

4.2   Turning Features into Scores

Now, to compute the score for each of the three possibilities (continue current run, continue best run so far, and start a new run) we compute the following function

$$p^k(f) \leftarrow \frac{1}{1 + \exp(w_0^k + \sum_i f_i w_i^k)},$$

whereby $k \in \{1, 2, 3\}$ marks whether the function marks the score for continuing the current run, continuing the best run, or starting a new run, and $f \in R^9$ is the feature vector that characterizes our search experience so far. Note that $p^k(f) \in (0, 1)$, whereby the function approaches 0 when the weighted sum in the denominators exponential function goes to infinity, and how the function approaches 1 when the same sum approaches minus infinity. Finally, note that we require a total of 30 weights to define the three functions. These weights will be learned later by a parameter tuner to achieve superior runtime behavior.

4.3   The Reactive Restart Framework

Given the weights w_i^k with $k \in \{1, 2, 3\}$ and $i \in \{0, \ldots, 9\}$, we can now define the framework within which we can embed any black-box optimization solver.[1]

---

[1] We say black-box because we do not need to know anything about the inner workings of the solver. However, we make two assumptions. First, that we can set a time limit to the solver where it stops, and that we can add more time and continue the interrupted computation later. Second, that whenever the solver stops it returns information when it found the first solution, when it found the best solution so far, and what the quality of the best solution found so far is.

Algorithm 1. Reactive Restart Framework Algorithm

1: **function** REACTIVERESTARTS $(S,\text{x},\text{timeout},k,r,w_{i\in\{0,...,9\}}^{k\in\{1,2,3\}})$
2:    (initTime, best) $\leftarrow S(\text{x}, newRun, stopAtFirstSolution)$
3:    interval $\leftarrow r \times$ initTime
4:    $b \leftarrow S(\text{x}, continueLastRun, \text{interval} - \text{initTime})$
5:    elapsedTime $\leftarrow$ interval
6:    UPDATE(best, b)
7:    **while** elapsedTime $\leq k \times$ timeout **do**
8:        $(a, b) \leftarrow S(\text{x}, \text{interval})$
9:        elapsedTime $\leftarrow$ elapsedTime + interval
10:       UPDATE(initTime, best, a, b)
11:       interval $\leftarrow r \times$ initTime
12:   INIT(F)
13:   **while** elapsedTime $\leq$ timeout **do**
14:       $p^k \leftarrow \frac{1}{1+\exp(w_0^k+\sum_i w_i^k F_i)}$ $\forall k \in \{1,2,3\}$
15:       $p^k \leftarrow \frac{p^k}{p^1+p^2+p^3}$ $\forall k \in \{1,2,3\}$
16:       pick random $x \in [0, 1]$
17:       **if** $x \leq p^1$ **then**
18:           $b \leftarrow S(\text{x}, continueBestRun, \text{interval})$
19:           elapsedTime $\leftarrow$ elapsedTime + interval
20:           UPDATE(best, b)
21:       **else if** $x \leq p^1 + p^2$ **then**
22:           $b \leftarrow S(\text{x}, continueLastRun, \text{interval})$
23:           elapsedTime $\leftarrow$ elapsedTime + interval
24:           UPDATE(best, b)
25:       **else**
26:           $(a, b) \leftarrow S(\text{x}, newRun, \text{interval})$
27:           elapsedTime $\leftarrow$ elapsedTime + interval
28:           UPDATE(initTime, best, a, b)
29:           interval $\leftarrow r \times$ initTime
30:       UPDATE(F)
31:   **return** best

We present a stylized version of our framework in Algorithm 1. Given are a randomized optimization algorithm S, an input x, a timeout, a fraction $k \in [0, 1]$, a factor $r \geq 1$, and weights w. The framework first runs S on x until a first solution is computed. It records the time to find this solution and sets the incremental time interval each run is given to r times this input-dependent value. Next, S' run on x is continued until this incremental time interval is reached. The best solution seen so far is recorded.

Now, the first phase begins, which lasts for the fraction of the total time allowed as specified by k. In this first phase, we start a new run on x every single time, whereby we update the best solution seen so far and the time it takes each time to find a first solution. The function UPDATE is assumed to record the best solution quality found so far as well as to maintain the running average of the time it took to compute a first solution for each new run.

After the first phase ends, we initialize the features based on the search experience so far. Then, we enter the main phase. Based on the given weights and the current features we compute scores for the three options how we can continue the computation at each step. We then choose randomly and proportionally to these scores whether we continue the best run so far, the last run, or whether we begin a new run.

No matter which choice we always keep the best solution found so far up to date. When we choose to start a new run, we also update the running average of the times it takes to find a first solution as well as the incremental time interval that results from this running average times the factor r. Finally, we update the features and continue until the time has run out.

The last ingredient needed to apply this framework in practice is a method for learning the weights w. Based on a training set of instances, we compute weights that result in superior performance using the gender-based genetic algorithm tuner GGA [19], following the same general approach for tuning hyper-parameterized search methods as introduced in [17].

5 Experimental Analysis

We now present our numerical analysis. First, we briefly introduce the combinatorial optimization problems, the solvers, and the instances used in our experiments. Second, we describe our comprehensive data collection, which allows us to conduct our investigations completely offline, that is, without the need of running any additional experiments. Third, we present the results of our investigations which show the effectiveness of our online method.

5.1 Problems and Benchmarks

First, we briefly introduce the two considered NP-complete problems, as well as the corresponding solvers and benchmarks used in our investigations.

Traveling Salesperson. The Traveling Salesperson Problem (TSP) considers an edge-weighted graph $G = (V, E, w)$, the vertices $V = \{1, \ldots, n\}$ are referred to as *cities*. It asks for a permutation π of V such that $\left(\sum_{i=1}^{n-1} w(\pi(i), \pi(i+1))\right) + w(\pi(n), \pi(1))$ (the cost of visiting the cities in the order of the permutation and then returning to the origin $\pi(1)$) is minimized.

Natural applications of the TSP are in areas like planning and logistics [21], but they are also encountered in a large number of other domains, such as genome sequencing, drilling problems, aiming telescopes, and data clustering [22]. TSP is one of the most important (and most studied) optimization problems.

We use the Chained-Lin-Kernighan (LINKERN) heuristic [23,24], a state-of-the-art incomplete solver for the Traveling Salesperson problem. Its stochastic behavior comes from random components during the creation of the initial tour.

The TSPlib is a classic repository of TSP instances [25]. For our investigations, we pick all 112 instances from TSPlib, and as additional challenging instances ch71009, mona-lisa100k, and usa115475.

Minimum Vertex Cover. Finding a minimum vertex cover of a graph is a classical NP-hard problem. Given an unweighted, undirected graph $G = (V, E)$, a vertex cover is defined as a subset of the vertices $S \subseteq V$, such that every edge of G has an endpoint in S, i.e. for all edges $\{u, v\} \in E$, $u \in S$ or $v \in S$. The decision problem k-vertex cover decides whether a vertex cover of size k exists. We consider the optimization variant to find a vertex cover of minimum size.

Applications arise in numerous areas such as network security, scheduling and VLSI design [26]. The vertex cover problem is also closely related to the problem of finding a maximum clique. This has a range of applications in bioinformatics and biology, such as identifying related protein sequences [27].

Numerous algorithms have been proposed for solving the vertex cover problem. We choose FASTVC [28] over the popular NUMVC [29] as a solver for the minimum vertex cover problem as it works better for massive graphs. FASTVC is based on two low-complexity heuristics, one for initial construction of a vertex cover, and one to choose the vertex to be removed in each exchanging step, which involves random draws from a set of candidates.

For our experimental investigations, we select all 86 instances used in [28]. Among these, the number of vertices ranges from about 1000 to over 4 million, and the number of edges ranges from about 2000 to over 56 million.

5.2   Data Collection

We recorded 10,000 independent, regular runs of the original solvers on each of the 115 TSP instances and on each of the 86 MVC instances. For TSP, the time limit per instance was 1 h. For MVC, we allowed 100 s. The runs were conducted on a compute cluster with Intel Xeon E5620 CPUs (2.4 GHz).

For each run, we make a record whenever a solver finds a better solution, together with the solution quality. Altogether, the records of our 20,000 runs take up over 8 GB when GZ-compressed with default settings. We plan to make these files publicly available (upon finding a suitable webserver) as a resource for studying the behaviour of these algorithms.

5.3   Training of Hyper

For each of the two benchmarks, we used two thirds of the respectively available instances for training. That is, we handed the parameterized framework to a recently improved version of GGA that uses surrogate models to predict where improved parameterizations may be found [30] We ran GGA for 70 generations with a population size of 100 individuals. The random replacement rate was set to 5%, the mutation rate was set to 5% as well.

5.4   Results

Following the training of HYPER on two thirds of the instances (per problem domain), we are left with 38 of the 115 TSP instances and 28 of the 86 MVC instances. We use these to compare the performance of the following investigated approaches:

1. SINGLE: the solver is run once with a random seed, allowing it to run for the total time given;
2. RESTARTS: the solver is restarted from scratch whenever a preset time limit is reached, and this loop is repeated until time is up;
3. LUBY: restarts based on the fixed Luby sequence [3], where one Luby time unit is based on five times the time the first run needs to produce the first solution;
4. BET-AND-RUN: the previously described bet-and-run strategy by Friedrich et al. [13];
5. HYPER: our trained hyper-parameterized bet-and-run restart strategy, as described above.

We will analyze the outcomes using several criteria. First, we compare the performance gaps achieved with respect to the optimal solution possible within the time budget.[2] Second, we consider the number of times an approach is able to find the best possible solution. Third, we compare the amount of time needed in order to compute the final results.

To start off, Tables 1 and 2 show the results of the individual solvers across the sets of 38 and 28 instances. Note that we are using the problem domain names TSP and MVC instead of the respective solvers to facilitate reading.

We observe that the number of times the best possible solution is found increases with increasing time budget. Note that this is not natural as the best possible solution is the best possible solution for the respective time limit! The fact that the relative gap decreases anyhow is therefore a reflection of the fact that the best restart can actually find the best solution rather quickly. With increasing time limits, the restarted approaches thus have more buffer to find this best quality solution as well.

Next, we find that SINGLE and RESTARTS are clearly outperformed by the other three approaches across both problem domains and across all total time budgets. On TSP, HYPER achieves less than half the performance gap of BET-AND-RUN when the total time budget is only 100 s. This advantage for HYPER becomes more and pronounced as the budget increases to 5,000 s. For this time limit, HYPER has a six-times lower average gap than BET-AND-RUN, which is marked improvement. At the same time, BET-AND-RUN can find the best solutions in only 67% of the runs, whereas HYPER's success rate is 84%.

MVC can be seen as a little bit more challenging in our setting, as the computation time budgets were rather short and FASTVC encountered significant initialization times on some of the large instances. As a consequence, the number of times where no solution has been produced by the various approaches is higher than for TSP, however, this number decreases with increasing time budget.

On MVC, HYPER and BET-AND-RUN are really close in terms of average performance gap, however, there is an advantage for HYPER in number of times the best possible solutions are found. In practice this is still a substantial improvement.

---

[2] This best possible solution is the best solution provided within the given time limit by any of the 10,000 runs we conducted.

Table 1. TSP results. Shown are time in seconds, and performance gap from the best possible solution within the respective time limit. "solutions" and "no solutions" refer to the number of times the approach has produced any solution at all. "best found" lists the number of times the best possible solution was found given 380 runs (38 instances ∗ 10 independent runs). Highlighted in dark blue and light blue are the best and second best average approaches.

			SINGLE		
time budget	solutions	no solutions	best found	average performance	average time
100	380	0	234	0.1415	12
200	378	2	239	0.1368	21
500	380	0	266	0.0885	95
1000	380	0	266	0.0877	105
2000	380	0	266	0.0762	165
5000	380	0	266	0.0596	290
			RESTARTS		
time budget	solutions	no solutions	best found	average performance	average time
100	380	0	252	0.0689	21
200	380	0	255	0.0618	35
500	380	0	259	0.0519	61
1000	380	0	261	0.0474	98
2000	380	0	261	0.0457	154
5000	380	0	258	0.0435	268
			LUBY		
time budget	solutions	no solutions	best found	average performance	average time
100	380	0	296	0.0274	19
200	380	0	299	0.0189	32
500	380	0	309	0.0135	75
1000	380	0	317	0.0108	127
2000	380	0	318	0.0090	229
5000	380	0	322	0.0070	476
			BET-AND-RUN		
time budget	solutions	no solutions	best found	average performance	average time
100	380	0	244	0.0487	5
200	380	0	245	0.0473	6
500	380	0	246	0.0444	8
1000	380	0	248	0.0436	13
2000	380	0	251	0.0429	22
5000	380	0	256	0.0419	49
			HYPER		
time budget	solutions	no solutions	best found	average performance	average time
100	380	0	295	0.0216	15
200	380	0	302	0.0142	26
500	380	0	307	0.0132	57
1000	380	0	307	0.0090	87
2000	380	0	319	0.0077	178
5000	380	0	321	0.0066	322

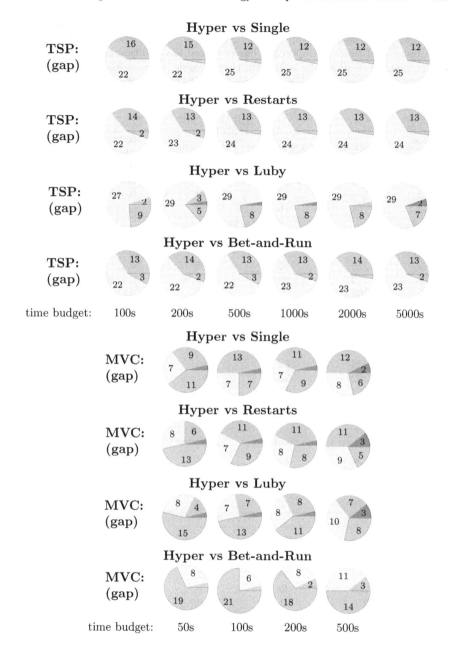

Fig. 1. Statistical comparison of HYPER with the other approaches using the Wilcoxon rank-sum test (significance level $p = 0.05$). The approaches are compared based on the quality gap to the best possible solution (smaller is better).

The colors have the following meaning: Green indicates that HYPER is statistically better, Red indicates that HYPER is statistically worse, Light gray indicates that both performed identical, Dark gray indicates that the differences were statistically insignificant. We have chosen pie charts on purpose because they allow for a quick qualitative comparison of results. (Color figure online)

Table 2. MVC results. Shown are time in seconds, and performance gap from the best possible solution within the respective time limit. "solutions" and "no solutions" refer to the number of times the approach has produced any solution at all. "best found" lists the number of times the respective best possible solution has been found given 280 runs (28 instances * 10 independent runs). Highlighted in dark blue and light blue are the best and second best average approaches.

time budget	solutions	no solutions	best found	average performance	average time
SINGLE					
5	211	69	74	0.1097	3
10	223	57	76	0.4558	5
20	254	26	80	0.6181	9
50	264	16	98	0.2273	19
RESTARTS					
5	228	52	76	0.1111	3
10	252	28	82	0.4140	6
20	268	12	88	0.6128	12
50	278	2	101	0.1802	25
LUBY					
5	228	52	80	0.1064	3
10	252	28	91	0.3907	6
20	268	12	91	0.5767	11
50	278	2	114	0.1032	23
BET-AND-RUN					
5	228	52	65	0.0800	3
10	252	28	79	0.3328	5
20	268	12	90	0.4721	9
50	278	2	105	0.0390	18
HYPER					
5	228	52	75	0.0781	3
10	252	28	87	0.3309	5
20	268	12	104	0.4710	9
50	278	2	119	0.0385	19

Interestingly, our results differ from [13], where Luby-based restarts performed not as well as RESTARTS, whereas in our study BET-AND-RUN is outperformed by LubyStat on TSP. This might be due to a different approach of setting t_{init} and because we use a larger instance set for TSP. Independent of this HYPER outperforms both.

Figure 1 adds to the results by showing the results of performing single-sided Wilcoxon rank-sum tests on the outcomes of 10 independent runs. For the two different problem domains, we observe the following. For TSP, HYPER dominates the field and is beaten five times by LUBY (as assessed by the statistical tests) in terms of quality gap to the optimum. For MVC, HYPER typically outperforms SINGLE, RESTARTS, and LUBY. In contrast to this, HYPER and BET-AND-RUN perform essentially comparably on MVC, and the differences are rarely significant.

Lastly, we summarize the investigations by testing whether the performance differences between HYPER and and the other approaches across all instances and time budgets are statistically significant. Again, we apply a single-sided Wilcoxon test to test the null hypothesis that two given distributions are identical. We compare the approaches based on the performance gap achieved, and across all time budgets and instances. In particular, we take the median of the 10 independent runs per instance, and then collect for each restart approach the medians across all instances and time limits. As a consequence, each approach has $38 * 6 = 228$ medians for TSP and $28 * 4 = 112$ medians for MVC.

Table 3 shows the results of these two tests. In summary, we can deduce from the outcome that HYPER performs no worse than existing approaches, and typically better. A closer inspection of the raw results of HYPER and BET-AND-RUN on MVC reveals that their performance is near-identical, despite the fact that the averages of HYPER are consistently lower than those of BET-AND-RUN (as seen in Table 2). In stark contrast to this, the performance comparisons on TSP are mostly highly significant and in an favor of HYPER.

Table 3. One-sided Wilcoxon rank-sum test to test whether the quality gap distribution of HYPER is shifted to the left of that of the other approaches. Shown are the p-values.

	SINGLE	RESTARTS	LUBY	BET-AND-RUN
TSP	0.000003	0.000015	0.478300	0.000002
MVC	0.060172	0.248689	0.354935	0.236808

6 Conclusion

We introduced the idea of learning reactive restart strategies for combinatorial search algorithms. We compared this new approach (HYPER) with other approaches, among them a very recent BET-AND-RUN approach that had been assessed comprehensively on TSP and MVC instances. Across both domains, HYPER resulted in markedly better average solution qualities, and it exhibited significantly increased rates of hitting the best possible solution.

As the investigated problem domains are structurally very different, we expect our approach to generalize to other problem domains as well, such as continuous and multi-objective optimization problems.

Future work will focus on the development of other runtime features as a basis for making restart decisions.

References

1. Gomes, C.P., Selman, B., Crato, N., Kautz, H.A.: Heavy-tailed phenomena in satisfiability and constraint satisfaction problems. J. Autom. Reason. **24**(1), 67–100 (2000)
2. Hoos, H.H.: Stochastic local search - methods, models, applications. Ph.D. thesis, TU Darmstadt (1998)
3. Luby, M., Sinclair, A., Zuckerman, D.: Optimal speedup of Las Vegas algorithms. Inf. Process. Lett. **47**(4), 173–180 (1993)
4. Wu, H., van Beek, P.: On universal restart strategies for backtracking search. In: Bessière, C. (ed.) CP 2007. LNCS, vol. 4741, pp. 681–695. Springer, Heidelberg (2007). doi:10.1007/978-3-540-74970-7_48
5. Audemard, G., Simon, L.: Refining restarts strategies for SAT and UNSAT. In: Milano, M. (ed.) CP 2012. LNCS, vol. 7514, pp. 118–126. Springer, Heidelberg (2012). doi:10.1007/978-3-642-33558-7_11
6. Marti, R.: Multi-start methods. In: Glover, F., Kochenberger, G.A. (eds.) Handbook of Metaheuristics, pp. 355–368 (2003)
7. Lourenço, H.R., Martin, O.C., Stützle, T.: Iterated local search: framework and applications. In: Gendreau, M., Potvin, J.Y. (eds.) Handbook of Metaheuristics. International Series in Operations Research & Management Science, vol. 146, pp. 363–397. Springer, Boston (2010). doi:10.1007/978-1-4419-1665-5_12
8. Schoenauer, M., Teytaud, F., Teytaud, O.: A rigorous runtime analysis for quasi-random restarts and decreasing stepsize. In: Hao, J.-K., Legrand, P., Collet, P., Monmarché, N., Lutton, E., Schoenauer, M. (eds.) EA 2011. LNCS, vol. 7401, pp. 37–48. Springer, Heidelberg (2012). doi:10.1007/978-3-642-35533-2_4
9. Biere, A.: Adaptive restart strategies for conflict driven SAT solvers. In: Kleine Büning, H., Zhao, X. (eds.) SAT 2008. LNCS, vol. 4996, pp. 28–33. Springer, Heidelberg (2008). doi:10.1007/978-3-540-79719-7_4
10. Huang, J.: The effect of restarts on the efficiency of clause learning. In: International Joint Conference on Artificial Intelligence (IJCAI), pp. 2318–2323 (2007)
11. Lalla-Ruiz, E., Voß, S.: Improving solver performance through redundancy. Syst. Sci. Syst. Eng. **25**(3), 303–325 (2016)
12. Fischetti, M., Monaci, M.: Exploiting erraticism in search. Oper. Res. **62**(1), 114–122 (2014)
13. Friedrich, T., Kötzing, T., Wagner, M.: A generic bet-and-run strategy for speeding up stochastic local search. In: Proceedings of the Thirty-First AAAI Conference on Artificial Intelligence, pp. 801–807 (2017)
14. Lissovoi, A., Sudholt, D., Wagner, M., Zarges, C.: Theoretical results on bet-and-run as an initialisation strategy. In: Genetic and Evolutionary Computation Conference (GECCO) (2017, accepted for publication)
15. Stützle, T., López-Ibáñez, M.: Automatic (offline) configuration of algorithms. In: Genetic and Evolutionary Computation Conference (GECCO), pp. 795–818 (2016)
16. Bezerra, L.C.T., López-Ibáñez, M., Stützle, T.: Automatic component-wise design of multiobjective evolutionary algorithms. IEEE Trans. Evol. Comput. **20**(3), 403–417 (2016)
17. Ansótegui, C., Pon, J., Tierney, K., Sellmann., M.: Reactive dialectic search portfolios for MaxSAT. In: AAAI Conference on Artificial Intelligence (2017, accepted for publication)
18. Kadioglu, S., Sellmann, M.: Dialectic search. In: Gent, I.P. (ed.) CP 2009. LNCS, vol. 5732, pp. 486–500. Springer, Heidelberg (2009). doi:10.1007/978-3-642-04244-7_39

19. Ansótegui, C., Sellmann, M., Tierney, K.: A gender-based genetic algorithm for the automatic configuration of algorithms. In: Gent, I.P. (ed.) CP 2009. LNCS, vol. 5732, pp. 142–157. Springer, Heidelberg (2009). doi:10.1007/978-3-642-04244-7_14

20. Hartigan, J.A.: Bounding the maximum of dependent random variables. Electron. J. Stat. **8**(2), 3126–3140 (2014)

21. Polacek, M., Doerner, K.F., Hartl, R.F., Kiechle, G., Reimann, M.: Scheduling periodic customer visits for a traveling salesperson. Eur. J. Oper. Res. **179**, 823–837 (2007)

22. Applegate, D.L., Bixby, R.E., Chvatal, V., Cook, W.J.: The Traveling Salesman Problem: A Computational Study. Princeton University Press, Princeton (2011)

23. Applegate, D.L., Cook, W.J., Rohe, A.: Chained Lin-Kernighan for large traveling salesman problems. INFORMS J. Comput. **15**(1), 82–92 (2003)

24. Cook, W.: The Traveling Salesperson Problem: Downloads (Website) (2003). http://www.math.uwaterloo.ca/tsp/concorde/downloads/downloads.htm. Accessed 21 Dec 2016

25. Reinelt, G.: TSPLIB - A traveling salesman problem library. ORSA J. Comput. **3**(4), 376–384 (1991). Instances: http://comopt.ifi.uni-heidelberg.de/software/ TSPLIB95/tsp/. Accessed 21 Dec 2016

26. Gomes, F.C., Meneses, C.N., Pardalos, P.M., Viana, G.V.R.: Experimental analysis of approximation algorithms for the vertex cover and set covering problems. Comput. Oper. Res. **33**(12), 3520–3534 (2006)

27. Abu-Khzam, F.N., Langston, M.A., Shanbhag, P., Symons, C.T.: Scalable parallel algorithms for FPT problems. Algorithmica **45**(3), 269–284 (2006)

28. Cai, S.: Balance between complexity and quality: local search for minimum vertex cover in massive graphs. In: International Joint Conference on Artificial Intelligence (IJCAI), pp. 747–753 (2015). Code: http://lcs.ios.ac.cn/caisw/MVC.html. Accessed 21 Dec 2016

29. Cai, S., Su, K., Luo, C., Sattar, A.: NuMVC: an efficient local search algorithm for minimum vertex cover. J. Artif. Intell. Res. **46**(1), 687–716 (2013)

30. Ansótegui, C., Malitsky, Y., Samulowitz, H., Sellmann, M., Tierney, K.: Model-based genetic algorithms for algorithm configuration. In: International Joint Conference on Artificial Intelligence (IJCAI), pp. 733–739 (2015)

Efficient Adaptive Implementation of the Serial Schedule Generation Scheme Using Preprocessing and Bloom Filters

Daniel Karapetyan[1]([⊠]) and Alexei Vernitski[2]

[1] Institute for Analytics and Data Science, University of Essex, Essex, UK
daniel.karapetyan@gmail.com
[2] Department of Mathematical Sciences, University of Essex, Essex, UK
asvern@essex.ac.uk

Abstract. The majority of scheduling metaheuristics use indirect representation of solutions as a way to efficiently explore the search space. Thus, a crucial part of such metaheuristics is a "schedule generation scheme" – procedure translating the indirect solution representation into a schedule. Schedule generation scheme is used every time a new candidate solution needs to be evaluated. Being relatively slow, it eats up most of the running time of the metaheuristic and, thus, its speed plays significant role in performance of the metaheuristic. Despite its importance, little attention has been paid in the literature to efficient implementation of schedule generation schemes. We give detailed description of serial schedule generation scheme, including new improvements, and propose a new approach for speeding it up, by using Bloom filters. The results are further strengthened by automated control of parameters. Finally, we employ online algorithm selection to dynamically choose which of the two implementations to use. This hybrid approach significantly outperforms conventional implementation on a wide range of instances.

Keywords: Resource-constrained project scheduling problem · Serial schedule generation scheme · Bloom filters · Online algorithm selection

1 Introduction

Resource Constrained Project Scheduling Problem (RCPSP) is to schedule a set of jobs J subject to precedence relationships and resource constraints. RCPSP is a powerful model generalising several classic scheduling problems such as job shop scheduling, flow shop scheduling and parallel machine scheduling.

In RCPSP, we are given a set of resources R and their capacities c_r, $r \in R$. In each time slot, c_r units of resource r are available and can be shared between jobs. Each job $j \in J$ has a prescribed consumption $v_{j,r}$ of each resource $r \in R$. We are also given the duration d_j of a job $j \in J$. A job consumes $v_{j,r}$ units of resource r in every time slot that it occupies. Once started, a job cannot be

© Springer International Publishing AG 2017
R. Battiti et al. (Eds.): LION 2017, LNCS 10556, pp. 124–138, 2017.
https://doi.org/10.1007/978-3-319-69404-7_9

interrupted (no preemption is allowed). Finally, each resource is assigned a set $pred_j \subset J$ of jobs that need to be completed before j can start.

There exist multiple extensions of RCPSP. In the multi-mode extension, each job can be executed in one of several modes, and then resource consumption and duration depend on the selected mode. In some applications, resource availability may vary with time. There could be set-up times associated with certain jobs. In multi-project extension, several projects run in parallel sharing some but not all resources. In this paper we focus on the basic version of RCPSP, however some of our results can be easily generalised to many of its extensions and variations.

Most of the real-world scheduling problems, including RCPSP, are NP-hard, and hence only problems of small size can be solved to optimality, whereas for larger problems (meta)heuristics are commonly used. Metaheuristics usually search in the space of feasible solutions; with a highly constrained problem such as RCPSP, browsing the space of feasible solutions is hard. Indeed, if a schedule is represented as a vector of job start times, then changing the start time of a single job is likely to cause constraint violations. Usual approach is to use indirect solution representation that could be conveniently handled by the metaheuristic but could also be efficiently translated into the direct representation.

Two translation procedures widely used in scheduling are *serial schedule generation scheme* (SSGS) and *parallel schedule generation scheme* [9]. Some studies conclude that SSGS gives better performance [7], while others suggest to employ both procedures within a metaheuristic [10]. Our research focuses on SSGS.

With SSGS, the indirect solution representation is a permutation π of jobs. The metaheuristic handles candidate solutions in indirect (permutation) form. Every time a candidate solution needs to be evaluated, SSGS is executed to translate the solution into a schedule $(t_j, \ j \in J)$, and only then the objective value can be computed, see Fig. 1.

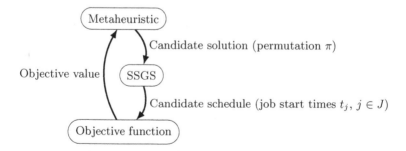

Fig. 1. Classic architecture of a scheduling metaheuristic. To obtain objective value of a candidate solution, metaheuristic uses SSGS to translate the candidate solution into a candidate schedule, which is then used by objective function.

SSGS is a simple procedure that iterates through J in the order given by π, and schedules one job at a time, choosing the earliest feasible slot for each

---

Algorithm 1. Serial Schedule Generation Scheme (SSGS). Here T is the upper bound on the makespan.

---

    input   : Solution π in permutation form, respecting precedence relations
    output : Schedule t_j, $j \in J$
1 $A_{t,r} \leftarrow c_r$ for every $t = 1, 2, \ldots, T$ and $r \in R$;
2 **for** $i = 1, 2, \ldots, |J|$ **do**
3      $j \leftarrow \pi(i)$;
4      $t^0 \leftarrow \max_{j' \in pred_j} t_{j'} + d_{j'}$;
5      $t_j \leftarrow find(j, t^0, A)$ (see Algorithm 2);
6      $update(j, t_j, A)$ (see Algorithm 3);
7 **return** t_j, $j \in J$;

Algorithm 2. Conventional implementation of $find(j, t^0, A)$ – a function to find the earliest feasible slot for job j.

---

    input   : Job $j \in J$ to be scheduled
    input   : Earliest start time t^0 as per precedence relations
    input   : Current availability A of resources
    output : Earliest feasible start time for job j
1 $t_j \leftarrow t^0$;
2 $t \leftarrow t_j$;
3 **while** $t < t_j + d_j$ **do**
4   **if** $A_{t,r} \geq v_{j,r}$ *for every* $r \in R$ **then**
5         $t \leftarrow t + 1$;
6   **else**
7         $t_j \leftarrow t + 1$;
8         $t \leftarrow t_j$;
9 **return** t_j;

---

job. The only requirement for π is to respect the precedence relations; otherwise SSGS produces a feasible schedule for any permutation of jobs. The pseudo-code of SSGS is given in Algorithm 1, and its two subroutines *find* and *update* in Algorithms 2 and 3.

Commonly, the objective of RCPSP is to find a schedule that minimises the makespan, i.e. the time required to complete all jobs; however other objective functions are also considered in the literature. We say that an objective function of a scheduling problem is *regular* if advancing the start time of a job cannot worsen the solution's objective value. Typical objective functions of RCPSP, including makespan, are regular. If the scheduling problem has a regular objective function, then SSGS guarantees to produce *active* solutions, i.e. solutions that cannot be improved by changing t_j for a single $j \in J$. Moreover, it was shown [8] that for any active schedule S there exists a permutation π for which SSGS will generate S. Since any optimal solution S is active, searching in the space of feasible permutations π is sufficient to solve the problem.

Algorithm 3. Procedure $update(j, t_j, A)$ to update resource availability A after scheduling job j at time t_j.

---

    input   : Job j and its start time t_j
    input   : Resource availability A
1 **for** $t \leftarrow t_j, t_j + 1, \ldots, t_j + d_j - 1$ **do**
2    $\lfloor$ $A_{j,r} \leftarrow A_{j,r} - v_{j,r}$ for every $r \in R$;

This is an important property of SSGS; the parallel schedule generation scheme, mentioned above, does not provide this guarantee [8] and, hence, may not in some circumstances allow a metaheuristic finding optimal or near-optimal solutions.

The runtime of a metaheuristic is divided between its search control mechanism that modifies solutions and makes decisions such as accepting or rejecting new solutions, and SSGS. While SSGS is a polynomial algorithm, in practice it eats up the majority of the metaheuristic runtime (over 98% as reported in [1]). In other words, by improving the speed of SSGS twofold, one will (almost) double the number of iterations a metaheuristic performs within the same time budget, and this increase in the number of iterations is likely to have a major effect on the quality of obtained solutions.

In our opinion, not enough attention was paid to SSGS – a crucial component of many scheduling algorithms, and this study is to close this gap. In this paper we discuss approaches to speed up the conventional implementation of SSGS. Main contributions of our paper are:

- A detailed description of SSGS including old and new speed-ups (Sect. 2).
- New implementation of SSGS employing Bloom filters for quick testing of resource availability (Sect. 3).
- A hybrid control mechanism that employs intelligent learning to dynamically select the best performing SSGS implementation (Sect. 4).

Empirical evaluation in Sect. 5 confirms that both of our implementations of SSGS perform significantly better than the conventional SSGS, and the hybrid control mechanism is capable of correctly choosing the best implementation while generating only negligible overheads.

2   SSGS Implementation Details

Before we proceed to introducing our main new contributions in Sects. 3 and 4, we describe what state-of-the-art implementation of SSGS we use, including some previously unpublished improvements.

2.1   Initialisation of A

The initialisation of A in line 1 of Algorithm 1 iterates through T slots, where T is the upper bound on the makespan. It was noted in [1] that instead of initialising

Algorithm 4. Enhanced implementation of $find(j, t^0, A)$

> **input** : Job $j \in J$ to be scheduled
> **input** : Earliest start time t^0 as per precedence relations
> **input** : Current availability A of resources
> **output** : Earliest feasible start time for job j

1 $t_j \leftarrow t^0$;
2 $t \leftarrow t_j + d_j - 1$;
3 $t^{test} \leftarrow t_j$;
4 **while** $t \geq t^{test}$ **do**
5   **if** $A_{t,r} \geq V_{j,r}$ *for every* $r \in R$ **then**
6         $t \leftarrow t - 1$;
7   **else**
8         $t^{test} \leftarrow t_j + d_j$;
9         $t_j \leftarrow t + 1$;
10        $t \leftarrow t_j + d_j - 1$;

11 **return** t_j;

A at every execution of SSGS, one can reuse this data structure between the executions. To correctly initialise A, at the end of SSGS we restore $A_{t,r}$ for each $r \in R$ and each slot where some job was scheduled: $A_{t,r} \leftarrow c_r$ for $r \in R$ and $t = 1, 2, \ldots, M$, where M is the makespan of the solution. Since $M \leq T$ and usually $M \ll T$, this notably improves the performance of SSGS [1].

2.2 Efficient Search of the Earliest Feasible Slot for a Job

The function $find(j, t^0, I, A)$ finds the earliest slot feasible for scheduling job j. Its conventional implementation (Algorithm 2) takes $O(T|R|)$ time, where T is the upper bound of the time horizon. Our enhanced implementation of $find$ (Algorithm 4), first proposed in [1], has the same worst case complexity but is more efficient in practice. It is inspired by the Knuth-Morris-Pratt substring search algorithm. Let t_j be the assumed starting time of job J. To verify if it is feasible, we need to test sufficiency of resources in slots $t_j, t_j + 1, \ldots, t_j + d_j - 1$. Unlike the conventional implementation, our enhanced version tests these slots in the reversed order. The order makes no difference if the slot is feasible, but otherwise testing in reversed order allows us to skip some slots; in particular, if slot t is found to have insufficient resources then we know that feasible t_j is at least $t + 1$.

A further speed up, which was not discussed in the literature before, is to avoid re-testing of slots with sufficient resources. Consider the point when we find that the resources in slot t are insufficient. By that time we know that the resources in $t+1, t+2, \ldots, t_j + d_j - 1$ are sufficient. Our heuristic is to remember that the earliest slot t^{test} to be tested in future iterations is $t_j + d_j$.

2.3 Preprocessing and Automated Parameter Control

We observe that in many applications, jobs are likely to require only a subset of resources. For example, in construction works, to dig a hole one does not need cranes or electricians, hence the 'dig a hole' job will not consume those resources. To exploit this observation, we pre-compute vector R_j of resources used by job j, and then iterate only through resources in R_j when testing resource sufficiency in *find* (Algorithm 4, line 5) and updating resource availability in *update* (Algorithm 3, line 2). Despite the simplicity of this idea, we are not aware of anyone using or mentioning it before.

We further observe that some jobs may consume only one resource. By creating specialised implementations of *find* and *update*, we can reduce the depth of nested loops. While this makes no difference from the theoretical point of view, in practice this leads to considerable improvement of performance. Correct implementations of *find* and *update* are identified during preprocessing and do not cause overheads during executions of SSGS.

Having individual vectors R_j of consumed resources for each job, we can also intelligently learn the order in which resource availability is tested (Algorithm 4, line 5). By doing this, we are aiming at minimising the expected number of iterations within the resource availability test. For example, if resource r is scarce and job j requires significant amount of r, then we are likely to place r at the beginning of R_j. More formally, we sort R_j in descending order of probability that the resource is found to be insufficient during the search. This probability is obtained empirically by a special implementation of SSGS which we call SSGS^{data}. SSGS^{data} is used once to count how many times each resource turned out to be insufficient during scheduling of a job. (To avoid bias, SSGS^{data} tests every resource in R_j even if the test could be terminated early.) After a single execution of SSGS^{data}, vectors R_j are optimised, and in further executions default implementation of SSGS is used.

One may notice that the data collected in the first execution of SSGS may get outdated after some time; this problem in addressed in Sect. 4.

Ordering of R_j is likely to be particularly effective on instances with asymmetric use of resources, i.e. on real instances. Nevertheless, we observed improvement of runtime even on pseudo-random instances as reported in Sect. 5.

3 SSGS Implementation Using Bloom Filters

Performance bottleneck of an algorithm is usually its innermost loop. Observe that the innermost loop of the *find* function is the test of resource sufficiency in a slot, see Algorithm 4, line 5. In this section we try to reduce average runtime of this test from $O(|R|)$ to $O(1)$ time. For this, we propose a novel way of using a data structure known as Bloom filter.

Bloom filters were introduced in [2] as a way of optimising dictionary lookups, and found many applications in computer science and electronic system engineering [3,13]. Bloom filters usually utilise pseudo-random hash functions to encode data, but in some applications [6] non-hash-based approaches are used. In our paper, we also use a non-hash-based approach, and to our knowledge, our paper is the first in which the structure of Bloom filters is chosen dynamically according to the statistical properties of the data, with the purpose of improving the speed of an optimisation algorithm.

In general, Bloom filters can be defined as a way of using data, and they are characterised by two aspects: first, all data is represented by short binary arrays of a fixed length (perhaps with a loss of accuracy); second, the process of querying data involves only bitwise comparison of binary arrays (which makes querying data very fast).

We represent both each job's resource consumption and resource availability at each time slot, by binary arrays of a fixed length; we call these binary arrays Bloom filters. Our Bloom filters will consist of bits which we call *resource level bits*. Each resource bit, denoted by $u_{r,k}$, $r \in R$, $k \in \{1, 2, \ldots, c_r\}$, means "$k$ units of resource r" (see details below). Let U be the set of all possible resource bits. A *Bloom filter structure* is an ordered subset $L \subseteq U$, see Fig. 2 for an example. Suppose that a certain Bloom filter structure L is fixed. Then we can introduce $B^L(j)$, the Bloom filter of job j, and $B^L(t)$, the Bloom filter of time slot t, for each j and t. Each $B^L(j)$ and $B^L(t)$ consists of $|L|$ bits defined as follows: if $u_{r,k}$ is the ith element of L then

$$B^L(j)_i = \begin{cases} 1 & \text{if } v_{j,r} \geq k, \\ 0 & \text{otherwise,} \end{cases} \quad \text{and} \quad B^L(t)_i = \begin{cases} 1 & \text{if } A_{t,r} \geq k, \\ 0 & \text{otherwise.} \end{cases}$$

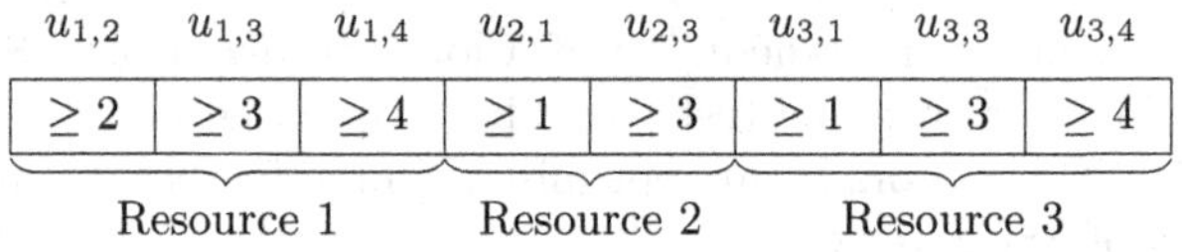

$u_{1,2}$	$u_{1,3}$	$u_{1,4}$	$u_{2,1}$	$u_{2,3}$	$u_{3,1}$	$u_{3,3}$	$u_{3,4}$
≥ 2	≥ 3	≥ 4	≥ 1	≥ 3	≥ 1	≥ 3	≥ 4

Resource 1   Resource 2   Resource 3

Fig. 2. Example of a Bloom filter structure for a problem with 3 resources, each having capacity 4.

To query if a job j can be scheduled in a time slot t, we compare Bloom filters $B^L(j)$ and $B^L(t)$ bitwise; then one of three situations is possible, as the following examples show. Consider a job j and three slots, t, t' and t'', with the following resource consumption/availabilities, and Bloom filter structure as in Fig. 2:

$v_{j,1} = 3$	$v_{j,2} = 2$	$v_{j,3} = 0$	$B^L(j) = (110\,10\,000)$
$A_{t,1} = 2$	$A_{t,2} = 3$	$A_{t,3} = 4$	$B^L(t) = (100\,11\,111)$
$A_{t',1} = 3$	$A_{t',2} = 1$	$A_{t',3} = 4$	$B^L(t') = (110\,10\,111)$
$A_{t'',1} = 3$	$A_{t'',2} = 2$	$A_{t'',3} = 2$	$B^L(t'') = (110\,10\,100)$

For two bit arrays of the same length, let notation '$\leq$' mean bitwise less or equal. By observing that $B^L(j) \not\leq B^L(t)$, we conclude that resources in slot t

are insufficient for j; this conclusion is guaranteed to be correct. By observing that $B^L(j) \leq B^L(t')$, we conclude tentatively that resources in slot t' may be sufficient for j; however, further verification of the complete data related to j and t' (that is, $v_{j,\cdot}$ and $A_{t',\cdot}$) is required to get a precise answer; one can see that $v_{j,2} \geq A_{t',2}$, hence this is what is called a *false positive*. Finally, we observe that $B^L(j) \leq B^L(t'')$, and a further test (comparing $v_{j,\cdot}$ and $A_{t'',\cdot}$) confirms that resources in t'' are indeed sufficient for j.

Values of $B^L(j)$, $j \in J$, are pre-computed when L is constructed, and $B^L(t)$, $t = 1, 2, \ldots, T$, are maintained by the algorithm.

3.1 Optimisation of Bloom Filter Structure

The length $|L|$ of a Bloom filter is limited to reduce space requirements and, more importantly for our application, speed up Bloom filter tests. Note that if $|L|$ is small (such as 32 or 64 bits) then we can exploit efficient bitwise operators implemented by all modern CPUs; then each Bloom filter test takes only one CPU operation. We set $|L| = 32$ in our implementation. While obeying this constraint, we aim at minimising the number of false positives, because false positives slow down the implementation.

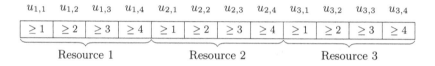

Fig. 3. Example of a Bloom filter structure for a problem with 3 resources, each having capacity 4, with $L = U$.

Our L building algorithm is as follows:

1. Start with $L = U$, such as in Fig. 3.
2. If $|L|$ is within the prescribed limit, stop.
3. Otherwise select $u_{r,k} \in L$ that is least important and delete it. Go to step 2.

By 'least important' we mean that the deletion of it is expected to have minimal impact of the expected number of false positives. Let $L = (\ldots, u_{r,q}, u_{r,k}, u_{r,m}, \ldots)$. Consider a job j such that $k \leq v_{j,r} < m$ and a slot t such that $q \leq A_{t,r} < k$. With L as defined above, Bloom filters correctly identify that resources in slot t are insufficient for job j: $B^L(j) \not\leq B^L(t)$. However, without the resource level bit $u_{r,k}$ we get a false positive: $B^L(j) \leq B^L(t)$. Thus, the probability of false positives caused by deleting $u_{r,k}$ from L in is as follows:

$$\left(\sum_{k=i}^{m-1} D_k^r \right) \cdot \left(\sum_{k=q}^{i-1} E_k^r \right),$$

where D_k^r is the probability that a randomly chosen job needs exactly k units of resource r, and E_k^r is the probability that a certain slot, when we examine it for scheduling a job, has exactly k units of resource r available. The probability distribution D^r is produced from the RCPSP instance data during pre-processing.[1] The probability distribution E^r is obtained empirically during the run of SSGS^{data} (see Sect. 2.2); each time resource sufficiency is tested within SSGS^{data}, its availability is recorded.

3.2   Additional Speed-ups

While positive result of a Bloom filter test generally requires further verification using full data, in some circumstances its correctness can be guaranteed. In particular, if for some $r \in R$ and $j \in J$ we have $u_{r,k} \in L$ and $v_{j,r} = k$, then the Bloom filter result, whether positive or negative, does not require verification.

Another observation is that updating $B^L(t)$ in *update* can be done in $O(|R_j|)$ operations instead of $O(|R|)$ operations. Indeed, instead of computing $B^L(t)$ from scratch, we can exploit our structure of Bloom filters. We update each bit related to resources $r \in R_j$, but we keep intact other bits. With some trivial pre-processing, this requires only $O(|R_j|)$ CPU operations.

We also note that if $|R_j| = 1$, i.e. job j uses only one resource, then Bloom filters will not speed up the *find* function for that job and, hence, in such cases we use the standard *find* function specialised for one resource (see Sect. 2.2).

4   Hybrid Control Mechanism

So far we have proposed two improved implementations of SSGS: one using Bloom filters (which we denote SSGS^{BF}), and the other one not using Bloom filters (which we denote SSGS^{NBF}). While it may look like SSGS^{BF} should always be superior to SSGS^{NBF}, in practice SSGS^{NBF} is often faster. Indeed, Bloom filters usually speed up the *find* function, but they also slow down *update*, as in SSGS^{BF} we need to update not only the values $A_{t,r}$ but also the Bloom filters $B^L(t)$ encoding resource availability. If, for example, the RCPSP instance has tight precedence relations and loose resource constraints then *find* may take only a few iterations, and then the gain in the speed of *find* may be less than the loss of speed of *update*. In such cases SSGS^{BF} is likely to be slower then SSGS^{NBF}.

In short, either of the two SSGS implementations can be superior in certain circumstances, and which one is faster mostly depends on the problem instance. In this section we discuss how to adaptively select the best SSGS implementation. Automated algorithm selection is commonly used in areas such as sorting, where multiple algorithms exist. A typical approach is then to extract easy to compute input data features and then apply off-line learning to develop a predictor of which algorithm is likely to perform best, see e.g. [5]. With sorting, this seems

---

[1] In multi-mode extension of RCPSP, this distribution depends on selected modes and hence needs to be obtained empirically, similarly to how we obtain E^r.

to be the most appropriate approach as the input data may vary significantly between executions of the algorithm. Our case is different in that the most crucial input data (the RCPSP instance) does not change between executions of SSGS. Thus, during the first few executions of SSGS, we can test how each implementation performs, and then select the faster one. This is a simple yet effective control mechanism which we call Hybrid.

Hybrid is entirely transparent for the metaheuristic; the metaheuristic simply calls SSGS whenever it needs to evaluate a candidate solution and/or generate a schedule. The Hybrid control mechanism is then intelligently deciding each time which implementation of SSGS to use based on information learnt during previous runs.

An example of how Hybrid performs is illustrated in Fig. 4. In the first execution, it uses $SSGS^{data}$ to collect data required for both $SSGS^{BF}$ and $SSGS^{NBF}$. For the next few executions, it alternates between $SSGS^{BF}$ and $SSGS^{NBF}$, measuring the time each of them takes. During this stage, Hybrid counts how many times $SSGS^{BF}$ was faster than the next execution of $SSGS^{NBF}$. Then we use the sign test [4] to compare the implementations. If the difference is significant (we use a 5% significance level for the sign test) then we stop alternating the implementations and in future use only the faster one. Otherwise we continue alternating the implementations, but for at most 100 executions. (Without such a limitation, there is a danger that the alternation will never stop – if the implementations perform similarly; since there are overheads associated with the alternation and time measurement, it is better to pick one of the implementations and use it in future executions.)

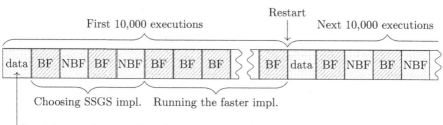

Fig. 4. Stages of the Hybrid control mechanism. Each square shows one execution of SSGS, and the text inside describes which implementation of SSGS is used. $SSGS^{data}$ is always used in the first execution. Further few executions (at most 100) alternate between $SSGS^{BF}$ and $SSGS^{NBF}$, with each execution being timed. Once sign test shows significant difference between the $SSGS^{BF}$ and $SSGS^{NBF}$ implementations, the faster one is used for the rest of executions. After 10,000 executions, previously collected data is erased and adaptation starts from scratch.

Our decision to use the sign test is based on two considerations: first, it is very fast, and second, it works for distributions which are not normal. This makes our approach different from [12] where the distributions of runtimes are assumed to be normal. (Note that in our experiments we observed that the distribution of running times of an SSGS implementation resembles Poisson distribution.)

As pointed out in this and previous sections, optimal choices of parameters of the SSGS implementations mostly depend on the RCPSP instance – which does not change throughout the metaheuristic run; however solution π also affects the performance. It should be noted though that metaheuristics usually apply only small changes to the solution at each iteration, and hence solution properties tend to change relatively slowly over time. Consequently, we assume that parameters chosen in one execution of SSGS are likely to remain efficient for some further executions. Thus, Hybrid 'restarts' every 10,000 executions, by which we mean that all the internal data collected by SSGS is erased, and learning starts from scratch, see Fig. 4. This periodicity of restarts is a compromise between accuracy of choices and overheads, and it was shown to be practical in our experiments.

5   Empirical Evaluation

In this Section we evaluate the implementations of SSGS discussed above. To replicate conditions within a metaheuristic, we designed a simplified version of Simulated Annealing. In each iteration of our metaheuristic, current solution is modified by moving a randomly selected job into a new randomly selected position (within the feasible range). If the new solution is not worse than the previous one, it is accepted. Otherwise the new solution is accepted with 50% probability. Our metaheuristic performs 1,000,000 iterations before terminating.

We evaluate SSGS^{BF}, SSGS^{NBF} and Hybrid. These implementations are compared to 'conventional' SSGS, denoted by SSGS^{conv}, which does not employ any preprocessing or Bloom filters and uses conventional implementation of *find* (Algorithm 2).

We found that instances in the standard RCPSP benchmark set PSPLIB [11] occupy a relatively small area of the feature space. For example, all the RCPSP instances in PSPLIB have exactly four resources, and the maximum job duration is always set to 10. Thus, we chose to use the PSPLIB instance generator, but with a wider range of settings. Note that for this study, we modified the PSPLIB instance generator by allowing jobs not to have any precedence relations. This was necessary to extend the range of network complexity parameter (to include instances with scarce precedence relations), and to speed up the generator, as the original implementation would not allow us to generate large instances within reasonable time.

All the experiments are conducted on a Windows Server 2012 machine based on Intel Xeon E5-2690 v4 2.60 GHz CPU. No concurrency is employed in any of the implementations or tests.

To see the effect of various instance features on SSGS performance, we select one feature at a time and plot average SSGS performance against the values of that feature. The rest of the features (or generator parameters) are then set as follows: number of jobs 120, number of resources 4, maximum duration of job 10, network complexity 1, resource factor 0.75, and resource strength 0.1. These values correspond to some typical settings used in PSPLIB. For formal definitions of the parameters we refer to [11].

For each combination of the instance generator settings, we produce 50 instances using different random generator seed values, and in each of our experiments the metaheuristic solves each instance once. Then the runtime of SSGS is said to be the overall time spent on solving those 50 instances, over 50,000,000 (which is the number of SSGS executions). The metaheuristic overheads are relatively small and are ignored.

From the results reported in Fig. 5 one can see that our implementations of SSGS are generally significantly faster than $SSGS^{conv}$, but performance of each implementation varies with the instance features. In some regions of the instance space $SSGS^{BF}$ outperforms $SSGS^{NBF}$, whereas in other regions $SSGS^{NBF}$ outperforms $SSGS^{BF}$. The difference in running times is significant, up to a factor of two in our experiments. At the same time, Hybrid is always close to the best of $SSGS^{BF}$ and $SSGS^{NBF}$, which shows efficiency of our algorithm selection approach. In fact, when $SSGS^{BF}$ and $SSGS^{NBF}$ perform similarly, Hybrid sometimes outperforms both; this behaviour is discussed below.

Another observation is that $SSGS^{NBF}$ is always faster than $SSGS^{conv}$ (always below the 100% mark) which is not surprising; indeed, $SSGS^{NBF}$ improves the performance of both *find* and *update*. In contrast, $SSGS^{BF}$ is sometimes slower than $SSGS^{conv}$; on some instances, the speed-up of the *find* function is overweighed by overheads in both *find* and *update*. Most important though is that Hybrid outperforms $SSGS^{conv}$ in each of our experiments by 8 to 68%, averaging at 43%. In other words, within a fixed time budget, an RCPSP metaheuristic employing Hybrid will be able to run around 1.8 times more iterations than if it used $SSGS^{conv}$.

To verify that Hybrid exhibits the adaptive behaviour and does not just stick to whichever implementation has been chosen initially, we recorded the implementation it used in every execution for several problems, see Fig. 6. For this experiment, we produced three instances: first instance has standard parameters except Resource Strength is 0.2; second instance has standard parameters except Resource Factor is 0.45; third instance has standard parameters except Maximum Job Duration is 20. These parameter values were selected such that the two SSGS implementations would be competitive and, therefore, switching between them would be a reasonable strategy. One can see that the switches occur several times throughout the run of the metaheuristic, indicating that Hybrid adapts to the changes of solution. For comparison, we disabled the adaptiveness and measured the performance if only implementation chosen initially is used throughout all iterations; the results are shown on Fig. 6. We conclude that Hybrid benefits from its adaptiveness.

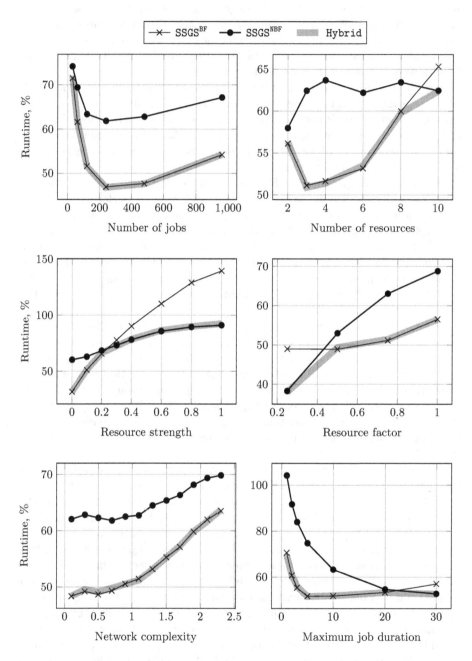

Fig. 5. These plots show how performance of the SSGS implementations depends on various instance features. Vertical axis gives the runtime of each implementation relative to $SSGS^{conv}$. ($SSGS^{conv}$ graph would be a horizontal line $y = 100\%$.)

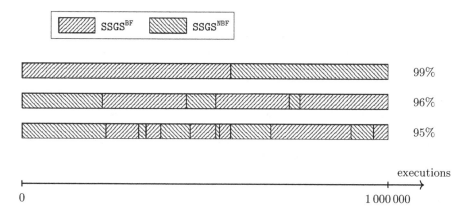

Fig. 6. This diagram shows how Hybrid switches between SSGS implementations while solving three problem instances. The number on the right shows the time spent by Hybrid compared with the time that would be needed if only the implementation chosen at the start would be used for all iterations.

6 Conclusions and Future Work

In this paper we discussed the crucial component of many scheduling metaheuristics, the serial schedule generation scheme (SSGS). SSGS eats up most of the runtime in a typical scheduling metaheuristic, therefore performance of SSGS is critical to the performance of the entire metaheuristic, and thus each speed-up of SSGS has significant impact. We described existing and some new speed-ups to SSGS, including preprocessing and automated parameter control. This implementation clearly outperformed the 'conventional' SSGS in our experiments. We further proposed a new implementation that uses Bloom filters, particularly efficient in certain regions of the instance space. To exploit strengths of both implementations, we proposed a hybrid control mechanism that learns the performance of each implementation and then adaptively chooses the SSGS version that is best for a particular problem instance and phase of the search. Experiments showed that this online algorithm selection mechanism is effective and makes Hybrid our clear choice. Note that Hybrid is entirely transparent for the metaheuristic which uses it as if it would be simple SSGS; all the learning and adaptation is hidden inside the implementation.

The idea behind online algorithm selection used in this project can be further developed by making the number of executions between restarts adaptable. To determine the point when the established relation between the SSGS implementations may have got outdated, we could treat the dynamics of the implementations' performance change as two random walks, and use the properties of these two random walks to predict when they may intersect.

All the implementations discussed in the paper, in C++, are available for downloading from http://csee.essex.ac.uk/staff/dkarap/rcpsp-ssgs.zip. The implementations are transparent and straightforward to use.

While we have only discussed SSGS for the simple RCPSP, our ideas can easily be applied in RCPSP extensions. We expect some of these ideas to work particularly well in multi-project RCPSP, where the overall number of resources is typically large but only a few of them are used by each job.

Acknowledgements. We would like to thank Prof. Rainer Kolisch for providing us with a C++ implementation of the PSPLIB generator. It should be noted that, although the C++ implementation was developed to reproduce the original Pascal implementation, the exact equivalence cannot be guaranteed; also, in our experiments we used a modification of the provided C++ code.

References

1. Asta, S., Karapetyan, D., Kheiri, A., Özcan, E., Parkes, A.J.: Combining Monte-Carlo and hyper-heuristic methods for the multi-mode resource-constrained multi-project scheduling problem. Inf. Sci. **373**, 476–498 (2016)
2. Bloom, B.H.: Space/time trade-offs in hash coding with allowable errors. Commun. ACM **13**(7), 422–426 (1970)
3. Broder, A., Mitzenmacher, M.: Network applications of bloom filters: a survey. Internet Math. **1**(4), 485–509 (2004)
4. Cohen, L., Holliday, M.: Practical Statistics for Students: An Introductory Text. Paul Chapman Publishing Ltd., London (1996)
5. Guo, H.: Algorithm selection for sorting and probabilistic inference: a machine learning-based approach. Ph.D. thesis, Kansas State University (2003)
6. Kayaturan, G.C., Vernitski, A.: A way of eliminating errors when using Bloom filters for routing in computer networks. In: Fifteenth International Conference on Networks, ICN 2016, pp. 52–57 (2016)
7. Kim, J.-L., Ellis Jr., R.D.: Comparing schedule generation schemes in resource-constrained project scheduling using elitist genetic algorithm. J. Constr. Eng. Manag. **136**(2), 160–169 (2010)
8. Kolisch, R.: Serial and parallel resource-constrained project scheduling methods revisited: theory and computation. Eur. J. Oper. Res. **90**(2), 320–333 (1996)
9. Kolisch, R., Hartmann, S.: Heuristic algorithms for the resource-constrained project scheduling problem: classification and computational analysis. In: Węglarz, J. (ed.) Project Scheduling, pp. 147–178. Springer, Boston (1999). doi:10.1007/978-1-4615-5533-9_7
10. Kolisch, R., Hartmann, S.: Experimental investigation of heuristics for resource-constrained project scheduling: an update. Eur. J. Oper. Res. **174**(1), 23–37 (2006)
11. Kolisch, R., Sprecher, A.: PSPLIB - a project scheduling problem library: OR software - ORSEP operations research software exchange program. Eur. J. Oper. Res. **96**(1), 205–216 (1997)
12. Lau, J., Arnold, M., Hind, M., Calder, B.: Online performance auditing: using hot optimizations without getting burned. SIGPLAN Not. **41**(6), 239–251 (2006)
13. Tarkoma, S., Rothenberg, C.E., Lagerspetz, E.: Theory and practice of bloom filters for distributed systems. IEEE Commun. Surv. Tutor. **14**(1), 131–155 (2012)

Interior Point and Newton Methods in Solving High Dimensional Flow Distribution Problems for Pipe Networks

Oleg O. Khamisov[1]([✉]) and Valery A. Stennikov[2]

[1] Skolkovo Institute of Science and Technology, Moscow, Russia
oleg.khamisov@skolkovotech.ru
[2] Melentiev Energy Systems Institute SB RAS, Irkutsk, Russia
sva@isem.irk.ru

Abstract. In this paper optimal flow distribution problem in pipe network is considered. The investigated problem is a convex sparse optimization problem with linear equality and inequality constrains. Newton method is used for problem with equality constrains only and obtains an approximate solution, which may not satisfy inequality constraints. Then Dikin Interior Point Method starts from the approximate solution and finds an optimal one. For problems of high dimension sparse matrix methods, namely Conjugate Gradient and Cholesky method with nested dissection, are applied. Since Dikin Interior Point Method works much slower then Newton Method on the matrices of big size, such approach allows us to obtain good starting point for this method by using comparatively fast Newton Method. Results of numerical experiments are presented.

Keywords: Pipe network · Convex optimization · Newton Method · Interior Point Method · Sparse matrix · Large-scale optimization

1 Introduction

In this paper optimal (steady-state) flow distribution problem in pipe network is considered. From mathematical point of view this problem is sparse large-scale convex optimization problem.

Several approaches to find steady-state solution exist. A comprehensive survey of methods is presented in [6]. Our paper uses approach similar to [2], but also includes sparse matrices techniques in order to efficiently solve high dimensional flow distribution problems.

The aim of this paper is application of new technology for solving the considered problem.

2 Problem Statement

Let us consider pipe network, which is given by its incidence matrix $\overline{A} \in \mathbb{R}^{(n+1) \times m}$, where m is the number of oriented edges and $n+1$ is the number

© Springer International Publishing AG 2017
R. Battiti et al. (Eds.): LION 2017, LNCS 10556, pp. 139–149, 2017.
https://doi.org/10.1007/978-3-319-69404-7_10

of nodes. Vector of nodal flow rates $\overline{Q} \in \mathbb{R}^{n+1}$, $\sum_{i=1}^{n} Q_i = 0$, with components $Q_i > 0$ for sources, $Q_i < 0$ for sinks and $Q_i = 0$ for the nodes of connection. The flow distribution problem has the following statement [5]:

$$F(x) \rightarrow \min, x \in \mathbb{R}^m, \tag{1}$$

$$\overline{A}x = \overline{Q}, \tag{2}$$

here x is the vector of unknown flows on the edges, that have to be found: $x_i > 0$ if flow coincides with the direction of the edge i, otherwise $x_i < 0$. Function F is a twice differentiable convex function, which will be described below.

System (2) describes the first Kirchhoff's law. Since $\overline{A}$ is an incidence matrix, we have rank $\overline{A} = n$, therefore it is possible to exclude one arbitrary row and get equivalent system:

$$Ax = Q, \tag{3}$$

here $A \in \mathbb{R}^{n \times m}$, and $Q \in \mathbb{R}^n$ are new matrix and right hand side vector, obtained by such exclusion. This is done to avoid working with singular matrix.

The objective function is defined in the following way [2,5]:

$$F(x) = \left(\sum_{i=1}^{m} \frac{S_i |x_i|^{\beta_i}}{\beta_i} - H_i x_i \right), \tag{4}$$

here S_i are edge resistance coefficients, $\beta_i \geq 3$ are integer numbers, H_i are pressure coefficients. Usually all β_i are assumed to be equal each other, in [2] all $\beta_i = \beta = 3$. Further calculations are done under this assumption, therefore objective function can be presented in the form

$$F(x) = \left(\sum_{i=1}^{m} \frac{S_i |x_i|^3}{3} - H_i x_i \right).$$

For the problem (1) with linear constraints (3) Newton Method can be applied with certain conditions, which are considered below.

If line contains a pump, x_i must be nonnegative. Let $I_1 \subseteq \{1, \ldots, m\}$ be a set of lines with pumps without flow limitations. Additionally, flow values on some lines $i \in I_2 \subseteq \{1, \ldots, m\}$, which can also contain pumps, must be limited from above by $\gamma_i > 0$. Note that $I_1 \cap I_2 = \emptyset$. Therefore, the following inequality constraints on the flows have to be taken into consideration

$$x_i \geq 0, \ i \in I_1, \ \gamma_i \geq x_i \geq \alpha_i, \ i \in I_2, \tag{5}$$

here, if pump is present on the edge $i \in I_2$ then $\alpha_i = 0$, otherwise $\alpha_i = -\gamma_i$. When (5) is present, the solution of problem (1) and (3) is found by Newton Method and then Dikin Interior Point method is applied to take into consideration inequality constraints (5).

3   Newton Method

Let us consider problem (1) and (3). The Lagrange function for this system has the following form:

$$L(x, \lambda) = F(x) + \lambda^T (Ax - Q), \lambda \in \mathbb{R}^n.$$

Due to the convexity of the objective function, equivalence of the Lagrange function gradient in x to zero is a sufficient minimum condition. Therefore Newton Method is applied to solve the equation

$$\frac{\partial L}{\partial x} = 0.$$

At $(k + 1)$ step of the Newton method vector x^{k+1} is calculated according to the following formula [8]

$$\begin{pmatrix} x^{k+1} \\ \lambda^{k+1} \end{pmatrix} = \begin{pmatrix} x^k \\ \lambda^k \end{pmatrix} - \left(\nabla^2 L(x^k, \lambda^k) \right)^{-1} \nabla L(x^k, \lambda^k),$$

where

$$\nabla L(x, \lambda) = \begin{pmatrix} \nabla F(x) + A^T \lambda \\ Ax - Q \end{pmatrix}, \nabla^2 L(x, \lambda) = \begin{pmatrix} \nabla^2 F(x) & A^T \\ A & 0 \end{pmatrix}.$$

Therefore, x^{k+1} is obtained from the following system of equations

$$\begin{pmatrix} \nabla^2 F(x^k) & A^T \\ A & 0 \end{pmatrix} \begin{pmatrix} x^{k+1} \\ \lambda^{k+1} \end{pmatrix} = \begin{pmatrix} \nabla^2 F(x^k) x^k - \nabla F(x^k) \\ Q \end{pmatrix}, \tag{6}$$

where $\nabla^2 F(x)$ is objective function Hessian.

$$\nabla^2 F(x) = 2\mathrm{diag}(S_1 |x_1|, \ldots, S_m |x_m|).$$

During each iteration it is necessary to find a solution of system of linear equations with symmetric indefinite matrix, therefore its solution can be found by Conjugate Gradient method (Algorithm 1) [9,10]. Denote by $y^k \in \mathbb{R}^{n+m}$ vector, consisting of x^k and λ^k. The pair x^k, λ^k obtained at the previous iteration of the Newton Method is used as starting approximation, to reduce the number of iterations in Conjugate Gradient method. Hessian $\nabla^2 L(x^k, \lambda^k)$ is not a positive definite matrix, moreover it can be singular, therefore the algorithm can fail if $q_i = 0$ on the line 3 of Algorithm 1. In this case Conjugate Gradient Method and Newton Methods stop without finding solution and, starting from the current approximation, we switch to Interior Point Method which is described in the next section. Otherwise Conjugate Gradient Method stops, when solution of linear system is found and Newton Method continues working.

Algorithm 1. Conjugate Gradient

1: $p^0 = r^0 = \nabla L(x^k, \lambda^k) - \nabla^2 L(x^k, \lambda^k) y^{k-1}$
2: **for** $i = 0, 1, \ldots,$ **until** $\|r^{i+1}\| < \epsilon$ **do**
3:     $q_i = (p^i)^T \nabla^2 L(x^k, \lambda^k) p^i$
4:     **if** $q_i = 0$ **then**
5:         exit   ▷ (Start interior point search)
6:     **end if**
7:     $\gamma_i = \frac{(r^i)^T r^i}{q_i}$
8:     $y^{k,i+1} = y^{k,i} + \gamma_i p^i$
9:     $r^{i+1} = r^i - \gamma_i \nabla^2 L(x^k, \lambda^k) p^i$
10:     $\psi_i = \frac{(r^{i+1})^T r^{i+1}}{(r^i)^T r^i}$
11:     $p^{i+1} = r^{i+1} + \psi_i p^i$
12: **end for**

Algorithm 2. Interior point search

1: Take initial point x^0 such that it strictly satisfies (5)
2: Take r^0 such that $\|r^0\| \geq \epsilon$
3: **for** $k = 1, 2 \ldots,$ **until** $\|r^k\| < \epsilon$ **do**

4:     $\sigma_i^k = \begin{cases} (x_i^k)^2, & i \in I_1, \\ \min\{(x^k - \alpha_i)^2, (x^k - \gamma_i)^2\}, & i \in I_2 \\ 1, & i \notin I_1 \cup I_2 \end{cases}$

5:     $D^k = diag(\sigma_1^k, \ldots, \sigma_m^k)$
6:     $B^k = AD^k A^T$
7:     $r^k = Ax^k - Q$
8:     Find u^k as the solution of the system $B^k u = r^k$
9:     $\delta^k = u^k A$
10:     $g^k = D^k \delta^k$
11:     $\lambda_k = \min\{1, \rho\nu_k\}$
12:     $x^{k+1} = x^k + \lambda_k g^k$
13: **end for**

---

4   Interior Point Method

Consider the problem (1), (3) and (5). In order to solve it by Dikin Interior Point Method [1,3,4] it is necessary to find interior point of the set given by constraints (3) and (5). Search for interior point is done by the Algorithm 2. On the line 11, ν^k is maximal feasible step length with respect to (5), $\rho \in (0,1)$ is a coefficient insuring that x^{k+1} is an interior point. Based on [1] it is taken equal to 0.8, ϵ is the given tolerance.

After interior point was found, Interior Point Method (Algorithm 3) which is similar to Algorithm 2, starts working. On the line 12, λ_k' is found as a result of exact one-dimensional minimization of convex function $\phi(\lambda) = F(x^k + \lambda g^k)$. Derivative of ϕ has the following form:

$$\phi'(\lambda) = \sum_{i=1}^{m} g_i^k \left(S_i |x_i^k + \lambda g_i^k|^2 sign(x_i^k + \lambda g_i^k) - H_i \right).$$

Algorithm 3. Interior Point Method

1: Take initial point x^0 so that it strictly satisfies constraints (3), (5)

2:

3: Take Φ_0 such that $\sqrt{\Phi_0} \geq \epsilon$

4: **for** $k = 1, 2 \ldots$, **until** $\sqrt{\Phi_k} < \epsilon$ **do**

5: $\qquad \sigma_i^k = \begin{cases} (x_i^k)^2, & i \in I_1, \\ \min\{(x^k - \alpha_i)^2, (x^k - \gamma_i)^2\}, & i \in I_2, \\ 1, & i \notin I_1 \cup I_2 \end{cases}$

6: $\qquad D^k = diag(\sigma_1^k, \ldots, \sigma_m^k)$

7: $\qquad W^k = AD^k, B^k = W^k A^T$

8: $\qquad c^k = \nabla F(x^k), d^k = W^k c^k$

9: $\qquad$ Find u^k as the solution of the system $B^k u = d^k$

10: $\qquad \delta^k = u^k A - c^k$

11: $\qquad g^k = D^k \delta^k$

12: $\qquad \lambda_k' = arg \min_{\lambda \in [0,1]} \{F(x^k + \lambda_k g^k)\}$

13: $\qquad \lambda_k = \min\{\lambda_k', \rho \nu_k\}$

14: $\qquad x^{k+1} = x^k + \lambda_k g^k$

15: $\qquad \Phi_k = \sum_{i=1}^n \sigma_i^k (\delta_i^k)^2$

16: **end for**

Let us use the following notation: θ_i^k is a root of the equation

$$S_i |x_i^k + \lambda g_i^k|^2 sign(x_i^k + \lambda g_i^k) - H_i = 0, \ i \in K = \{i : g_i^k \neq 0\}.$$

As can be seen, θ_i^k have the following form:

$$\theta_i^k = \begin{cases} \left[\left(\frac{H_i}{S_i} \right)^{\frac{1}{2}} - x_i^k \right] \Big/ g_i^k, & \frac{H_i}{S_i} \geq 0, \\ \left[-\left(-\frac{H_i}{S_i} \right)^{\frac{1}{2}} - x_i^k \right] \Big/ g_i^k, & \frac{H_i}{S_i} < 0. \end{cases}$$

Function $\phi'(\lambda)$ is monotonously increasing, therefore solution λ_k' of the equation $\phi'(\lambda) = 0$ belongs to the interval $[\min_{i \in K} \theta_i^k, \max_{i \in K} \theta_i^k]$. Search for λ_k' on this interval is done by cubic interpolation method [7].

On the line 13, ν_k is defined in the same way as is in Algorithm 2.

As it can be seen, the most computationally intensive part in the both algorithms is the solution of linear systems. Since B^k is always positive definite, sparse Cholesky method is applicable here. As it is shown in [11–13] the sparse Cholesky method consists of the following steps:

1. Fill-in reduction (permutation of the matrix B aimed to reduce amount of nonzero elements in Cholesky matrix). In this paper nested dissection [10, 11] is used as the fill-in reduction technique.
2. Symbolic factorization (search for indexes of nonzero elements in Cholesky matrix).
3. Numerical factorization (calculation of the Cholesky matrix nonzero elements).

In the both algorithms we have $B^k = AD^kA^T$, so at each iteration only matrix D^k is changing. It does not affect the nonzero structure of B^k, which is the same as the structure of AA^T, therefore fill-in reduction and symbolic factorization can be done before the cycle. Additionally numerical factorization can be done not at every iteration, but only if change of x^k becomes small or when change x^k results in constrains violation. Moreover, if fill-in reduction and symbolic factorization are done during the interior point search, this information can be used in Interior Point Method, as well as numerical factorization during several first iterations. Optimized algorithm, that unifies interior point search and Interior Point method is given by the Algorithm 4. Computational effect of such approach is described in the section Numerical result.

Algorithm 4. Interior Point Method optimized

1: Take initial point x^0 so that it strictly satisfies (5)
2: Obtain permutation, granted by fill-in reduction, based on the nonzero structure of AA^T
3: Do symbolic factorization.
4:   ▷ (Start interior point search)
5: Take Φ_0 such that $\sqrt{\Phi_0} \geq \epsilon$
6: Choose g^0 such that $\|g^0\| > \epsilon$, $\lambda_0 = 1$.
7: **for** $k = 1, 2, \ldots$ **until** $\sqrt{\Phi_k} < \epsilon$ **do**
8:     **if** $\lambda_{k-1}\|g^{k-1}\| \leq \epsilon$ **then**
9:         Calculate D^k
10:        Refresh B^k and do numerical factorization
11:    **end if**
12:    Calculate u^k using already obtained Cholesky decomposition
13:    Calculate x^{k+1} and r^{k+1}
14:    **if** x^{k+1} does not satisfy inequality constrains (5) **then**
15:        Go to line 9
16:    **end if**
17:    Calculate Φ_k
18: **end for**
19:   ▷ (Start Interior Point method)
20: Set $g^0 = 0$
21: Take Φ_0 such that $\sqrt{\Phi_0} \geq \epsilon$
22: **for** $k = 1, 2, \ldots$ **until** $\sqrt{\Phi_k} < \epsilon$ **do**
23:    **if** $\lambda_{k-1}\|g^{k-1}\| \leq \epsilon$ **then**
24:        Calculate D^k
25:        Refresh B^k and do numerical factorization
26:    **end if**
27:    Calculate u^k using already obtained Cholesky decomposition
28:    Calculate x^{k+1}
29:    **if** x^{k+1} does not satisfy inequality constrains (5) **then**
30:        Go to line 24
31:    **end if**
32:    Calculate Φ_k
33: **end for**

---

5   Matrices Multiplication in Interior Point Method

Approach used in Algorithm 4 allows us to reduce the amount of matrix multiplications $AD^k A^T$, which is the most expensive operation after Cholesky decomposition. Additionally, since matrices D^k are diagonal without zero entries, and A is an incidence matrix, multiplication can be simplified in the following way.

Let us define $\overline{B}^k = \overline{A}D^k\overline{A}^T$, then if $i \neq j$ and exists t such that $\overline{A}_{ti} \neq 0$ and $\overline{A}_{tj} \neq 0$, then $\overline{B}_{ij}^k = -D_t^k$. Otherwise, if $i \neq j$ and such t does not exist, $\overline{B}_{ij}^k = 0$. $\overline{B}_{ii}^k = -\sum_{j=1, j \neq i}^n \overline{B}_{ij}^k$.

Matrix B^k is obtained from $\overline{B}^k$ by excluding column and row with the same index as the index of the row excluded from $\overline{A}$. Without loss of generality, we can assume that the last row is excluded from $\overline{A}$. Explicit computation of $\overline{B}^k$ is unnecessary.

If columns of A are sorted in lexicographic order depending on the row indexes of nonzero elements, we can say that j_i is column index, starting form which columns cannot have nonzero elements with row indexes less then i. Then index t corresponding to non-diagonal entry B_{ij}^k is j_i plus amount of non-diagonal nonzero entries of the i-th column of B^k before B_{ij}^k. Therefore diagonal entries are calculated by the following formula

$$B_{ii}^k = \sum_{j=i, j \neq i}^n B_{ij}^k + \begin{cases} D_{j_{i+1}-1, j_{i+1}-1}, & \text{if column } j_{i+1} - 1 \text{ has } 1 \\ & \text{nonzero element,} \\ 0, & \text{otherwise.} \end{cases}$$

6   Acceleration by Constant Multiplication

Since constants S_i, H_i and Q_i are of order 10^{-4}, 10^3 and 10^2 respectively [2], in practice it can lead to computational difficulties. To improve performance, scaled vector x^s is used instead of x:

$$x^s = \nu x, \ \nu > 0. \tag{7}$$

Then problem (1), (3) and (5) is reformulated in the following form:

$$F^s(x^s) = \left(\sum_{i=1}^m \frac{S_i |x_i^s|^3}{3} - \frac{H_i}{\nu^2} x_i^s \right) \to \min, \ x^s \in \mathbb{R}^m,$$

$$Ax^s = \frac{1}{\nu}Q,$$

$$x_i^s \geq \frac{\alpha_i}{\nu}, \ i \in I_1, \ \frac{\alpha_i}{\nu} \geq x_i^s \geq \frac{\gamma_i}{\nu}, \ i \in I_2,$$

here $F^s(x^s) = \frac{F(x^s)}{\nu^3}$. Numerical experiments show the following. When

$$\nu = \sqrt[\beta]{0.001} = 0.1,$$

all constants S_i, $\frac{H_i}{\nu^2}$ and $\frac{Q_i}{\nu}$ are approximately of the same order. In this case we obtain essential acceleration in computations, that can be seen in the Table 1.

7   Combined Method for Constrained Problem

As it can be seen from Table 2, Newton Method works much faster, then Interior Point Method, but it cannot work with inequality constraints (5) efficiently. Therefore in the case, when inequality constraints are present, the following scheme is used. Firstly Newton Method is used for solving (1) and (3). Its solution is projected on the set

$$\Omega = \{x \mid x_i \geq \delta_i^1, \, i \in I_1, \, x_i \in [\alpha_i + \xi_i, \gamma_i - \xi_i], \xi_i = \delta_i^2(\alpha_i - \gamma_i), \, i \in I_2\}, \quad (8)$$

here $\delta_i^1 > 0$, $i \in I_1$, $\delta_i^2 > 0$, $i \in I_2$. This set is chosen so that any its point satisfies strictly (5). Obtained projection is taken as initial for interior point search and Interior Point Method. Formalization of the algorithm has form 5. Results of the algorithm work are presented in Table 3, its comparison with usage of Interior Point Method is given in Table 4.

Algorithm 5. Combined method

1: Run Newton Method for the problem (1), (3)
2: Do fill-in reduction and symbolic factorization for the matrix AA^T
3: Run Algorithm 4 without fill-in reduction and symbolic factorization starting form the closest point to the one obtained by the Newton Method, that belongs to Ω (8).

---

8   Numerical Results

The algorithms were coded in C++. Results of numerical experiments are given in Tables 1, 2, 3 and 4. Computations were made in PC with Intel Core i7 / 2.4 GHz / 16 GB.

Here the following notations are used: n — amount of nodes in system (rows in A), m amount of edges in system (columns in A), column Scaling describes whether x is scaled according to (7) or not, Iter. — number of iterations, CG iter. — average among all Newton Method steps number of iterations for Conjugate Gradient, IPS iter — number of iterations of interior (and feasible) point search (Algorithm 2), Time — time in seconds, NZ (%) — number of nonzero entries relative to all entries in the matrices A and L^k, L^k — Cholesky matrix, obtained from B^k (nonzero structures of L^k and B^k is always same on any iteration, since only diagonal matrix D^k changes), B_c^k — amount of calculations of values B^k and numerical factorizations, Total time — full computational time.

In Table 1 problem without inequality constrains (5) is considered. Therefore Newton Method by itself is sufficient to find optimal solution (theoretically it can fail because of matrix being indefinite or possibly singular, but such cases did not happen in numerical experiments). Additionally Cholesky factorization can be done only once, due to the fact that matrices D^k and B^k are same for

Table 1. Problem without inequality constraints (5) with and without scaling

Matrix size		Scaling	Newton Method			Interior Point Method		
n	m		CG iter.	Iter.	Time	IPS iter.	Iter.	Time
1000	1500	No	1327	5	0.16	11	17	0.1
		Yes	338	5	0.1	10	6	0.1
5000	7500	No	1406	6	1.3	11	19	2
		Yes	398	5	0.35	9	19	2
10000	15000	No	1594	6	2.9	11	22	10.5
		Yes	574	7	1.2	9	19	10.1

Table 2. Problem without inequality constraints (5). Newton Method and Dikin Interior Point Method

Matrix size		NZ(%)		Newton Method			Interior Point Method		
N	M	A	L^k	CG iter.	Iter.	Time	IPS iter.	Iter.	Time
1000	1500	0.2	7.6	338	5	0.1	10	6	0.1
5000	7500	0.04	7.3	398	5	0.4	10	6	1.4
10000	15000	0.02	7.3	396	6	0.8	10	7	10
20000	30000	0.01	7.2	419	6	1.7	10	8	73
30000	45000	0.0067	7.1	414	6	2.6	10	8	225
50000	75000	0.004	7.9	443	6	5.2	10	8	1254
100000	150000	0.002	5.6	479	6	14	10	10	6220
1000000	1500000	0.0002	—	837	6	327	—	—	>10000

all iterations. As can be seen, scaling (7) allows to reduce number of iterations for both Newton Method and Interior Point Method.

In Table 2 Newton Method and Interior Point Method with scaling are compared. As can be seen, Interior Point Method works much slower, because density of L^k is higher, than density of A, therefore Cholesky decomposition together with solution of triangular system with L^k require more computations, than computations with A.

In Table 3 problem (1), (3) and (5) is considered, combined method 5 is used. Firstly Newton Method is applied to find solution of problem (1) and (3). Then obtained point is projected to the shrunk area, as described in Sect. 7 (here $\delta_i^1 = 10$, $\delta_i^2 = 0.3$). Since in interior point search and in Interior Point Method matrix B^k changes every several iterations, amount of calculations of B^k is presented in column B^k calc. for interior point search and for Interior Point Method.

In Table 4 problem (1), (3) and (5) is considered. Combined method 5 is compared with usage of Interior Point Method only. As can be seen, with growth of the problem size, performance difference increases in favor of method 5.

Table 3. Problem with inequality constraints (5)

Matrix size		Newton Method			Interior Point Method					Total time
					IPS					
n	m	CG iter.	Iter.	Time	Iter.	B_c^k	Iter.	B_c^k	Time	
1000	1500	478	10	0.3	10	1	30	2	0.2	0.5
5000	7500	1383	12	2.5	10	1	110	3	7.8	8.2
10000	15000	1476	12	5.4	10	1	737	6	85	95
20000	30000	2631	12	20	10	1	954	7	693	713
30000	45000	3776	12	46	10	1	8351	14	4667	4713

Table 4. Problem with inequality constraints (5). Solution with Newton method and without it

Matrix size		Newton Method	Interior Point Method				Total time
			IPS				
n	m		Iter.	B^k calc.	Iter.	B^k calc.	
1000	1500	Used	10	1	30	2	0.5
		Not used	11	2	36	2	0.2
5000	7500	Used	10	1	110	3	8.2
		Not used	20	3	61	3	7.9
10000	15000	Used	10	1	737	6	95
		Not used	183	3	1199	6	111
20000	30000	Used	10	1	954	7	713
		Not used	18	3	3608	6	1198

9   Conclusion

A method, aimed to solve high dimensional flow distribution problem is presented in this work. Combination of Newton Method and Dikin Interior Point Method based on sparse matrices techniques is used, to provide solution for such problems. Numerical experiments show, that it allows to obtain solution for systems consisting of 10000 — 30000 nodes in reasonable time on a personal computer. Additionally presented algorithm provides opportunity for parallel computations.

Main advantage of the suggested combination of Newton Method and Interior Point Method consists in the following. This combination grants more precise solution, than Interior Point Method alone. Since Newton Method works fast, its usage worsen computational time negligibly. Reduced running time is obtained due to the good starting point for Interior Point Method provided by Newton Method.

References

1. Dikin, I.I.: Interior Point Method in Linear and Nonlinear programming. Moscow, Krasand (2010). (in Russian)
2. Novitskiy, N.N., Dikin, I.I.: Calculation of feasible pipeline network operating conditions by the interior-point method. Bull. Russ. Acad. Sci. Energy. (5) (2003). (in Russian)
3. Dikin, I.I.: Iterative solution of problems of linear and quadratic programming. Sov. Math. Dokl. **8**, 674–675 (1967)
4. Vanderbei, R.J.: Linear Programming Foundations and Extentions, 4th edn. Springer, Heidelberg (2014)
5. Merenkov, A.P., Khasilev, V.Y.: Theory of Hydralic Networks. Moscow, Nauka (1985). (in Russian)
6. Farhat, I.A., Al-Hawary, M.E.: Optimization methods applied for solving the short-term hydrothermal coordination problem. Electr. Power Syst. Res. **79**, 1308–1320 (2009)
7. Nocedal, J., Wright, S.: Numerical Optimization. Springer, Heidelberg (2006)
8. Boyd, S., Vandenberghe, L.: Convex Optimization. Cambridge University Press, Cambridge (2004)
9. Van der Vorst, H.A.: Iterative Krylov Methods for Large Linear Systems. Cambridge Monographs on Applied and Computational Mathematics, vol. 13, 2nd edn. Cambridge University Press, Cambridge (2003). Ciarlet, P.G., Iserles, A., Kohn, R.V., Wright M.H. (eds.)
10. Saad, Y.: Iterative Methods for Sparce Linear Systems, 2nd edn. SIAM, Philadelphia (2003)
11. Pissanetsky, S.: Sparce Matrix Technology. Academic Press, New York (1984)
12. Davis, T.A.: Direct Methods for Sparce Linear Systems. SIAM, Philadelphia (2006)
13. Gilbert, J.R., Ng, E.G., Peyton B.W.: An Efficient Algorythm to Compute Row and Column Counts for Sparce Cholesky Factorization. Oak Ridge National Laboratory (1992)

Hierarchical Clustering and Multilevel Refinement for the Bike-Sharing Station Planning Problem

Christian Kloimüllner$^{(\boxtimes)}$ and Günther R. Raidl

Institute of Computer Graphics and Algorithms, TU Wien,
Favoritenstraße 9–11/1861, 1040 Vienna, Austria
{kloimuellner,raidl}@ac.tuwien.ac.at

Abstract. We investigate the Bike-Sharing Station Planning Problem
(BSSPP). A bike-sharing system consists of a set of rental stations, each
with a certain number of parking slots, distributed over a geographical
region. Customers can rent available bikes at any station and return them
at any other station with free parking slots. The initial decision process
where to build stations of which size or how to extend an existing system
by new stations and/or changing existing station configurations is crucial
as it actually determines the satisfiable customer demand, costs, as well
as the rebalancing effort arising by the need to regularly move bikes from
some stations tending to run full to stations tending to run empty. We
consider as objective the maximization of the satisfied customer demand
under budget constraints for fixed and variable costs, including the costs
for rebalancing. As bike-sharing stations are usually implemented within
larger cities and the potential station locations are manifold, the size
of practical instances of the underlying optimization problem is rather
large, which makes a manual decision process a hardly comprehensible
and understandable task but also a computational optimization very
challenging. We therefore propose to state the BSSPP on the basis of
a hierarchical clustering of the considered underlying geographical cells
with potential customers and possible stations. In this way the estimated
existing demand can be more compactly expressed by a relatively sparse
weighted graph instead of a complete matrix with mostly small non-zero
entries. For this advanced problem formulation we describe an efficient
linear programming approach for evaluating candidate solutions, and for
solving the problem a first multilevel refinement heuristic based on mixed
integer linear programming. Our experiments show that it is possible
to approach instances with up to 2000 geographical cells in reasonable
computation times.

Keywords: Bike-Sharing Station Planning Problem · Hierarchical clustering · Multilevel refinement · Facility location problem

1 Introduction

Many large cities around the world have already built bike sharing systems
(BSS), and many more are considering to introduce one or extend an existing

© Springer International Publishing AG 2017
R. Battiti et al. (Eds.): LION 2017, LNCS 10556, pp. 150–165, 2017.
https://doi.org/10.1007/978-3-319-69404-7_11

one. These systems consist of rental stations around the city or a certain part of it where customers can rent and return bikes. A rental station has a specific number of parking slots where a bike can be taken from or returned to. On the contrary to bike-rental systems, BSSs encourage a short-term usage of bikes. As bikes are typically returned at a different station than they have been taken from, a need for active rebalancing arises as the demand for bikes to rent and parking slots to return bikes is not equally distributed among the stations.

Finding a good combination of station locations and building these stations in the right size is crucial when planning a BSS as these stations obviously directly determine the satisfied customer demand in terms of bike trips, the arising rebalancing effort, and the resulting fixed and variable costs. Stations close to public transport, business parks, or large housing developments will likely face a high demand whereas stations in sparser inhabited areas will probably face a lower demand. However, also the station density and connectedness of the actual regions to be covered play crucial roles. Some solitary station that is far from any other station will most likely not fulfill much demand. Moreover, a clever choice of station locations might also exploit the natural demands and customer flows in order to keep the rebalancing effort and associated costs reasonable.

As BSSs are usually implemented in rather large cities the problem of finding optimal locations for rental stations and sizing these stations appropriately is challenging and manually hardly comprehensible. Thus, there is the need for computational techniques supporting this decision-making. Besides fixed costs for building the system, an integrated approach should also estimate maintenance and rebalancing costs over a certain time horizon such that overall costs for the operator can be approximated more precisely. It is further important to consider the customer demands in a time-dependent way because there usually exists a morning peak and an afternoon peak which is due to commuters, people going to work, and students. Between these peaks, the demand of the system is usually a bit lower. We refer to this problem as *Bike Sharing Station Planning Problem* (BSSPP). The objective we consider here is to determine for a specified total-cost budget and a separate fixed-cost budget a selection of locations where rental stations of an also to be determined size should be erected in order to maximize the actually fulfilled customer demand.

In this work, we first concentrate on how to efficiently model the BSSPP such that we can also deal with very large instances with thousands of considered geographical cells for customers and potential station locations. To this end we propose to utilize a hierarchical clustering to express the estimated potential customer demand on it. We will then describe a *linear programming* (LP) based method to evaluate candidate solutions, and finally present a first novel *multilevel refinement heuristic* (MLR), based on mixed integer linear programming (MIP), to approach the optimization problem.

In Sect. 2 we discuss related work. Section 3 defines the BSSPP formally, also introducing the hierarchical clustering. Sections 3.3 and 3.4 describe LP models for determining the actually fulfilled customer demands for a candidate solution and estimating the required rebalancing effort, respectively. The MLR is then described in Sect. 4. First computational results on randomly generated instances are shown in Sect. 5, and finally, conclusions are drawn in Sect. 6.

2   Related Work

There already exists some work which tries to find optimal station locations for BSSs, although mostly considering different aspects. To the best of our knowledge, Yang et al. [12] were the first who considered the problem in 2010. They relate the problem to *hub location problems*, a special variant of the well-known *facility location problem*, and propose a mathematical model for it. The considered objective is to minimize the walking distance by prospective customers, fixed costs, and, a penalty for uncovered demands. The authors solve the problem by a heuristic approach in which a first part of the algorithm tries to identify the location of rental stations and a second, inner part tries to find shortest paths between origin and destination pairs. The authors illustrate their approach by a small example consisting of 11 candidate cells for bike stations.

Lin et al. [6] propose a mixed integer non-linear programming model and solve a small example instance with 11 candidate stations by the commercial solver LINGO, and furthermore provide a sensitivity analysis. Martinez et al. [8] develop approaches for a case study within Lisbon having 565 prospective candidate stations. They propose a hybrid approach consisting of a heuristic part utilizing a *mixed integer linear programming* (MIP) formulation. Locations as well as the fleet dimension are optimized, e-bikes are also considered, and rebalancing requirements are estimated.

Lin et al. [7] propose a heuristic algorithm for solving the hub location inventory problem arising in BSSPP. They do not only optimize station locations but their algorithm also identifies where to build bike lanes. As a subproblem they have to determine the travel patterns of the customers, i.e., solve a flow problem for a given configuration. They illustrate their approach on a small example consisting of 11 candidate locations for stations. Saharidis et al. [9] propose a MIP formulation which minimizes unmet demands and walking distance for prospective customers. They test their approach in a case study for the city center of Athens having 50 candidate cells for stations. Chen et al. [1] provide a mathematical non-linear programming model and solve the problem utilizing an improved immune algorithm. They define three different types of rental stations depending on their location (e.g., near a metro station, supermarkets). Their aim is that stations in the residential area have enough bikes available such that the morning peak can be managed and that stations near metro lines or important places have enough free parking slots available to manage incoming bikes during the morning peak. They provide a case study for a particular metro line of Nianjing city including 10 district stations and 31 residential stations. In [2] Chen and Sun aim at satisfying a given demand and minimizing travel times of the users. The authors propose an integer programming model which they solve with the LINGO solver. A computational analysis is provided on a small example. Frade et al. [3] describe an approach for a case study of the city of

Coimbra, Portugal. They present a compact MIP model which they solve using the XPRESS solver. Their objective is to maximize the demand covered by the BSS under budget constraints. They also include the net revenue in their mathematical model which reduces the costs incurred by building the BSS. Their single test instance consists only of 29 cells or *traffic zones*, how they call it. Hu et al. [5] also present a case study for a BSS along a metro line. They aim at minimizing total costs incurred by building particular BSS stations. In their computational study they consider three scenarios, each consisting of ten possible station candidates. They solve the proposed MIP model by the LINGO solver. Last but not least, Gavalas et al. [4] summarized diverse algorithmic approaches for the design and management of vehicle-sharing systems.

We conclude that all previous works on computational optimization approaches for designing BSS only consider rather small scenarios. Most previous work accomplishes the optimization with compact mathematical models that are directly solved by a MIP solver. Such methods, however, are clearly unsuited for tackling large realistic scenarios of cities with up to 2000 cells or more. In the following, we therefore propose a novel multilevel refinement heuristic based on a hierarchical clustering of the demand data.

3   Problem Formalization

The considered geographical area is partitioned into cells. Let S be the set of cells where a BSS station may potentially be located (*station cells*), and let V be the set of cells where some positive travel demand (outgoing, ingoing, or both) from prospective customers of the BSS exists (*customer cells*).

To handle such a large number of cells effectively, we consider a hierarchical abstraction as crucial in order to represent and model the further data in a meaningful and relatively compact form. To this end, we are expecting a hierarchical clustering of all customer cells V as input.

This hierarchical clustering is given in the form of a rooted tree with the inner nodes corresponding to clusters and the leafs corresponding to the cells. All cells have the same depth which is equal to the height of the tree, denoted by h. Let $C = C_0 \cup \ldots \cup C_h$ be the set of all tree nodes, with C_d corresponding to the subset of nodes at depth $d = 0, \ldots, h$. $C_0 = \{0\}$ contains only the root node 0 representing the single cluster with all cells, while $C_h = V$. Let super$(p) \in C$ be the immediate predecessor (parent cluster) of some node $p \in C \setminus C_0$ and sub$(p) \subset C$ be the set of immediate successors (children) of a cluster $p \in C \setminus C_h$.

As the travel demand of potential users varies over time we are given a (small) set of periods $T = \{1, \ldots, \tau\}$ for a "typical" day for which the planning shall be done. The estimated existing travel demand occurring in each period $t \in T$ from/to any cell $v \in V$ is given by a weighted directed graph $G^t = (C^t, A^t)$. All relevant outgoing travel demand at a cell v is represented by outgoing arcs $(v, p) \in A^t$ with $p \in C$ and corresponding values (weights) $d_{v,p}^t > 0$, i.e., (v, p) represents all expected trips from v to any cell represented by p in period t that might ideally be satisfied, and $d_{v,p}^t$ indicates the expected number of these trips.

Moreover, for each time period $t \in T$ we are given its duration denoted by δ_t^{period} and we are given a global parameter δ^{rent} which defines the average duration of a single trip performed by some user of the BSS.

The following conditions must hold to keep this graph as compact and meaningful as possible: the target node p of an arc (v, p) must not be a predecessor of v in the cluster tree. Self-loops (v, v), however, are allowed and important to model demand where the destination corresponds to the origin, arcs representing a neglectable demand, i.e., below a certain threshold, shall be avoided. Consequently, if there is an arc (v, p) no further arc (v, q) is allowed to any node q being a successor or a predecessor of p.

All estimated ingoing travel demand for each cell $v \in V$ is given correspondingly by arcs $(p, v) \in A^t$ with $p \in C$ with demand values $d_{p,v}^t \geq 0$, and corresponding conditions must hold.

Furthermore, it is an important property, that ingoing and outgoing demands have to be consistent: Let us denote by $V(p)$ the subset of all cells from V contained in cluster $p \in C$, i.e., the leafs of the subtree rooted in p, and by $C(p)$ the subset of all the nodes $q \in C$ that are part of the subtree rooted in p, including p and $V(p)$. For any $p \in C \setminus V$ it must hold that

$$\sum_{(v,q)\in A^t | v \in V(p), q \notin C(p)} d_{v,q}^t \geq \sum_{(q,v)\in A^t | q \in C(p), v \notin V(p)} d_{q,v}^t \tag{1}$$

and

$$\sum_{(q,v)\in A^t | q \notin C(p), v \in V(p)} d_{q,v}^t \geq \sum_{(v,q)\in A^t | v \notin V(p), q \in C(p)} d_{v,q}^t. \tag{2}$$

Condition (1) ensures that the total demand originating at the leafs of the subtree rooted at p and leading to a destination outside of the tree is never less than the total ingoing demand at all the cells outside the tree originating from some cluster inside the tree. Condition (2) provides a symmetric condition for the total ingoing demand at all the leafs of the tree. Furthermore, for the root node $p = 0$, inequalities (1) and (2) must hold with equality.

For each customer cell $v \in V$, we are given a (typically small) set $S(v) \subseteq S$ of station cells in the vicinity by which v's demand may be (partly) fulfilled. Furthermore, let $a_{v,s} \in (0, 1]$, $\forall v \in V$, $s \in S(v)$, be an attractiveness value indicating the expected proportion of demand from v (ingoing as well as outgoing) that can at most be fulfilled with a sufficiently sized station at s. These attractiveness values will be determined primarily based on the walking distances among the stations (the value will typically roughly exponentially decrease with the distance), but can be in general an arbitrary distance decay model. If there is a one-to-one correspondence of cells in V and S, for each $v \in V$, $v \in S(v)$, $a_{v,v} = 1$ will typically hold.

For the costs of building a station we consider here only a (strongly) simplified linear model, but we distinguish fixed costs for building the station and initially buying the bikes, variable costs for maintaining the station and the respective bikes, and costs for performing the rebalancing. Let b^{fix} and b^{var} be the average fixed and variable costs per bike slot, and let b^{reb} be the average costs for rebalancing one bike per day over the whole planning horizon. The fixed costs for a station in cell $s \in S$ with x_s slots are then $\text{fixcost}(s) = b^{\text{fix}} \cdot x_s$ and the total costs are $\text{totalcost}(s) = b^{\text{fix}} \cdot x_s + b^{\text{var}} \cdot x_s + b^{\text{reb}} \cdot Q_x(s)$, where $Q_x(s)$ denotes an estimation for the number of bikes that need to be redistributed from station s to some other station. We assume here that the size of each station, i.e., the number of its slots, can be freely chosen from 0 (i.e., no station is built) up to some maximum cell-dependent capacity $z_s \in \mathbb{N}$. The determination of the rebalancing effort for a given candidate solution will be described in Sect. 3.4. We remark that this cost model only is a first very rough estimate. Considering location dependent costs, costs for a station to be built that are independent of the number of slots, and a more restricted selection of station sizes is left for future research.

We assume that a total budget $B_{\text{max}}^{\text{tot}}$ is given as well as a budget for only the sum of all fixed costs $B_{\text{max}}^{\text{fix}} < B_{\text{max}}^{\text{tot}}$, and both must not be exceeded in a feasible solution.

3.1   Solution Representation

A solution $x = \{x_s \in \mathbb{N} \mid s \in S\}$ assigns each station cell $s \in S$ an amount of parking slots to be built, possibly also 0 which would mean that no station is going to be built in cell s.

3.2   Objective

The goal is to maximize the expected total number of journeys in the system, i.e., the total demand that actually can be fulfilled at each day over all time periods, considering the available budgets $B_{\text{max}}^{\text{tot}}$ and $B_{\text{max}}^{\text{fix}}$.

Let $D(x,t)$ be the total demand fulfilled by solution x in time period $t \in T$, and let $Q_x(s)$ be the required rebalancing effort arising at each station $s \in S \mid x_s \neq 0$ in terms of the number of bikes to be moved to some other station. The calculation of these values will be considered separately in Sects. 3.3 and 3.4. The BSSPP can then be stated as the following MIP.

$$\max \sum_{t \in T} D(x,t) \tag{3}$$

$$\sum_{s \in S} (b^{\text{fix}} \cdot x_s + b^{\text{var}} \cdot x_s + b^{\text{reb}} \cdot Q_x(s)) \leq B_{\text{max}}^{\text{tot}} \tag{4}$$

$$\sum_{s \in S} b^{\text{fix}} \cdot x_s \leq B_{\text{max}}^{\text{fix}} \tag{5}$$

$$x_s \in \{0, \ldots, z_s\} \qquad\qquad s \in S \tag{6}$$

Inequality (4) calculates the total costs over all stations and ensures that the total budget is not exceeded, while inequality (5) restricts only the fixed costs over all stations by the respective budget.

3.3 Calculation of Fulfilled Customer Demand

To determine the overall fulfilled demand for a specific, given solution x and a certain time slot $t \in T$, we first make the following local definitions. Let $S' = \{s \in S \mid x_s \neq 0\}$ correspond to the set of cells where a station actually is located, $V' = \{v \in V \mid S(v) \cap S' \neq \emptyset\}$ be the set of customer cells whose demand can possibly (partly) be fulfilled as at least one station exists in the neighborhood. Moreover, let $C' = \{p \in C \mid V(p) \cap V' \neq \emptyset\}$ be the set of all nodes in the hierarchical clustering representing relevant customer cells, i.e., cells whose demand can possibly be fulfilled. The set $S'(v) = S(v) \cap V'$, $\forall v \in V'$ refers to the existing stations that might fulfill part of v's demand, and $V'(p) = V(p) \cap V'$, $\forall p \in C'$ denotes the existing customer cells contained in cluster p. $C'(p)$ refers to the subset of all the nodes $q \in C'$ that are part of the subtree rooted at p, including p and $V'(p)$, and $G' = (C', A')$ with $A' = \{(p,q) \in A^t \mid p,q \in C'\}$ is then the correspondingly reduced demand graph.

In the following we use variables u, v, w for referencing customer cells in V', variables p, q for referencing cluster nodes in C' (which might possibly also be customer cells), variable s for station cells in S', and α, β for arbitrary nodes in $C' \cup S$.

We further define for each arc in A' corresponding to a specific demand an individual flow network depending on the kind of the arc:

- Arcs $(u,v) \in A'$ with $u, v \in V'$, including the case $u = v$:
  $G_f^{u,v} = (V_f^{u,v}, A_f^{u,v})$ with node set $V_f^{u,v} = \{u\} \cup S'(u) \cup S'(v) \cup \{v\}$ and arc set $A_f^{u,v} = (\{u\} \times S'(u)) \cup (S'(u) \times S'(v)) \cup (S'(v) \times \{v\})$.
- Arcs $(v,p) \in A'$ with $v \in V', p \in C' \setminus V'$:
  $G_f^{v,p} = (V_f^{v,p}, A_f^{v,p})$ with node set $V_f^{v,p} = \{v\} \cup S'(v) \cup \{p\}$ and arc set $A_f^{v,p} = (\{v\} \times S'(v)) \cup (S'(v) \times \{p\})$.
- Arcs $(p,v) \in A'$ with $p \in C' \setminus V', v \in V'$:
  $G_f^{p,v} = (V_f^{p,v}, A_f^{p,v})$ with node set $V_f^{p,v} = \{p\} \cup S'(v) \cup \{v\}$ and arc set $A_f^{p,v} = (\{p\} \times S'(v)) \cup (S'(v) \times \{v\})$.

All arcs $(\alpha, \beta) \in A_f^{p,q}$ of all flow networks have associated corresponding flow variables $0 \leq f_{\alpha,\beta}^{p,q} \leq d_{p,q}^t$. The fulfilled demands can be modeled within these networks as maximum flows. Furthermore, we utilize variables H_p^{in}, H_p^{out} $\forall p \in C' \setminus V'$, for the total inflow/outflow at all customer cells $V'(p)$ originating at/targeted to cluster nodes from outside cluster p, i.e., $C' \setminus C'(p) \setminus V'$. Variables F_p^{in}, F_p^{out}, $\forall p \in C' \setminus V'$, represent the total ingoing/outgoing flows at all cluster nodes q within cluster p originating at/targeted to customer cells outside cluster p, i.e., $V' \setminus V'(p)$, respectively. The flow variables, however, depend on each other and the stations' capacities. A weighting factor ω is used to adjust the number of trips which can be performed in time period t by using only a single bike. The following LP is used to compute the total satisfied demand

$$D(x,t) = \max \sum_{(v,p)\in A' \mid v\in V'} \sum_{(v,s)\in A_f^{v,p}} f_{v,s}^{v,p} \tag{7}$$

$$\text{s.t.} \sum_{(v,s)\in A_f^{v,p}} f_{v,s}^{v,p} \le d_{v,p}^t \qquad\qquad (v,p)\in A' \mid v\in V' \tag{8}$$

$$\sum_{(s,v)\in A_f^{p,v}} f_{s,v}^{p,v} \le d_{p,v}^t \qquad\qquad (p,v)\in A' \mid v\in V' \tag{9}$$

$$f_{u,s}^{u,v} = \sum_{s'\in S'(v)} f_{s,s'}^{u,v} \qquad\qquad \begin{array}{l}(u,v)\in A' \mid u,v\in V',\\ s\in S'(u)\end{array} \tag{10}$$

$$\sum_{s'\in S'(u)} f_{s',s}^{u,v} = f_{s,v}^{u,v} \qquad\qquad \begin{array}{l}(u,v)\in A' \mid u,v\in V',\\ s\in S'(v)\end{array} \tag{11}$$

$$f_{v,s}^{v,p} = f_{s,p}^{v,p} \qquad\qquad \begin{array}{l}(v,p)\in A' \mid v\in V',\\ p\in C'\setminus V', s\in S'(v)\end{array} \tag{12}$$

$$f_{p,s}^{p,v} = f_{s,v}^{p,v} \qquad\qquad \begin{array}{l}(p,v)\in A' \mid v\in V',\\ p\in C'\setminus V', s\in S'(v)\end{array} \tag{13}$$

$$-x_s \le \sum_{(p,q)\in A'} \sum_{(\alpha,s)\in A_f^{p,q}} f_{\alpha,s}^{p,q}$$
$$-\sum_{(p,q)\in A'} \sum_{(s,\alpha)\in A_f^{p,q}} f_{s,\alpha}^{p,q} \qquad\qquad s\in S' \tag{14}$$
$$-\omega\cdot \frac{\delta^{\text{rent}}\cdot \sum_{(p,q)\in A'} \sum_{(\alpha,s)\in A_f^{p,q}} f_{\alpha,s}^{p,q}}{\delta_t^{\text{period}}}$$

$$x_s \ge \sum_{(p,q)\in A'} \sum_{(\alpha,s)\in A_f^{p,q}} f_{\alpha,s}^{p,q}$$
$$-\sum_{(p,q)\in A'} \sum_{(s,\alpha)\in A_f^{p,q}} f_{s,\alpha}^{p,q} \qquad\qquad s\in S' \tag{15}$$
$$+\omega\cdot \frac{\delta^{\text{rent}}\cdot \sum_{(p,q)\in A'} \sum_{(s,\alpha)\in A_f^{p,q}} f_{s,\alpha}^{p,q}}{\delta_t^{\text{period}}}$$

$$H_p^{\text{in}} = \sum_{(q,v)\in A' \mid q\notin C'(p)\cup V', v\in V'(p)} \qquad\qquad p\in C'\setminus V' \tag{16}$$
$$\sum_{(s,q)\in A_f^{q,v}} f_{s,q}^{q,v}$$

$$F_p^{\text{in}} = \sum_{(v,q)\in A' \mid v\notin V'(p), q\in C'(p)\setminus V'} \qquad\qquad p\in C'\setminus V' \tag{17}$$
$$\sum_{(s,q)\in A_f^{v,q}} f_{s,q}^{v,p}$$

$$H_p^{in} \geq F_p^{in} \qquad\qquad p \in C' \setminus V' \setminus \{0\} \qquad (18)$$

$$H_0^{in} = F_0^{in} \qquad\qquad (19)$$

$$H_p^{out} = \sum_{(v,q)\in A' \mid v \in V'(p), q \notin C'(p)\cup V'} \qquad\qquad p \in C' \setminus V' \qquad (20)$$
$$\sum_{(q,s)\in A_f^{q,v}} f_{q,s}^{q,v}$$

$$F_p^{out} = \sum_{(q,v)\in A' \mid q \in C'(p)\setminus V', v \notin V'(p)} \qquad\qquad p \in C' \setminus V' \qquad (21)$$
$$\sum_{(p,s)\in A_f^{q,v}} f_{q,s}^{q,v}$$

$$H_p^{out} \geq F_p^{out} \qquad\qquad p \in C' \setminus V' \setminus \{0\} \qquad (22)$$

$$H_0^{out} = F_0^{out} \qquad\qquad (23)$$

$$0 \leq f_{v,s}^{v,p} \leq a_{v,s} \cdot d_{v,p}^t \qquad\qquad (v,p) \in A' \mid v \in V', \qquad (24)$$
$$(v,s) \in A_f^{v,p}$$

$$0 \leq f_{s,v}^{p,v} \leq a_{s,v} \cdot d_{p,v}^t \qquad\qquad (p,v) \in A' \mid v \in V', \qquad (25)$$
$$(s,v) \in A_f^{p,v}$$

$$0 \leq f_{\alpha,\beta}^{p,q} \leq d_{p,q}^t \qquad\qquad (p,q) \in A', \qquad (26)$$
$$(\alpha,\beta) \in A_f^{p,q} \mid \alpha, \beta \notin V'$$

$$F_p^{in}, F_p^{out} \geq 0 \qquad\qquad p \in C' \setminus V' \qquad (27)$$

$$H_p^{in}, H_p^{out} \geq 0 \qquad\qquad p \in C' \setminus V' \qquad (28)$$

Objective function (7) maximizes the total outgoing flow over all $v \in V'$, i.e., the fulfilled demand. Note that this also corresponds to the total ingoing flow over all v. Inequalities (8) limit the total flow leaving $v \in V'$, for each demand $(v,p) \in A' \mid v \in V'$ to $d_{v,p}^t$. Inequalities (9) do the same w.r.t. ingoing demands. Equalities (10) and (11) provide the flow conservation at source and destination stations s for $(u,v) \in A'$ with $u, v \in V'$. Equalities (12) provide the flow conservation at the source station in case of an arc $(v,p) \in A'$ towards a cluster node p, while (13) provide the flow conservation at the destination station in case of an arc $(p,v) \in A'$ originating at a cluster node p. Inequalities (14) and (15) provide the capacity limitations at each station $v \in V'$. It is the accumulated demand occurring at the particular station including a "compensation term" for large values of ingoing as well as outgoing demand. The fraction $\delta_t^{period}/\delta^{rent}$ represents the number of trips which can ideally be performed in period t using a single bike. The weighting factor ω is used to adjust this value such that it better reflects reality as the bike trips are not likely to be performed "optimally" with respect to the distribution over the whole time period in real world. Equalities (16) compute the total outgoing flow for the leafs of the subtree rooted at p to any cluster which is not part of the subtree rooted at p. Equalities (17) compute the total ingoing flow for each cluster node p by considering the ingoing flow from any $v \in V$ for which p is not a predecessor to every cluster of the subtree rooted at p. Inequalities (18) ensure that there must not be more ingoing flow to clusters of the subtree rooted at p as there is outgoing flow from the leafs contained in the subtree rooted at p. Equality (19) ensures that at the top level, i.e., at the root node 0, the outgoing flow from leaf nodes to cluster nodes and the ingoing flow from cluster nodes to leaf nodes is balanced, i.e., the same amount. Inequalities (21)–(23) state the corresponding constraints for the outgoing flow instead of the ingoing flow. Equations (24) and (25) provide the domain definitions for the flow variables from/to a cell v to/from a neighboring station s by considering the demand weighted by factor $a_{v,s}$. For all remaining flow variables, (26) provide the domain definitions based on the demands. The remaining variables are just restricted to be non-negative in (27) and (28).

3.4   Calculation of Rebalancing Costs

We state an LP for minimizing the total rebalancing effort over all time periods T at each station $s \in S'$ by choosing an appropriate initial fill level for each period, ensuring that the whole prospective customer demand is fulfilled. We estimate the rebalancing effort by considering the necessary changes in the fill levels inbetween the time periods. The LP uses the following decision variables. By $y_{t,s}$ we refer to the initial fill level of station $s \in S'$ at the beginning of time period $t \in T$, and by $r_{t,s}^+$ and $r_{t,s}^-$ we denote the number of bikes which need to be delivered to, respectively picked up from, station $s \in S'$ at the end of period $t \in T$ to achieve the fill levels $y_{t+1,s}$ (or $y_{1,s}$ in case of $t = \tau$).

The accumulated demand $D_{t,v}^{\mathrm{acc}}$ can be calculated by utilizing the solution of the previous model from Sect. 3.3, c.f. inequalities (14) and (15). The following LP is solved for each station $s \in S' \mid x_s \neq 0$ independently. For station cells $s \in S \setminus S'$, i.e., where no station is actually built in solution x, $Q_x(s) = 0$.

$$Q_x(s) = \min \quad \sum_{t \in T} r_{t,s}^+ + r_{t,s}^- \tag{29}$$

$$\text{s.t.} \quad y_{t,s} + r_{t,s}^+ \geq D_{t,s}^{\mathrm{acc}} \qquad\qquad t \in T \tag{30}$$

$$x_s - y_{t,s} + r_{t,s}^- \geq -D_{t,s}^{\mathrm{acc}} \qquad\qquad t \in T \tag{31}$$

$$y_{t+1,s} = y_{t,s} - D_{t,s}^{\mathrm{acc}} + r_{t,s}^+ - r_{t,s}^- \qquad t \in T \setminus \{\tau\} \tag{32}$$

$$y_{1,s} = y_{\tau,s} - D_{\tau,s}^{\mathrm{acc}} + r_{\tau,s}^+ - r_{\tau,s}^- \tag{33}$$

$$0 \leq y_{t,s} \leq x_s \qquad\qquad t \in T \tag{34}$$

$$0 \leq r_{t,s}^+ \leq D_{t,s}^{\mathrm{acc}} \qquad\qquad t \in T \tag{35}$$

$$0 \leq r_{t,s}^- \leq -D_{t,s}^{\mathrm{acc}} \qquad\qquad t \in T \tag{36}$$

Objective function (29) minimizes the number of rebalanced bikes, i.e., number of bikes that have to be delivered $r_{t,s}^+$ and number of bikes that have to be picked up $r_{t,s}^-$. Inequalities (30) compute the number of bikes that have to be delivered to the corresponding station in order to meet the given demand. Inequalities (31) compute the number of bikes that have to be picked up from the corresponding station in order to meet the given demand. Inequalities (32) state a recursion in order to compute the fill level for the next time period. Inequalities (33) state that for each station the fill level for the next day has to be again the initial fill level of the first period. Inequalities (34)–(36) are the domain definitions for the number of bikes to be moved and the fill level for each time period.

4   Multilevel Refinement Approach

Clearly, practical instances of the problem are far too large to be approached by a direct exact MIP approach. However, also basic constructive techniques or metaheuristics with simple, classical neighborhoods are unlikely to yield reasonable results when making decisions on a low level without considering crucial relationships on higher abstraction levels, i.e., a more global view. Classical local search techniques on the natural variable domains concerning decisions for individual stations may only fine-tune a solution but are hardly able to overcome bad solutions in which larger regions need to be either supplied with new stations or where many stations need to be removed. We therefore have the strong need of some technique that exploits also a higher-level view, deciding for larger areas about the supply of stations in principle. Multilevel refinement strategies can provide this point-of-view.

In multilevel refinement strategies [11] the whole problem is iteratively coarsened (aggregated) until a certain problem size is reached that can be reasonably handled by some exact or heuristic optimization technique. After obtaining a solution at this highest abstraction level, the solution is iteratively extended to the previous lower level problem instance and possibly refined by some local search, until a solution to the original problem at the lowest level, i.e., the original problem instance, is obtained. For a general discussion and the generic framework we refer to the work of Walshaw [10].

To apply multilevel refinement to BSSPP we essentially have to decide how to realize the procedures for coarsening an instance for the next higher level, solving a reasonably small instance, and extending a solution to a solution at the next lower level. In the following, we denote all problem instance data at level l by an additional superscript l. By P_l we generally refer to the problem at level l of the MLR algorithm described here.

4.1   Coarsening

We have to derive the more abstract problem instance P_{l+1} from a given instance P_l. Naturally, we can exploit the already existing customer cell cluster hierarchy for the coarsening. Remember that all customer cells appear in the cluster hierarchy always at the same level. We coarsen the problem by considering the customer cells and the station cells separately.

Coarsening of Customer Cells. The main strategy for coarsening the customer cells is to merge cells having the same parent cluster together with their parent. This means $V^{l+1} = C^l_{h^l-1}$ or simply $V^{l+1} = C_{h-l-1}$, i.e., each cluster node at depth $h - l - 1$ corresponds to a customer cell at level $l + 1$ representing the merged set of customer nodes contained in C_{h-l-1}. The hierarchical clustering of P_l becomes $C^{l+1} = C_0 \cup \ldots \cup C_{h-l}$. Remember that we already defined the function super(p) to return the parent cluster of some node p, and therefore super$(p^l) : C^l \to C^{l+1}$ also returns the cluster from C^{l+1} in which cluster $p^l \in C^l$ is merged into. The new demand graph $G^{t,l+1} = (C^{t,l+1}, A^{t,l+1})$ consists of the arc set $A^{t,l+1} = \bigcup_{(p^l,q^l) \in A^{t,l}}(\text{super}(p^l), \text{super}(q^l))$. This demand graph may again contain self-loops, but it is still simple, i.e., multiple arcs from $A^{t,l}$ may map to the same single arc in $A^{t,l+1}$ and the respective demand values are merged. Considering an arc $(p^{l+1}, q^{l+1}) \in A^{t,l+1}$, its associated demand is thus

$$d^{t,l+1}_{p^{l+1},q^{l+1}} = \sum_{(p^l,q^l) \in A^{t,l}\,|\,p^{l+1}=\text{super}(p^l),\,q^{l+1}=\text{super}(q^l)} d^{t,l}_{p^l,q^l}. \tag{37}$$

Note that the conditions for a valid demand graph and valid demand values stated in inequalities (1) and (2) will still hold when aggregating in this way, since the total ingoing and outgoing demand at each cluster $p \in C^{l+1}$ (including the demands from and to all existing subnodes) stays the same.

Coarsening of Station Cells. To coarsen the station cells we need to define a hierarchical clustering for them as well. For simplicity we assume from now on that $S = V$ holds, i.e., there is a one-to-one correspondence of considered station cells and customer cells. This also appears reasonable in a practical setting. We can then apply the hierarchical clustering defined for the customer cells also to the station cells. Maximum station capacities for aggregated stations $s^{l+1} \in S^{l+1}$ are naturally calculated by the sum of the respective maximum capacities of the underlying station cells, i.e., $z^{l+1}_{s^{l+1}} = \sum_{s^l \in \text{sub}(s^{l+1})} z^l_{s^l}$.

Coarsening of Neighborhoods. A coarsened neighborhood mapping $S^{l+1}(v^{l+1})$ for each customer cell $v^{l+1} \in V^{l+1}$ and respective attractiveness values $a_{v^{l+1},s^{l+1}}$ for station cells $s^{l+1} \in S^{l+1}(v^{l+1})$ are determined as follows. The neighborhood mapping is retained as long as the attractiveness value in the coarsened problem instance does not fall below a certain threshold $\lambda \in (0, 1)$:

$$S^{l+1}(v^{l+1}) = \left\{ s^{l+1} \in \bigcup_{v^l \in \text{sub}(v^{l+1})} \text{super}(S^l(v^l)) \mid a_{v^{l+1},s^{l+1}} \geq \lambda \right\} \tag{38}$$

with the aggregated attractiveness values being

$$a_{v^{l+1},s^{l+1}} = \begin{cases} 1 & \text{if } v^{l+1} = s^{l+1} \\ \dfrac{\sum_{v^l \in \text{sub}(v^{l+1})} \sum_{s^l \in \text{sub}(s^{l+1}) \cap S^l(v^l)} (a_{v^l,s^l})}{|\text{sub}(v^{l+1})| \cdot |\text{sub}(s^{l+1})|} & \text{if } v^{l+1} \neq s^{l+1}. \end{cases} \tag{39}$$

4.2   Initialization

The initial problem becomes coarsened until we reach some level l where it can be reasonably solved as it is then small enough. In our experiments with binary clustering trees here we are stopping the coarsening when the clustering tree has no more than $2^5 = 32$ leaf nodes, or in other words, at a height of five. For initializing the solution at the coarsest level we utilize a MIP model. In this model, the objective stated in Sect. 3.2, the demand calculation for every time period stated in Sect. 3.3, and the rebalancing LP model stated in Sect. 3.4 are put together. By solving this model we obtain an optimal solution for the coarsest level, which forms the basis for proceeding with the next step of the algorithm, the *extension* to derive step-by-step a more detailed solutions.

4.3   Extension

In the extension step we derive from a solution x^{l+1} at level $l+1$ a solution x^l at level l, i.e., we have to decide for each aggregated station $s^{l+1} \in S^{l+1}$ with $x^{l+1}_{s^{l+1}} > 0$ slots how they should be realized by the respective underlying station cells $\mathrm{sub}(s^{l+1})$ at level l. We do this in a way so that the globally fulfilled demand is again maximized by solving the following MIP.

$$\max \quad \sum_{t \in T} D(x^l, t) \tag{40}$$

$$\text{s.t.} \quad \sum_{s^l \in S^l} \left(b^{\mathrm{fix}} \cdot x^l_{s^l} + b^{\mathrm{var}} \cdot x^l_{s^l} + b^{\mathrm{reb}} \cdot Q_{x^l}(s^l) \right) \leq B^{\mathrm{tot}}_{\max} \tag{41}$$

$$\sum_{s^l \in S^l} b^{\mathrm{fix}} \cdot x_{s^l} \leq B^{\mathrm{fix}}_{\max} \tag{42}$$

$$\sum_{s^l \in \mathrm{sub}(s^{l+1})} x^l_{s^l} \leq x^{l+1}_{s^{l+1}} \qquad s^{l+1} \in S^{l+1} \tag{43}$$

$$x^l_{s^l} \in \{0, \dots, z^l_s\} \qquad s^l \in S^l \tag{44}$$

The objective (40) maximizes the total satisfiable demand. Inequalities (41) restrict the maximum total budget whereas inequalities (42) restrict the maximum fixed budget. Inequalities (43) are the bounds on the total number of slots for the station nodes $s^l \in \mathrm{sub}(s^{l+1})$. The number of parking slots in each cell $x^l_{s^l}$ is restricted by the maximum number of parking slots allowed in this cell (44).

5   Computational Results

For our experiments we created seven different benchmark sets[1], each one containing 20 different, random instances. We consider instances with 200, 300, 500, 800, 1000, 1500, and 2000 customer cells, where each customer cell is also a possible location for a station to be built. Customer cells are aligned on a grid in

---

[1] https://www.ac.tuwien.ac.at/files/resources/instances/bsspp/lion17.bz2.

the plane and euclidean distances have been calculated based on which a hierarchical clustering with the complete-linkage method was computed. Demands among the leaf nodes were chosen randomly, considering the pairwise distance between customer cells, and demands below a certain threshold have been aggregated upwards in the clustering tree such that the demand graphs get sparser. Only cells within 200 m walking distance are considered to be in the vicinity of a customer cell and respective attractiveness values are chosen randomly but in correlation with the distances. We set the maximum station size to $z_s = 40$ for all cells in all test cases. For slot costs we set $b^{\text{fix}} = 1750$ €, and $b^{\text{var}} = 1000$ €, which are reasonable estimates in the Vienna area gathered from real BSSs. The costs for rebalancing a single bike for one day have been estimated with 3 € per bike and per day. When projecting this cost to the optimization horizon, e.g., 1 year, we get $b^{\text{reb}} = 365 \cdot 3 = 1095$ €. For coarsening of attractiveness values, we set the corresponding parameter $\lambda = 0$ and for adjusting the number of trips which can be performed in a particular time period $t \in T$ by using only a single bike we set $\omega = 1.2$. Each instance contains four time periods which we selected as follows: 4:30 am to 8:00 am, 8:00 am to 12:00 Noon, 12:00 Noon to 6:15 pm, and 6:15 pm to 4:30 am. The duration for each time period $t \in T$ has been set accordingly and the average trip duration has been set to $t^{\text{rent}} = 10$ min.

All algorithms are implemented in C++ and have been compiled with gcc 4.8. For solving the LPs and MIPs we used Gurobi 7.0. All experiments were executed as single threads on an Intel Xeon E5540 2.53 GHz Quad Core processor.

Table 1 summarizes obtained results. For every instance set we state the name containing the number of nodes, the number of different instances we have tested on (#runs), the maximum total budget ($B_{\text{max}}^{\text{tot}}$), and the maximum fixed budget ($B_{\text{max}}^{\text{fix}}$). For the proposed MLR, we list the average objective value ($\overline{\text{obj}}$), i.e., the expected fulfilled demand in terms of the number of journeys, the average number of coarsening levels ($\overline{\text{#coarsen}}$), the median time ($\widetilde{\text{time}}$), and the average total costs ($\overline{\text{totcost}}$) as well as the average fixed costs ($\overline{\text{fixcost}}$) for building the number of slots in the solution. Most importantly, it can be seen that the proposed MLR scales very well to large instances up to 2000 customer cells.

Table 1. Results for the multilevel refinement heuristic (MLR).

Instance				MLR				
Name	#runs	$B_{\text{max}}^{\text{tot}}$ [€]	$B_{\text{max}}^{\text{fix}}$ [€]	$\overline{\text{obj}}$	$\overline{\text{#coarsen}}$	$\widetilde{\text{time}}$ [s]	$\overline{\text{totcost}}$ [€]	$\overline{\text{fixcost}}$ [€]
BSSPP_200	20	200,000.00	130,000.00	9,651.98	3	46.2	198,000.00	126,000.00
BSSPP_300	20	350,000.00	250,000.00	10,951.79	5	60.8	349,250.00	222,250.00
BSSPP_500	20	500,000.00	350,000.00	16,057.78	6	121.6	497,750.00	316,750.00
BSSPP_800	20	850,000.00	550,000.00	28,862.21	6	263.9	849,750.00	540,750.00
BSSPP_1000	20	1,000,000.00	700,000.00	28,967.58	8	346.7	998,250.00	635,250.00
BSSPP_1500	20	1,500,000.00	1,000,000.00	41,208.19	8	574.5	1,498,475.00	953,575.00
BSSPP_2000	20	2,000,000.00	1,300,000.00	55,892.06	8	803.4	1,999,250.00	1,272,250.00
Average				27,370.22	6.3		912,960.71	580,975.00

6   Conclusion and Future Work

We presented an innovative approach to the BSSPP. Previous work only considers very small instances and case studies to small parts of a city whereas we aim at solving more realistic large-scale scenarios arising in large cities. As we have to cope with thousands of customer cells and potential station cells it is most fundamental to model the potential demands efficiently. To this end, we proposed to use a hierarchical clustering and defining the demand graph on it. This approach can drastically reduce the data in comparison to a complete demand matrix with only a very reasonable information loss. Moreover, we provided MIP formulations to compute the satisfiable demand by given configurations and to compute the prospective rebalancing costs. Putting them together under the objective of maximizing the expected satisfied total demand and adding further constraints for complying with given monetary budget constraints, we obtained a MIP model that solves our definition of the BSSPP exactly. Because this MIP model can in practice still only be solved for rather small instances, we further suggested a multilevel refinement heuristic utilizing the same hierarchical clustering we are given as input. Using this approach we have shown to be able to solve instances with up to 2000 nodes in reasonable computation times.

In future work it is important to make the cost model more realistic and to test on more realistic benchmark instances. In particular, we aim at considering also fixed costs for building a station which are independent of the number of slots. Furthermore, in practice also only a small, restricted set of different station configurations is possible per station cell. These extensions introduce interesting research questions especially in relation to the multilevel refinement procedure.

Acknowledgements. We thank the LOGISTIKUM Steyr, the Austrian Institute of Technology, and Rosinak & Partner for the collaboration on this topic. This work is supported by the Austrian Research Promotion Agency (FFG) under contract 849028.

References

1. Chen, J., Chen, X., Jiang, H., Zhu, S., Li, X., Li, Z.: Determining the optimal layout design for public bicycle system within the attractive scope of a metro station. Math. Probl. Eng. Article ID 456013, 8 p. (2015)
2. Chen, Q., Sun, T.: A model for the layout of bike stations in public bike-sharing systems. J. Adv. Transport. **49**(8), 884–900 (2015)
3. Frade, I., Ribeiro, A.: Bike-sharing stations: a maximal covering location approach. Transport. Res. A-Pol. **82**, 216–227 (2015)
4. Gavalas, D., Konstantopoulos, C., Pantziou, G.: Design & management of vehicle sharing systems: a survey of algorithmic approaches. In: Obaidat, M.S., Nicopolitidis, P. (eds.) Smart Cities and Homes: Key Enabling Technologies, pp. 261–289. Elsevier Science, Amsterdam (2016). Chap. 13
5. Hu, S.R., Liu, C.T.: An optimal location model for a bicycle sharing program with truck dispatching consideration. In: IEEE 17th International Conference on Intelligent Transportation Systems (ITSC), pp. 1775–1780. IEEE (2014)

6. Lin, J.R., Yang, T.H.: Strategic design of public bicycle sharing systems with service level constraints. Transport. Res. E-Log. **47**(2), 284–294 (2011)
7. Lin, J.R., Yang, T.H., Chang, Y.C.: A hub location inventory model for bicycle sharing system design: formulation and solution. Comput. Ind. Eng. **65**(1), 77–86 (2013)
8. Martinez, L.M., Caetano, L., Eiró, T., Cruz, F.: An optimisation algorithm to establish the location of stations of a mixed fleet biking system: an application to the city of Lisbon. Procedia Soc. Behav. Sci. **54**, 513–524 (2012)
9. Saharidis, G., Fragkogios, A., Zygouri, E.: A multi-periodic optimization modeling approach for the establishment of a bike sharing network: a case study of the city of Athens. In: Proceedings of the International Multi Conference of Engineers and Computer Scientists 2014. LNECS, vol. II, No. 2210, pp. 1226–1231. Newswood Limited (2014)
10. Walshaw, C.: A multilevel approach to the travelling salesman problem. Oper. Res. **50**(5), 862–877 (2002)
11. Walshaw, C.: Multilevel refinement for combinatorial optimisation problems. Ann. Oper. Res. **131**(1), 325–372 (2004)
12. Yang, T.H., Lin, J.R., Chang, Y.C.: Strategic design of public bicycle sharing systems incorporating with bicycle stocks considerations. In: 40th International Conference on Computers and Industrial Engineering (CIE), pp. 1–6. IEEE (2010)

Decomposition Descent Method for Limit Optimization Problems

Igor Konnov[✉]

Institute of Computational Mathematics and Information Technologies,
Kazan Federal University, Kremlevskaya St. 18, 420008 Kazan, Russia
konn-igor@ya.ru
http://kpfu.ru

Abstract. We consider a general limit optimization problem whose goal function need not be smooth in general and only approximation sequences are known instead of exact values of this function. We suggest to apply a two-level approach where approximate solutions of a sequence of mixed variational inequality problems are inserted in the iterative scheme of a selective decomposition descent method. Its convergence is attained under coercivity type conditions.

Keywords: Optimization problems · Limit problems · Non-smooth functions · Mixed variational inequality · Decomposition descent method · Coercivity conditions

1 Introduction

We first consider the general optimization problem, which consists in finding the minimal value of some function p over the corresponding feasible set X. For brevity, we write this problem as

$$\min_{\mathbf{x} \in X} \to p(\mathbf{x}). \tag{1}$$

Its solution set will be denoted by X^* and the optimal value of the function by p^*, i.e.

$$p^* = \inf_{\mathbf{x} \in X} p(\mathbf{x}).$$

In order to develop efficient solution methods for this problem we should exploit certain additional information about its properties, which are related to some classes of applications.

In what follows, we denote by $\mathbb{R}^s$ the real s-dimensional Euclidean space, all elements of such spaces being column vectors represented by a lower case Roman alphabet in boldface, e.g. $\mathbf{x}$. For any vectors $\mathbf{x}$ and $\mathbf{y}$ of $\mathbb{R}^s$, we denote by $\langle \mathbf{x}, \mathbf{y} \rangle$ their scalar product, i.e.,

$$\langle \mathbf{x}, \mathbf{y} \rangle = \mathbf{x}^\top \mathbf{y} = \sum_{i=1}^{s} x_i y_i,$$

© Springer International Publishing AG 2017
R. Battiti et al. (Eds.): LION 2017, LNCS 10556, pp. 166–179, 2017.
https://doi.org/10.1007/978-3-319-69404-7_12

and by $\|\mathbf{x}\|$ the Euclidean norm of $\mathbf{x}$, i.e., $\|\mathbf{x}\| = \sqrt{\langle \mathbf{x}, \mathbf{x}\rangle}$. Next, we define for brevity $M = \{1, \ldots, n\}$, $|\mathcal{A}|$ will denote the cardinality of a finite set $\mathcal{A}$. As usual, $\mathbb{R}$ will denote the set of real numbers, $\bar{\mathbb{R}} = \mathbb{R} \bigcup \{+\infty\}$.

Let us consider a partition of the N-dimensional space

$$\mathbb{R}^N = \mathbb{R}^{N_1} \times \ldots \times \mathbb{R}^{N_n}, \tag{2}$$

i.e.

$$\mathcal{N} = \bigcup_{i=1}^{n} \mathcal{N}_i,$$

where $\mathcal{N} = \{1, \ldots, N\}$, $N = |\mathcal{N}|$, $N_i = |\mathcal{N}_i|$, and $\mathcal{N}_i \bigcap \mathcal{N}_j = \varnothing$ if $i \neq j$. This means that any point $\mathbf{x} = (x_1, \ldots, x_N)^\top \in \mathbb{R}^N$ is represented by $\mathbf{x} = (\mathbf{x}_1, \ldots, \mathbf{x}_n)^\top$ where $\mathbf{x}_i = (x_j)_{j \in \mathcal{N}_i} \in \mathbb{R}^{N_i}$ for $i \in M$. The simplest case where $n_i = 1$ for all $i \in M$ and $n = N$ corresponds to the scalar coordinate partition.

Rather recently, partially decomposable optimization problems were paid significant attention due to their various big data applications; see e.g. [1–3] and the references therein. In these problems, the cost function and feasible set are specialized as follows:

$$p(\mathbf{x}) = f(\mathbf{x}) + h(\mathbf{x}), \tag{3}$$

$$h(\mathbf{x}) = \sum_{i=1}^{n} h_i(\mathbf{x}_i), \tag{4}$$

$$X = X_1 \times \ldots \times X_n = \prod_{i=1}^{n} X_i, \tag{5}$$

where $f : \mathbb{R}^N \to \bar{\mathbb{R}}$ is a function, which is continuous on X, $h_i : \mathbb{R}^{N_i} \to \bar{\mathbb{R}}$ is a convex function, and X_i is a convex set in $\mathbb{R}^{N_i}$ for $i = 1, \ldots, n$. Note that the function $f : \mathbb{R}^N \to \bar{\mathbb{R}}$ is not supposed to be convex in general. That is, we have to solve a non-convex and non-differentiable optimization problem, which appears too difficult for solution with usual subgradient type methods. Nevertheless, one can develop efficient coordinate-wise decomposition methods for finding stationary points of problem (1), (3)–(5) for a smooth f; see e.g. [3–6] and the references therein. Then the stationary points can be defined as solutions of the following mixed variational inequality (MVI for short): Find a point $\mathbf{x}^* \in X$ such that

$$\langle f'(\mathbf{x}^*), \mathbf{y} - \mathbf{x}^* \rangle + \sum_{i=1}^{n} [h_i(\mathbf{y}_i) - h_i(\mathbf{x}_i^*)] \geq 0 \tag{6}$$
$$\forall \mathbf{y}_i \in X_i, \quad \text{for } i = 1, \ldots, n.$$

Here also $\mathbf{y} = (\mathbf{y}_1, \ldots, \mathbf{y}_n)^\top$ as above.

We observe that all these solution methods also require exact values of the cost function and parameters of the feasible set. However, this is often impossible for big data problems due to the calculation errors and incompleteness of the necessary information. In addition, the same situation arises if we find it useful to replace the initial problem by a sequence of auxiliary ones with better properties. For instance, we can suppose that f in (3) is a general non-smooth function and also replace it with a sequence of its smooth approximations. In other words, we have to develop methods for *limit (or non-stationary)* problems.

There exist a number of methods for such limit optimization and variational inequality problems, but they are based essentially upon convexity assumptions and restrictive concordance rules for accuracy, approximation, and iteration parameters, which creates serious difficulties for their implementation; see e.g. [7–9] and the references therein. For instance, if iteration parameters are dependent of attained accuracy or approximation, they can not be evaluated properly. Besides, the mutual dependence of these parameters usually leads to very restrictive control rules and slow convergence.

In this paper, we intend to suggest a descent coordinate-wise decomposition method for the following problem: Find a point $\mathbf{x}^* \in X$ such that

$$\exists \mathbf{g}^* \in \mathbf{G}(\mathbf{x}^*), \langle \mathbf{g}^*, \mathbf{y} - \mathbf{x}^* \rangle + \sum_{i=1}^{n} [h_i(\mathbf{y}_i) - h_i(\mathbf{x}_i^*)] \geq 0 \tag{7}$$
$$\forall \mathbf{y}_i \in X_i, \quad \text{for } i = 1, \dots, n;$$

where $\mathbf{G} : X \to \Pi(\mathbb{R}^N)$ is a point-to-set mapping whose values are considered as generalized gradient sets of the function f; cf. (6). Here $\Pi(A)$ denotes the family of all nonempty subsets of a set A. For instance, if f is a locally Lipschitz function, we can set $\mathbf{G}$ to be its Clarke subdifferential mapping. If f is a convex function, we simply set $\mathbf{G}(\mathbf{x})$ to be the usual subdifferential $\partial f(\mathbf{x})$ of f at $\mathbf{x}$, then each solution of (7) clearly solves (1), (3)–(5). Next, we suppose that only sequences of approximations are known instead of the exact values of $\mathbf{G}$ and h.

In creating the desired solution method for the limit (or non-stationary) problem (7), we combine the selective descent splitting method from [6] with changing the tolerance parameters corresponding to a sequence of mixed variational inequality problems, which does not require evaluation of accuracy of approximate solutions, and utilization of some coercivity conditions; see e.g. [10]. This approach allows us to prove convergence without special concordance rules for all the parameters and tolerances.

2   Auxiliary Problem Properties

Let us consider a partially partitionable optimization problem of the form

$$\min_{\mathbf{x} \in X} \to \varphi(\mathbf{x}) = \{\mu(\mathbf{x}) + \eta(\mathbf{x})\}, \tag{8}$$

where the function $\mu : \mathbb{R}^N \to \bar{\mathbb{R}}$ is smooth on X, but not necessary convex. This problem will serve as approximation of the basic problem (1), (3)–(5). We will use the same partition (2) of the space $\mathbb{R}^N$ and fix the assumption on the feasible set.

(A1) It holds that (5) where X_i are non-empty, convex, and closed sets in $\mathbb{R}^{N_i}$ for $i = 1, \ldots, n$.

Also, we suppose that

$$\eta(\mathbf{x}) = \sum_{i=1}^{n} \eta_i(\mathbf{x}_i), \tag{9}$$

where $\eta_i : \mathbb{R}^{N_i} \to \bar{\mathbb{R}}$ is convex and has the non-empty subdifferential $\partial\eta_i(\mathbf{x}_i)$ at each point $\mathbf{x}_i \in X_i$, for $i \in M$. Then each function η_i is lower semi-continuous, the function η is lower semi-continuous, and

$$\partial\eta(\mathbf{x}) = \partial\eta_1(\mathbf{x}_1) \times \ldots \times \partial\eta_n(\mathbf{x}_n).$$

So, our problem (8) and (9) is rewritten as

$$\min_{\mathbf{x} \in X_1 \times \ldots \times X_n} \to \varphi(\mathbf{x}) = \left\{ \mu(\mathbf{x}) + \sum_{i=1}^{n} \eta_i(\mathbf{x}_i) \right\}. \tag{10}$$

Set $\mathbf{g}(\mathbf{x}) = \mu'(\mathbf{x})$, then

$$\mathbf{g}(\mathbf{x}) = (\mathbf{g}_1(\mathbf{x}), \ldots, \mathbf{g}_n(\mathbf{x}))^\top, \text{ where } \mathbf{g}_i(\mathbf{x}) = \left(\frac{\partial\mu(\mathbf{x})}{\partial x_j} \right)_{j \in \mathcal{N}_i} \in \mathbb{R}^{N_i}, \ i = 1, \ldots, n.$$

Given a point $\mathbf{x} \in X$, we say that a vector $\mathbf{d}$ is feasible for $\mathbf{x}$ if $\mathbf{x} + \alpha\mathbf{d} \in X$ for some $\alpha > 0$. From the assumptions above it follows that the function φ is directionally differentiable at each point $\mathbf{x} \in X$, that is, its directional derivative with respect to any feasible vector $\mathbf{d}$ is defined by the formula:

$$\varphi'(\mathbf{x}; \mathbf{d}) = \langle \mathbf{g}(\mathbf{x}), \mathbf{d} \rangle + \eta'(\mathbf{x}; \mathbf{d}), \text{ with } \eta'(\mathbf{x}; \mathbf{d}) = \sum_{i=1}^{n} \max_{\mathbf{b}_i \in \partial\eta_i(\mathbf{x}_i)} \langle \mathbf{b}_i, \mathbf{d}_i \rangle;$$

see e.g. [11].

We recall that a function $f : \mathbb{R}^s \to \bar{\mathbb{R}}$ is said to be *coercive* on a set $D \subset \mathbb{R}^s$ if $\{f(\mathbf{u}^k)\} \to +\infty$ for any sequence $\{\mathbf{u}^k\} \subset D$, $\|\mathbf{u}^k\| \to \infty$. We will in addition suppose that the function $\varphi : \mathbb{R}^N \to \bar{\mathbb{R}}$ is coercive on X, then problem (8) and (9) (or (10)) has a solution.

Problem (10) was considered in particular in [6]. We will utilize some properties obtained there and the descent splitting method from [6] will serve as a basic element of the two-level method for the general limit optimization problem. For this reason, we give some properties from [6] without proofs. We start our considerations from the optimality condition.

Lemma 1 [6, *Proposition 2.1*].

(a) *Each solution of problem (10) is a solution of the following MVI: Find a point $\mathbf{x}^* \in X = X_1 \times \cdots \times X_n$ such that*

$$\sum_{i=1}^{n} \langle \mathbf{g}_i(\mathbf{x}^*), \mathbf{y}_i - \mathbf{x}_i^* \rangle + \sum_{i=1}^{n} [\eta_i(\mathbf{y}_i) - \eta_i(\mathbf{x}_i^*)] \geq 0 \tag{11}$$

$$\forall \mathbf{y}_i \in X_i, \quad for \ i = 1, \ldots, n.$$

(b) If μ is convex, then each solution of MVI (11) solves problem (10).

In what follows, we denote by $\tilde{X}^0$ the solution set of MVI (11) and call it the set of *stationary points* of problem (10); cf. (6).

Fix $\alpha > 0$. For each point $\mathbf{x} \in X$ we can define $\mathbf{y}(\mathbf{x}) = (\mathbf{y}_1(\mathbf{x}), \dots, \mathbf{y}_n(\mathbf{x}))^\top \in X$ such that

$$\sum_{i=1}^{n} \langle \mathbf{g}_i(\mathbf{x}) + \alpha(\mathbf{y}_i(\mathbf{x}) - \mathbf{x}_i), \mathbf{y}_i - \mathbf{y}_i(\mathbf{x}) \rangle + \sum_{i=1}^{n} [\eta_i(\mathbf{y}_i) - \eta_i(\mathbf{y}_i(\mathbf{x}))] \geq 0 \qquad (12)$$
$$\forall \mathbf{y}_i \in X_i, \quad \text{for } i = 1, \dots, n.$$

This MVI gives a necessary and sufficient optimality condition for the optimization problem:

$$\min_{\mathbf{y} \in X_1 \times \dots \times X_n} \rightarrow \sum_{i=1}^{n} \Phi_i(\mathbf{x}, \mathbf{y}_i), \qquad (13)$$

where

$$\Phi_i(\mathbf{x}, \mathbf{y}_i) = \langle \mathbf{g}_i(\mathbf{x}), \mathbf{y}_i \rangle + 0.5\alpha \|\mathbf{x}_i - \mathbf{y}_i\|^2 + \eta_i(\mathbf{y}_i) \qquad (14)$$

for $i = 1, \dots, n$. Under the above assumptions each $\Phi_i(\mathbf{x}, \cdot)$ is strongly convex, hence problem (13) and (14) (or (12)) has the unique solution $\mathbf{y}(\mathbf{x})$, thus defining the single-valued mapping $\mathbf{x} \mapsto \mathbf{y}(\mathbf{x})$. Observe that all the components of $\mathbf{y}(\mathbf{x})$ can be found independently, i.e. (13) and (14) is equivalent to n independent optimization problems of the form

$$\min_{\mathbf{y}_i \in X_i} \rightarrow \Phi_i(\mathbf{x}, \mathbf{y}_i), \qquad (15)$$

for $i = 1, \dots, n$ and $\mathbf{y}_i(\mathbf{x})$ just solves (15).

Lemma 2 [6, *Proposition 2.2*].

(a) $\mathbf{x} = \mathbf{y}(\mathbf{x}) \Longleftrightarrow \mathbf{x} \in \tilde{X}^0$;
(b) The mapping $\mathbf{x} \mapsto \mathbf{y}(\mathbf{x})$ is continuous on X.

Set $\Delta(\mathbf{x}) = \|\mathbf{x} - \mathbf{y}(\mathbf{x})\|$, then $\Delta^2(\mathbf{x}) = \sum_{i=1}^{n} \Delta_i^2(\mathbf{x})$ where $\Delta_i(\mathbf{x}) = \|\mathbf{x}_i - \mathbf{y}_i(\mathbf{x})\|$. From Lemma 2 we conclude that the value $\Delta(\mathbf{x})$ can serve as accuracy measure for MVI (11).

We need also a descent property from [6, Lemma 2.1].

Lemma 3. *Take any point $\mathbf{x} \in X$ and an index $i \in M$. If*

$$\mathbf{d}_s = \begin{cases} \mathbf{y}_i(\mathbf{x}) - \mathbf{x}_i & \text{if } s = i, \\ \mathbf{0} & \text{if } s \neq i; \end{cases}$$

then

$$\varphi'(\mathbf{x}; \mathbf{d}) \leq -\alpha \|\mathbf{y}_i(\mathbf{x}) - \mathbf{x}_i\|^2.$$

Denote by $\mathbb{Z}_+$ the set of non-negative integers. Following [6] we describe the basic algorithm for MVI (11) as follows.

Algorithm (DDS). *Input:* A point $\mathbf{x}^0 \in X$. *Output:* A point $\mathbf{z}$. *Parameters:* Numbers $\alpha > 0$, $\delta > 0$, $\beta \in (0, \alpha)$, $\theta \in (0, 1)$.

At the k-th iteration, $k = 0, 1, \ldots$, we have a point $\mathbf{x}^k \in X$.

Step 1: Choose an index $i \in M$ such that $\Delta_i(\mathbf{x}^k) \geq \delta$, set $i_k = i$,

$$\mathbf{d}_s^k = \begin{cases} \mathbf{y}_s(\mathbf{x}^k) - \mathbf{x}_s^k & \text{if } s = i_k, \\ 0 & \text{if } s \neq i_k; \end{cases}$$

and go to Step 3. Otherwise (i.e. when $\Delta_s(\mathbf{x}^k) < \delta$ for all $s \in M$) go to Step 2.

Step 2: Set $\mathbf{z} = \mathbf{x}^k$ and stop.

Step 3: Determine m as the smallest number in $\mathbb{Z}_+$ such that

$$\varphi(\mathbf{x}^k + \theta^m \mathbf{d}^k) \leq \varphi(\mathbf{x}^k) - \beta \theta^m \Delta_i^2(\mathbf{x}^k), \tag{16}$$

set $\lambda_k = \theta^m$, $\mathbf{x}^{k+1} = \mathbf{x}^k + \lambda_k \mathbf{d}^k$, and $k = k + 1$. The iteration is complete.

Although its properties are similar to those in [6], we give their proofs here for more clarity of exposition.

Lemma 4. *The line-search procedure in Step 3 is always finite.*

Proof. If we suppose that the line-search procedure is infinite, then

$$\theta^{-m}(\varphi(\mathbf{x}^k + \theta^m \mathbf{d}^k) - \varphi(\mathbf{x}^k)) > -\beta \Delta_i^2(\mathbf{x}^k),$$

for $m \to \infty$, hence, by taking the limit we have $\varphi'(\mathbf{x}^k; \mathbf{d}^k) \geq -\beta \Delta_i^2(\mathbf{x}^k)$, but Lemma 3 gives $\varphi'(\mathbf{x}^k; \mathbf{d}^k) \leq -\alpha \Delta_i^2(\mathbf{x}^k)$, hence $\alpha \leq \beta$, a contradiction. □

We obtain the main property of the basic cycle.

Proposition 1. *The number of iterations in Algorithm (DDS) is finite.*

Proof. By construction, we have $-\infty < \varphi^* \leq \varphi(\mathbf{x}^k)$ and $\varphi(\mathbf{x}^{k+1}) \leq \varphi(\mathbf{x}^k) - \beta \delta^2 \lambda_k$, hence the sequence $\{\mathbf{x}^k\}$ is bounded and has limit points, besides,

$$\lim_{k \to \infty} \lambda_k = 0.$$

Suppose that the sequence $\{\mathbf{x}^k\}$ is infinite. Since the set M is finite, there is an index $i_k = i$, which is repeated infinitely. Take the corresponding subsequence $\{k_s\}$, then, without loss of generality, we can suppose that the subsequence $\{\mathbf{x}^{k_s}\}$ converges to a point $\bar{\mathbf{x}}$, besides, $\Delta_{i_{k_s}}(\mathbf{x}^{k_s}) = \|\mathbf{d}_i^{k_s}\|$, and we have

$$(\lambda_{k_s}/\theta)^{-1}(\varphi(\mathbf{x}^{k_s} + (\lambda_{k_s}/\theta)\mathbf{d}^{k_s}) - \varphi(\mathbf{x}^{k_s})) > -\beta \|\mathbf{d}_i^{k_s}\|^2.$$

Using the mean value theorem (see e.g. [11, Theorem 2.3.7]), we obtain

$$\langle \mathbf{g}_i^{k_s} + \mathbf{t}_i^{k_s}, \mathbf{d}_i^{k_s} \rangle = \langle \mathbf{g}^{k_s} + \mathbf{t}^{k_s}, \mathbf{d}^{k_s} \rangle > -\beta \|\mathbf{d}_i^{k_s}\|^2,$$

for some $\mathbf{g}^{k_s} = \mu'(\mathbf{x}^{k_s} + (\lambda_{k_s}/\theta)\xi_{k_s}\mathbf{d}^{k_s})$, $\mathbf{t}^{k_s} \in \partial\eta(\mathbf{x}^{k_s} + (\lambda_{k_s}/\theta)\xi_{k_s}\mathbf{d}^{k_s})$, $\xi_{k_s} \in (0,1)$. By taking the limit $s \to \infty$ we have

$$\langle \mu'(\bar{\mathbf{x}}) + \bar{\mathbf{t}}, \bar{\mathbf{d}} \rangle = \langle \mathbf{g}_i(\bar{\mathbf{x}}) + \bar{\mathbf{t}}_i, \bar{\mathbf{d}}_i \rangle \geq -\beta\|\bar{\mathbf{d}}_i\|^2,$$

for some $\bar{\mathbf{t}} \in \partial\eta(\bar{\mathbf{x}})$, where

$$\bar{\mathbf{d}}_s = \begin{cases} \mathbf{y}_i(\bar{\mathbf{x}}) - \bar{\mathbf{x}}_i & \text{if } s = i, \\ \mathbf{0} & \text{if } s \neq i; \end{cases}$$

due to Lemma 2. On the other hand, using Lemma 3 gives

$$\langle \mu'(\bar{\mathbf{x}}) + \bar{\mathbf{t}}, \bar{\mathbf{d}} \rangle \leq \varphi'(\bar{\mathbf{x}}; \bar{\mathbf{d}}) \leq -\alpha\|\bar{\mathbf{d}}_i\|^2,$$

besides, by construction, we have $\|\mathbf{d}_i^{k_s}\| \geq \delta$, hence $\|\bar{\mathbf{d}}_i\| \geq \delta > 0$ and $\alpha \leq \beta$, which is a contradiction. $\qquad\square$

3   Limit Decomposition Method and its Convergence

We now intend to describe a general iterative method for the limit MVI (7). First we introduce the approximation assumptions.

(A2) There exists a sequence of continuous mappings $\mathbf{G}_l : X \to \mathbb{R}^N$, which are the gradients of functions $f_l : \mathbb{R}^N \to \bar{\mathbb{R}}$, $l = 1, 2, \ldots$, such that the relations $\{\mathbf{y}^l\} \to \bar{\mathbf{y}}$ and $\mathbf{y}^l \in X$ imply $\{\mathbf{G}_l(\mathbf{y}^l)\} \to \bar{\mathbf{g}} \in \mathbf{G}(\bar{\mathbf{y}})$.

(A3) For each $i = 1, \ldots, n$ there exists a sequence of convex functions $h_{l,i} : \mathbb{R}^{N_i} \to \bar{\mathbb{R}}$, such that each of them is subdifferentiable on X_i and that the relations $\{\mathbf{u}^l\} \to \bar{\mathbf{u}}$ and $\mathbf{u}^l \in X_i$ imply $\{h_{l,i}(\mathbf{u}^l)\} \to h_i(\bar{\mathbf{u}})$.

Condition **(A2)** means that the limit set-valued mapping $\mathbf{G}$ at any point is approximated by a sequence of gradients $\{\mathbf{G}_l\}$. In fact, if $\mathbf{G}$ is the Clarke subdifferential of a locally Lipschitz function f, it can be always approximated by a sequence of gradients within condition **(A2)**; see [12,13]. Observe also that if there is a subsequence $\mathbf{y}^{l_s} \in X$ with $\{\mathbf{y}^{l_s}\} \to \bar{\mathbf{y}}$, then **(A2)** implies $\{\mathbf{G}_{l_s}(\mathbf{y}^{l_s})\} \to \bar{\mathbf{g}} \in \mathbf{G}(\bar{\mathbf{y}})$ and the same is true for **(A3)**. At the same time, the non-differentiability of the functions f or h is not obligatory, the main property is the existence of the approximation sequences indicated in **(A2)** and **(A3)**.

So, we replace MVI (7) with a sequence of MVIs: Find a point $\bar{\mathbf{z}}^l \in X = X_1 \times \ldots \times X_n$ such that

$$\sum_{i=1}^n \langle \mathbf{G}_{l,i}(\bar{\mathbf{z}}^l), \mathbf{y}_i - \bar{\mathbf{z}}_i^l \rangle + \sum_{i=1}^n [h_{l,i}(\mathbf{y}_i) - h_{l,i}(\bar{\mathbf{z}}_i^l)] \geq 0 \tag{17}$$
$$\forall \mathbf{y}_i \in X_i, \quad \text{for } i = 1, \ldots, n;$$

where we use the partition of $\mathbf{G}_l$ which corresponds to that of the space $\mathbb{R}^N$, i.e.

$$\mathbf{G}_l(\mathbf{x}) = (\mathbf{G}_{l,1}(\mathbf{x}), \ldots, \mathbf{G}_{l,n}(\mathbf{x}))^\top, \text{ where } \mathbf{G}_{l,i}(\mathbf{x}) \in \mathbb{R}^{N_i}, \; i = 1, \ldots, n.$$

Similarly, we set

$$h_l(\mathbf{x}) = \sum_{i=1}^{n} h_{l,i}(\mathbf{x}_i).$$

Since the feasible set X may be unbounded, we introduce also coercivity conditions.

(C1) For each fixed $l = 1, 2, \ldots$, the function $f_l(\mathbf{x}) + h_l(\mathbf{x})$ is coercive on the set X, that is, $\{f_l(\mathbf{w}^k) + h_l(\mathbf{w}^k)\} \to +\infty$ if $\{\mathbf{w}^k\} \subset X$, $\|\mathbf{w}^k\| \to \infty$ as $k \to \infty$.

(C2) There exist a number $\sigma > 0$ and a point $\bar{\mathbf{v}} \in X$ such that for any sequences $\{\mathbf{u}^l\}$ and $\{\mathbf{d}^l\}$ satisfying the conditions:

$$\mathbf{u}^l \in X, \{\|\mathbf{u}^l\|\} \to +\infty, \{\mathbf{d}^l\} \to \mathbf{0};$$

it holds that

$$\liminf_{l \to \infty} \left\{ \langle G_l(\mathbf{u}^l) + \tau \mathbf{d}^l, \bar{\mathbf{v}} - \mathbf{u}^l - \mathbf{d}^l \rangle + [h_l(\bar{\mathbf{v}}) - h_l(\mathbf{u}^l - \mathbf{d}^l)] \right\} \leq -\sigma \text{ if } \tau > 0.$$

Clearly, **(C1)** gives a custom coercivity condition for each function $f_l(\mathbf{x}) + h_l(\mathbf{x})$, which provides existence of solutions of each particular problem (17). Obviously, **(C1)** holds if X is bounded. At the same time, **(C2)** gives a similar coercivity condition for the whole sequence of these problems approximating the limit MVI (7). It also holds if X is bounded. In the unbounded case **(C2)** is weaker than the following coercivity condition:

$$\|\bar{\mathbf{v}} - \mathbf{u}^l - \mathbf{d}^l\|^{-1} \left\{ \langle G_l(\mathbf{u}^l) + \tau \mathbf{d}^l, \bar{\mathbf{v}} - \mathbf{u}^l - \mathbf{d}^l \rangle + [h_l(\bar{\mathbf{v}}) - h_l(\mathbf{u}^l - \mathbf{d}^l)] \right\} \to -\infty \text{ as } l \to \infty.$$

Similar conditions are also usual for penalty type methods; see e.g. [14,15]. We therefore conclude that conditions **(C1)** and **(C2)** are not restrictive.

The whole decomposition method for the non-stationary MVI (7) has a two-level iteration scheme where each stage of the upper level invokes Algorithm (DDS) with different parameters.

Method (DNS). Choose a point $\mathbf{z}^0 \in X$ and a sequence $\{\delta_l\} \to +0$.

At the l-th stage, $l = 1, 2, \ldots$, we have a point $\mathbf{z}^{l-1} \in X$ and a number δ_l. Set

$$\mu(\mathbf{x}) = f_l(\mathbf{x}), \quad \eta(\mathbf{x}) = h_l(\mathbf{x}),$$

apply Algorithm (DDS) with $\mathbf{x}^0 = \mathbf{z}^{l-1}$, $\delta = \delta_l$ and obtain a point $\mathbf{z}^l = \mathbf{z}$ as its output.

We now establish the main convergence result.

Theorem 1. *Suppose that assumptions* **(A1)**–**(A3)** *and* **(C1)**–**(C2)** *are fulfilled, besides,* $\{\delta_l\} \to +0$. *Then:*

(i) problem (17) has a solution;
(ii) the number of iterations at each stage of Method (DNS) is finite;

(iii) the sequence $\{\mathbf{z}^l\}$ generated by Method (DNS) has limit points and all these limit points are solutions of MVI (7);

(iv) if f is convex, then all the limit points of $\{\mathbf{z}^l\}$ belong to X^*.

Proof. We first observe that **(C1)** implies that each problem (17) has a solution since the cost function

$$\mu(\mathbf{x}) = f_l(\mathbf{x}) + h_l(\mathbf{x})$$

is coercive, hence the set

$$X_l(\mathbf{x}^0) = \{\mathbf{y} \in X \mid \mu(\mathbf{y}) \le \mu(\mathbf{x}^0)\}$$

is bounded. It follows that the optimization problem

$$\min_{\mathbf{x} \in X} \to \mu(\mathbf{x})$$

has a solution and so is MVI (17) due to Lemma 1. Hence, assertion (i) is true. Next, from Proposition 1 we now have that assertion (ii) is also true.

By (ii), the sequence $\{\mathbf{z}^l\}$ is well-defined and (12) implies

$$\langle \mathbf{G}_l(\mathbf{z}^l) + \alpha(\mathbf{y}^l(\mathbf{z}^l) - \mathbf{z}^l), \mathbf{y} - \mathbf{y}^l(\mathbf{z}^l) \rangle + [h_l(\mathbf{y}) - h_l(\mathbf{y}^l(\mathbf{z}^l))] \ge 0 \quad \forall \mathbf{y} \in X. \quad (18)$$

Besides, the stopping rule in Algorithm (DDS) gives

$$\alpha \delta_l \sqrt{n} \ge \alpha \|\mathbf{z}^l - \mathbf{y}^l(\mathbf{z}^l)\|. \quad (19)$$

We now proceed to show that $\{\mathbf{z}^l\}$ is bounded. Conversely, suppose that $\{\|\mathbf{z}^l\|\} \to +\infty$. Applying (18) with $\mathbf{y} = \bar{\mathbf{v}}$, we have

$$0 \le \langle \mathbf{g}^l + \mathbf{d}^l, \bar{\mathbf{v}} - \tilde{\mathbf{z}}^l \rangle + [h_l(\bar{\mathbf{v}}) - h_l(\tilde{\mathbf{z}}^l)].$$

Here and below, for brevity we set $\mathbf{g}^l = \mathbf{G}_l(\mathbf{z}^l)$, $\tilde{\mathbf{z}}^l = \mathbf{y}^l(\mathbf{z}^l)$, and $\mathbf{d}^l = \alpha(\mathbf{y}(\mathbf{z}^l) - \mathbf{z}^l)$. Take a subsequence $\{l_s\}$ such that

$$\lim_{s \to \infty} \left\{ \langle \mathbf{g}^{l_s} + \mathbf{d}^{l_s}, \bar{\mathbf{v}} - \tilde{\mathbf{z}}^{l_s} \rangle + [h_{l_s}(\bar{\mathbf{v}}) - h_{l_s}(\tilde{\mathbf{z}}^{l_s})] \right\}$$
$$= \liminf_{l \to \infty} \left\{ \langle \mathbf{g}^l + \mathbf{d}^l, \bar{\mathbf{v}} - \tilde{\mathbf{z}}^l \rangle + [h_l(\bar{\mathbf{v}}) - h_l(\tilde{\mathbf{z}}^l)] \right\},$$

then, by **(C2)**, we have

$$0 \le \lim_{s \to \infty} \left\{ \langle \mathbf{g}^{l_s} + \mathbf{d}^{l_s}, \bar{\mathbf{v}} - \tilde{\mathbf{z}}^{l_s} \rangle + [h_{l_s}(\bar{\mathbf{v}}) - h_{l_s}(\tilde{\mathbf{z}}^{l_s})] \right\} \le -\sigma < 0,$$

a contradiction. Therefore, the sequence $\{\mathbf{z}^l\}$ is bounded and has limit points. Let $\bar{\mathbf{z}}$ be an arbitrary limit point for $\{\mathbf{z}^l\}$, i.e.

$$\bar{\mathbf{z}} = \lim_{s \to \infty} \mathbf{z}^{l_s}.$$

Since $\mathbf{z}^l \in X$, we have $\bar{\mathbf{z}} \in X$. It now follows from **(A2)** that $\lim_{s \to \infty} \mathbf{g}^{l_s} = \bar{\mathbf{g}} \in \mathbf{G}(\bar{\mathbf{z}})$.

Fix an arbitrary point $\mathbf{y} \in X$, then, using (18) and (19) and **(A3)**, we have

$$\langle \bar{\mathbf{g}}, \mathbf{y} - \bar{\mathbf{z}} \rangle + [h(\mathbf{y}) - h(\bar{\mathbf{z}})] = \lim_{s \to \infty} \left\{ \langle \mathbf{g}^{l_s}, \mathbf{y} - \mathbf{z}^{l_s} \rangle + [h_{l_s}(\mathbf{y}) - h_{l_s}(\mathbf{z}^{l_s})] \right\}$$
$$= \lim_{s \to \infty} \left\{ \langle \mathbf{g}^{l_s} + \mathbf{d}^{l_s}, \mathbf{y} - \tilde{\mathbf{z}}^{l_s} \rangle + [h_{l_s}(\mathbf{y}) - h_{l_s}(\tilde{\mathbf{z}}^{l_s})] \right\} \geq 0,$$

therefore $\bar{\mathbf{z}}$ solves MVI (7) and assertion (iii) holds true.

Next, if f is convex, then so is p and each limit point of $\{\mathbf{z}^l\}$ belongs to X^*, which gives assertion (iv). $\square$

We observe that the above proof implies that MVI (7) has a solution.

4   Modifications and Applications

Method (DNS) admits various modifications. In particular, we can take the exact one-dimensional minimization rule instead of the current Armijo line-search (16) in Algorithm (DDS), then the assertions of Theorem 1 remain true. Next, if the function μ (i.e. each function f_l) is convex, we can replace (16) with the following:

$$\langle \mathbf{g}_i(\mathbf{x}^k + \theta^m \mathbf{d}^k), \mathbf{d}_i^k \rangle + \theta^{-m} \left\{ \eta_i(\mathbf{x}_i^k + \theta^m \mathbf{d}_i^k) - \eta_i(\mathbf{x}_i^k) \right\} \leq -\beta \alpha^{-1} \Delta_i^2(\mathbf{x}^k).$$

Moreover, if the gradient of the function μ is Lipschitz continuous, we have an explicit lower bound for the step-size and utilize the fixed step-size version of Algorithm (DDS), which leads to further reduction of computational expenses. It can be applied if the partial gradients of μ is Lipschitz continuous; see [6] for more details.

We give now only two instances in order to illustrate possible applications.

The first instance is the linear inverse problem that arises very often in signal and image processing; see e.g. [16] for more examples. The problem consists in solving a linear system of equations

$$A\mathbf{x} = \mathbf{b},$$

where A is a $m \times n$ matrix, $\mathbf{b}$ is a vector in $\mathbb{R}^m$, whose exact values may be unknown or admit some noise perturbations. If $A^\top A$ is ill-conditioned, the custom approach based on the least squares minimization problem

$$\min_{\mathbf{x}} \to \| A\mathbf{x} - \mathbf{b} \|^2$$

may give very inexact approximations. In order to enhance its properties, one can utilize a family of regularized problems of the form

$$\min_{\mathbf{x}} \to \| A\mathbf{x} - \mathbf{b} \|^2 + \varepsilon h(\mathbf{x}), \tag{20}$$

where $h(\mathbf{x}) = \|\mathbf{x}\|^2$ or $h(\mathbf{x}) = \|\mathbf{x}\|_1 \triangleq \sum_{i=1}^{n} |x_i|$, $\varepsilon > 0$ is a parameter. Note that the non-smooth regularization term yields additionally sparse solutions with rather small number of non-zero components; see e.g. [2,17].

The second instance is the basic machine learning problem, which is called the linear support vector machine. It consists in finding the optimal partition of the feature space $\mathbb{R}^n$ by using some given training sequence $\mathbf{x}^i$, $i = 1, \ldots, l$ where each point $\mathbf{x}^i$ has a binary label $y_i \in \{-1, +1\}$ indicating the class. We have to find a separating hyperplane. Usually, its parameters are found from the solution of the optimization problem

$$\min_{\mathbf{w} \in \mathbb{R}^n} \rightarrow (1/p)\|\mathbf{w}\|_p^p + C \sum_{i=1}^{l} L(\langle \mathbf{w}, \mathbf{x}^i \rangle; y_i), \tag{21}$$

where L is a loss function and $C > 0$ is a penalty parameter. The usual choice is $L(z; y) = \max\{0; 1 - yz\}$ and p is either 1 or 2; see e.g. [1,5] for more details. Observe that the data of the observation points $\mathbf{x}^i$ can be again inexact or even non-stationary.

Next, taking $p = 2$, we can rewrite this problem as

$$\min_{\mathbf{w}, \xi} \rightarrow 0.5\|\mathbf{w}\|^2 + C \sum_{i=1}^{l} \xi_i,$$

subject to

$$1 - y_i \langle \mathbf{w}, \mathbf{x}^i \rangle \leq \xi_i, \ \xi_i \geq 0, \ i = 1, \ldots, l.$$

Its dual has the quadratic programming format:

$$\max_{0 \leq \alpha_i \leq C, i=1,\ldots,l} \rightarrow \sum_{i=1}^{l} \alpha_i - 0.5 \sum_{s=1}^{l} \sum_{t=1}^{l} (\alpha_s y_s)(\alpha_t y_t)\langle \mathbf{x}^s, \mathbf{x}^t \rangle. \tag{22}$$

Observe that all these problems fall into format (1), (3)–(5) and that they can be treated as limit problems.

5   Computational Experiments

In order to evaluate the computational properties of the proposed method we carried out preliminary series of test experiments. For simplicity, we took only unconstrained test problems of form (1), (3)–(5) where $X = \mathbb{R}^N$ with the single-dimensional (coordinate) partition of the space, i.e., set $N = n$ and $N_i = 1$ for $i = 1, \ldots, n$. In all the experiments, we took the limit function f to be convex and quadratic, namely,

$$f(\mathbf{x}) = 0.5\|A\mathbf{x} - \mathbf{b}\|^2 + 0.5\|\mathbf{x}\|^2,$$

where A was an $n \times n$ matrix, $\mathbf{b}$ a fixed vector whose elements were defined by trigonometric functions, whereas the limit function h was defined either as zero or a non-smooth and convex one. In view of examples (20)–(22) from Sect. 4, these limit problems were approximated by the perturbed sequence of the following optimization problems

$$\min_{\mathbf{x} \in X} \rightarrow \{f_l(\mathbf{x}) + h(\mathbf{x})\}, \tag{23}$$

i.e. we utilized only the perturbation for f, where

$$f_l(\mathbf{x}) = 0.5\|A(\varepsilon_l)\mathbf{x} - \mathbf{b}(\varepsilon_l)\|^2 + 0.5\|\mathbf{x}\|^2,$$

$$a_{ij}(\varepsilon) = a_{ij} + 0.1\varepsilon\sin(ij) \quad \text{and} \quad b_i(\varepsilon) = b_i + 0.2\varepsilon\sin(i), \quad \text{for } i, j = 1, \ldots, n.$$

The main goal was to compare (DNS) with the usual (splitting) gradient descent method (GNS for short); see [18]. It calculates all the components for the direction finding procedure. Both the methods used the same line-search strategy and were applied sequentially to each non-stationary problem (23) with the following rule $\varepsilon_{l+1} = \nu\varepsilon_l$ for changing the perturbation. In (GNS), this change occurs after satisfying the inequality

$$\Delta_l(\mathbf{x}) = \|\mathbf{x} - \mathbf{y}^l(\mathbf{x})\| \le \delta_l,$$

where $\mathbf{y}^l(\mathbf{x})$ is a unique solution of the problem

$$\min_{\mathbf{y}} \to \left\{ \langle f_l'(\mathbf{x}), \mathbf{y} \rangle + 0.5\alpha\|\mathbf{x} - \mathbf{y}\|^2 + h(\mathbf{y}) \right\}.$$

In both the methods, we chose the rule $\delta_{l+1} = \nu\delta_l$ with $\nu = 0.5$. Similarly, for the limit problem, we set

$$\Delta(\mathbf{x}) = \|\mathbf{x} - \mathbf{y}(\mathbf{x})\|,$$

where $\mathbf{y}(\mathbf{x})$ is a unique solution of the problem

$$\min_{\mathbf{y}} \to \left\{ \langle f'(\mathbf{x}), \mathbf{y} \rangle + 0.5\alpha\|\mathbf{x} - \mathbf{y}\|^2 + h(\mathbf{y}) \right\}.$$

We took $\Delta(\mathbf{x}^k)$ as accuracy measure for solving the limit problem, chose the accuracy 0.1, took the same starting point $z_j^0 = j|\sin(j)|$ for $j = 1, \ldots, n$, and set $\alpha = 1$ for both the methods. The methods were implemented in Delphi with double precision arithmetic.

In the first two series, we set $h \equiv 0$ and took versions with exact line-search. In the first series, we took the elements $a_{ij} = \sin(i/j)\cos(ij)$ and $b_i = (1/i)\sin(i)$. The results are given in Table 1. In the second series, we took the elements $a_{ij} = 1/(i+j) + 2\sin(i/j)\cos(ij)/j$ and $b_i = n\sin(i)$. The results are given in Table 2. In the third series, we took the elements $a_{ij} = 1/(i+j) + 2\sin(i/j)\cos(ij)/j$ and $b_i = n\sin(i)$ as above, but also chose

$$h(\mathbf{x}) = \sum_{i=1}^{N} |x_i|.$$

So, the cost function is non-smooth. Here we took versions with the Armijo line-search. The results are given in Table 3, where (cl) now denotes the total number of calculations of partial derivatives of f_l. Therefore, (DNS) showed rather stable and rapid convergence, and the explicit preference over (GNS) if the dimensionality was greater than 20.

Table 1. The numbers of iterations (it) and partial derivatives calculations (cl)

	(GNS)		(DNS)	
	it	cl	it	cl
$N = 2$	2	4	2	7
$N = 5$	10	50	19	64
$N = 10$	17	170	67	235
$N = 20$	36	720	194	688
$N = 40$	105	4200	734	2935
$N = 80$	228	18240	3500	11241
$N = 100$	201	20100	4641	16473

Table 2. The numbers of iterations (it) and partial derivatives calculations (cl)

	(GNS)		(DNS)	
	it	cl	it	cl
$N = 2$	6	12	2	7
$N = 5$	10	50	12	49
$N = 10$	19	190	25	139
$N = 20$	38	760	55	349
$N = 40$	72	2880	119	871
$N = 80$	164	13120	309	2461
$N = 100$	252	25200	414	3279

Table 3. The numbers of iterations (it) and partial derivatives calculations (cl)

	(GNS)		(DNS)	
	it	cl	it	cl
$N = 2$	13	26	7	23
$N = 5$	16	80	32	84
$N = 10$	16	160	76	245
$N = 20$	40	800	257	982
$N = 40$	72	2880	485	1923
$N = 80$	135	10800	1127	4286
$N = 100$	188	18800	1374	6075

6   Conclusions

We described a new class of coordinate-wise descent splitting methods for limit decomposable composite optimization problems involving set-valued mappings and non-smooth functions. The method is based on selective coordinate variations together with changing the tolerance parameters corresponding to a sequence of mixed variational inequality problems without explicit evaluation of accuracy of approximate solutions. We proved convergence without special concordance rules for all the parameters and tolerances. Series of computational tests confirmed rather satisfactory convergence.

Acknowledgement. The results of this work were obtained within the state assignment of the Ministry of Science and Education of Russia, project No. 1.460.2016/1.4. In this work, the author was also supported by Russian Foundation for Basic Research, project No. 16-01-00109 and by grant No. 297689 from Academy of Finland.

References

1. Burges, C.J.C.: A tutorial on support vector machines for pattern recognition. Data Mining Know. Disc. **2**, 121–167 (1998)
2. Cevher, V., Becker, S., Schmidt, M.: Convex optimization for big data. Signal Process. Magaz. **31**, 32–43 (2014)
3. Facchinei, F., Scutari, G., Sagratella, S.: Parallel selective algorithms for nonconvex big data optimization. IEEE Trans. Sig. Process. **63**, 1874–1889 (2015)
4. Tseng, P., Yun, S.: A coordinate gradient descent method for nonsmooth separable minimization. Math. Progr. **117**, 387–423 (2010)
5. Richtárik, P., Takáč, M.: Parallel coordinate descent methods for big data optimization. Math. Program. **156**, 433–484 (2016)
6. Konnov, I.V.: Sequential threshold control in descent splitting methods for decomposable optimization problems. Optim. Meth. Softw. **30**, 1238–1254 (2015)
7. Alart, P., Lemaire, B.: Penalization in non-classical convex programming via variational convergence. Math. Program. **51**, 307–331 (1991)
8. Cominetti, R.: Coupling the proximal point algorithm with approximation methods. J. Optim. Theor. Appl. **95**, 581–600 (1997)
9. Salmon, G., Nguyen, V.H., Strodiot, J.J.: Coupling the auxiliary problem principle and epiconvergence theory for solving general variational inequalities. J. Optim. Theor. Appl. **104**, 629–657 (2000)
10. Konnov, I.V.: An inexact penalty method for non stationary generalized variational inequalities. Set-Valued Variat. Anal. **23**, 239–248 (2015)
11. Clarke, F.H.: Optimization and Nonsmooth Analysis. Wiley, New York (1983)
12. Ermoliev, Y.M., Norkin, V.I., Wets, R.J.B.: The minimization of semicontinuous functions: mollifier subgradient. SIAM J. Contr. Optim. **33**, 149–167 (1995)
13. Czarnecki, M.-O., Rifford, L.: Approximation and regularization of lipschitz functions: convergence of the gradients. Trans. Amer. Math. Soc. **358**, 4467–4520 (2006)
14. Gwinner, J.: On the penalty method for constrained variational inequalities. In: Hiriart-Urruty, J.-B., Oettli, W., Stoer, J. (eds.) Optimization: Theory and Algorithms, pp. 197–211. Marcel Dekker, New York (1981)
15. Blum, E., Oettli, W.: From optimization and variational inequalities to equilibrium problems. The Math. Stud. **63**, 127–149 (1994)
16. Engl, H.W., Hanke, M., Neubauer, A.: Regularization of Inverse Problems. Kluwer Academic Publishers, Dordrecht (1996)
17. Tibshirani, R.: Regression shrinkage and selection via the lasso. J. Royal Stat. Soc. Ser. B. **58**, 267–288 (1996)
18. Fukushima, M., Mine, H.: A generalized proximal point algorithm for certain nonconvex minimization problems. Int. J. Syst. Sci. **12**, 989–1000 (1981)

RAMBO: Resource-Aware Model-Based Optimization with Scheduling for Heterogeneous Runtimes and a Comparison with Asynchronous Model-Based Optimization

Helena Kotthaus[1]([⊠]), Jakob Richter[2], Andreas Lang[1], Janek Thomas[3],
Bernd Bischl[3], Peter Marwedel[1], Jörg Rahnenführer[2], and Michel Lang[2]

[1] Department of Computer Science 12, TU Dortmund University,
Dortmund, Germany
helena.kotthaus@tu-dortmund.de
[2] Department of Statistics, TU Dortmund University, Dortmund, Germany
[3] Department of Statistics, LMU München, Munich, Germany

Abstract. Sequential model-based optimization is a popular technique for global optimization of expensive black-box functions. It uses a regression model to approximate the objective function and iteratively proposes new interesting points. Deviating from the original formulation, it is often indispensable to apply parallelization to speed up the computation. This is usually achieved by evaluating as many points per iteration as there are workers available. However, if runtimes of the objective function are heterogeneous, resources might be wasted by idle workers. Our new knapsack-based scheduling approach aims at increasing the effectiveness of parallel optimization by efficient resource utilization. Derived from an extra regression model we use runtime predictions of point evaluations to efficiently map evaluations to workers and reduce idling. We compare our approach to five established parallelization strategies on a set of continuous functions with heterogeneous runtimes. Our benchmark covers comparisons of synchronous and asynchronous model-based approaches and investigates the scalability.

Keywords: Black-box optimization · Model-based optimization · Global optimization · Resource-aware scheduling · Performance management · Parallelization

1 Introduction

Efficient global optimization of expensive black-box functions is of interest to many fields of research. In the engineering industry, computationally expensive models have to be optimized; for machine learning hyperparameters have to be tuned; and for computer experiments in general, expensive algorithms have parameters that have to be optimized to obtain a well-performing algorithm configuration. The problems of global optimization can usually be modeled by a

R. Battiti et al. (Eds.): LION 2017, LNCS 10556, pp. 180–195, 2017.
https://doi.org/10.1007/978-3-319-69404-7_13

real-valued objective function f with a d-dimensional domain space. The challenge is to find the best point possible within a very limited time budget.

Together with [1, 22, 23], Model-based optimization (MBO) [15] is a state-of-the-art algorithm for *expensive* black-box functions. Starting on an initial design of already evaluated configurations, a regression model guides the search to new configurations by predicting the outcome of the black-box on yet unseen configurations. Based on this prediction an infill criterion (also called acquisition function) proposes a new promising configuration for evaluation. In each iteration the regression model is updated on the evaluated configurations of the previous iteration until the budget is exhausted. Jones et al. [15] proposed this now popular *Efficient Global Optimization* (EGO) algorithm. EGO sequentially adds points to the design using Kriging as a surrogate and the *Expected Improvement* (EI) as an infill criterion. Following, other infill criteria [14], specializations e.g. for categorical search spaces like in SMAC [10] and noisy optimization [20] have been introduced.

For computer experiments, parallelization has become of increasing interest to reduce the overall computation time. Originally, the EGO algorithm iteratively proposes one point to be evaluated after another. To allow for parallelization, infill criteria and techniques (constant liar, Kriging believer, qEI [9], qLCB [11], MOI-MBO [4]) have been suggested that propose multiple points in each iteration. Usually, the number of proposed points equals the number of available CPUs. However, these methods still do use the available resources inefficiently if the runtime of the black-box is heterogeneous. Before new proposals can be generated, the results of all evaluations within one iteration are gathered to update the model. Consequently all CPUs have to wait for the slowest function evaluation before receiving a new point proposal. This can lead to idling CPUs that are not contributing to the optimization. The goal in general is to use all available CPU time to solve the optimization problem.

One approach to avoid idling is to desynchronize the model update. Here, the model is updated each time an evaluation has finished, letting each parallel worker propose the next point for evaluation itself. Such asynchronous techniques have been suggested and discussed by [8, 12]. The main challenge is to modify the infill criterion to deal with points that are currently under evaluation to avoid evaluations of very similar configurations. The *Expected Expected Improvement* (EEI) [13] is one possibility for such a modification.

Another strategy is to keep the synchronous model update and schedule the evaluations of the proposed points in such a way that idling is reduced. Such an approach was presented in [19]. Here, a second regression model is used to predict runtimes for the proposed points which are used as an input for scheduling.

Our article contains the following contributions: First, we extended the parallel, resource-aware synchronous model-based optimization strategy proposed in [19] with an improved resource-aware scheduling algorithm. This algorithm, which replaces the original simple first fit heuristic, is based on a knapsack solver to better handle heterogeneous runtimes. Furthermore we use a clustering-based refinement strategy to ensure improved spatial diversity of the evaluated points.

Second, we compare our algorithm to three asynchronous MBO strategies that also aim at using all available CPU time to solve the optimization problem in parallel. Two of them [8,12] use Kriging as a surrogate and the third is included in SMAC [10] which uses a random forest surrogate.

Third, we benchmark the MBO algorithms on a set of established continuous test functions combined with simulated runtimes. For each function we use a 2- and a 5-dimensional version each of which is optimized using 4 and 16 CPUs in parallel to investigate scalability.

Compared to the considered asynchronous approaches, our new approach converges faster to the optima if the runtime estimates used as input for scheduling are reliable.

2   Model-Based Global Optimization

The aim of global optimization is to find the global minimum of a given function $f : \mathcal{X} \rightarrow \mathbb{R}$, $f(\boldsymbol{x}) = y, \boldsymbol{x} = (x_1, \ldots, x_d)^T$. Here, we assume $\mathcal{X} \subset \mathbb{R}^d$, usually expressed by simple box constraints. The optimization is guided by a surrogate model which estimates the response surface of the black-box function f (see also [22,23]). The surrogate is comparably inexpensive to evaluate and is utilized to propose new promising points $\boldsymbol{x}^*$ in an iterative fashion. A promising point $\boldsymbol{x}^*$ is determined by optimizing some infill criterion. After, $f(\boldsymbol{x}^*)$ is evaluated to obtain the corresponding objective value y, the surrogate model is refitted and a new point is proposed. The infill criterion quantifies the potential improvement based on an exploitation-exploration trade-off where a low (good) expected value of the solution $\hat{\mu}(\boldsymbol{x})$ is rewarded, and low estimated uncertainty $\hat{s}(\boldsymbol{x})$ is penalized. A popular infill criterion, especially for Kriging surrogate models, is the expected improvement

$$
\begin{aligned}
\mathrm{EI}(\boldsymbol{x}) &= \mathbb{E}(\max(y_{\min} - \hat{\mu}(\boldsymbol{x}), 0)) \\
&= (y_{\min} - \hat{\mu}(\boldsymbol{x})) \, \Phi\left(\frac{y_{\min} - \hat{\mu}(\boldsymbol{x})}{\hat{s}(\boldsymbol{x})}\right) + \hat{s}(\boldsymbol{x})\phi\left(\frac{y_{\min} - \hat{\mu}(\boldsymbol{x})}{\hat{s}(\boldsymbol{x})}\right),
\end{aligned}
$$

where Φ is the distribution and ϕ is the density function of the standard normal distribution and $y_{\min}$ is the best observed function value so far. Alternatively, the comparably simpler lower confidence bound criterion

$$
\mathrm{LCB}(\boldsymbol{x}, \lambda) = \hat{\mu}(\boldsymbol{x}) - \lambda \hat{s}(\boldsymbol{x}), \quad \lambda \in \mathbb{R}
$$

is used, where $\hat{\mu}(\boldsymbol{x})$ denotes the posterior mean and $\hat{s}(\boldsymbol{x})$ the posterior standard deviation of the regression model at point $\boldsymbol{x}$. Before entering the iterative process, initially some points have to be pre-evaluated. These points are generally chosen in a space-filling manner to uniformly cover the input space. The optimization usually stops after a target objective value is reached or a predefined budget is exhausted [22,23].

2.1   Parallel MBO

Ordinary MBO is sequential by design. However, applications like hyperparameter optimization for machine learning or computer simulations have driven the rapid development of extensions for parallel execution of multiple point evaluations. The parallel extensions either focus on a synchronous model update using infill criteria with multi-point proposals or implement an asynchronous evaluation where each worker generates one new point proposal individually.

Multi-Point proposals derive not only one single point x^* from a surrogate model, but q points $x_1^*, \ldots, x_q^*$ simultaneously. The q proposed points must be sufficiently different from each other to avoid multiple evaluations with the same configuration. For this reason Hutter et al. [11] introduced the criterion

$$\text{qLCB}(x, \lambda_j) = \hat{\mu}(x) - \lambda_j \hat{s}(x) \text{ with } \lambda_j \sim \text{Exp}(\lambda) \qquad (1)$$

as an intuitive extension of the LCB criterion using an exponentially distributed random variable. Since λ guides the trade-off between exploration and exploitation, sampling multiple different λ_j might result in different "best" points by varying the impact of the standard deviation. The qLCB criterion was implemented in a distributed version of SMAC [11]. An extension of the EI criterion is the qEI criterion [9] which directly optimizes the expected improvement over q points. A closed form solution to calculate qEI exists for $q = 2$ and useful approximations can be applied for $q \leq 10$ [7]. However, as the computation is using Monte Carlo sampling, it is quite expensive. A less expensive and popular alternative is *Kriging believer* approach [9]. Here, the first point is proposed using the single-point EI criterion. Its posterior mean value is treated as a real value of f to refit the surrogate, effectively penalizing the surrounding region with a lower standard deviation for the next point proposal using EI again. This is repeated until q proposals are generated.

In combination with parallel synchronous execution the above described multi-point infill approaches can lead to underutilized systems because a new batch of points can only be proposed as soon as the slowest function evaluation is terminated. Snoek et al. [21] introduce the EI per second to address heterogeneous runtimes. The runtime of a configuration is estimated using a second surrogate model and a combined infill criterion can be constructed which favors less expensive configurations.

We also use surrogate models to estimate resource requirements but instead of adapting the infill criterion, we use them for the scheduling of parallel function evaluations. Our goal is to guide MBO to interesting regions in a faster and resource-aware way without directly favoring less expensive configurations.

Asynchronous Execution approaches the problem of parallelizing MBO from a different angle. Instead of evaluating multiple points in batches to synchronously refit the model, the model is refitted after each function evaluation to increase CPU utilization workers. Here, each worker propose the next point for evaluation itself, even when configurations x_{busy} are currently under evaluation on other processing units. The busy evaluations have to be taken into account by the surrogate model to avoid that new point proposals are identical or very similar to pending evaluations. Here, the Kriging believer approach [9] can be applied to block these regions. Another theoretically well-founded way to impute pending values is the *expected* EI (EEI) [8,13,21]. The unknown value of $f(x_{busy})$ is integrated out by calculating the expected value of y_{busy} via Monte Carlo sampling, which is, similar to qEI, computationally demanding. For each Monte Carlo iteration values $y_{1,busy}, \ldots, y_{\mu,busy}$ are drawn from the posterior distribution of the surrogate regression model at $x_{1,busy}, \ldots, x_{\mu,busy}$, with μ denoting the number of pending evaluations. These values are combined with the set of already known evaluations and used to fit the surrogate model. The EEI can then simply be calculated by averaging the individual expected improvement values that are formed by each Monte Carlo sample:

$$\widehat{\mathrm{EEI}(x)} = \frac{1}{n_{\mathrm{sim}}} \sum_{i=1}^{n_{\mathrm{sim}}} \mathrm{EI}_i(x) \tag{2}$$

whereas n_{sim} denotes the number of Monte Carlo iterations.

Besides the advantage of an increased CPU utilization, asynchronous execution can also potentially cause additional runtime overhead due to the higher number of model refits, especially when the number of workers increases. Therefore our experiments include a comparison with most of the above described approaches to investigate the advantages and disadvantages.

Instead of using asynchronous execution to efficiently utilize parallel computer architectures, our new approach uses the synchronous execution combined with resource-aware scheduling and is presented in the following section.

3   Resource-Aware Scheduling with Synchronous Model Update

The goal of our new scheduling strategy is to guide MBO to interesting regions in a faster and resource-aware way. To efficiently map jobs (proposed points) to available resources our strategy needs to know the resource demands of jobs before execution. Therefore, we estimate the runtime of each job using a regression model. Additionally, we calculate an execution priority for each job based on the multi-point infill criterion. In the following, we will describe these inputs.

3.1   Infill Criterion - Priority

The priorities of the proposed points should reflect their usefulness for optimization. In our setup we opt for the qLCB (1) to generate a set of job proposals by optimizing the LCB for q randomly drawn values of $\lambda_j \sim \mathrm{Exp}(\frac{1}{2})$, as in Richter et al. [19]. qLCB is suitable because the proposals are independent of each other. There is no direct order of the set of obtained candidates x_j^* in terms of how promising or important one candidate is in comparison to each other. Therefore, we introduce an order that steers the search more towards promising areas. We give the highest priority to the point x_j that was proposed using the smallest value of λ_j. We define the priority for each point as $p_j := -\lambda_j$.

3.2   Resource Estimation

To estimate the resource demands of proposed candidates, we use a separate regression model. To adapt to the domain space of the objective function, we choose the same regression method used for the surrogate. In the same fashion as for the MBO algorithm, runtimes are predicted in each MBO iteration based on all previously evaluated jobs and measured runtimes.

3.3   Resource-Aware Knapsack Scheduling

The goal of our scheduling strategy is to reduce the CPU idle time on the workers while acquiring the feedback of the workers in the shortest possible time to avoid model update delay. The set of points proposed by the multi-point infill criterion forms the set of jobs $J = \{1, \ldots, q\}$ that we want to execute on the available CPUs $K = \{1, \ldots, m\}$. For each job the estimated runtime is given by $\hat{t}_j$ and the corresponding priority is given by p_j. To reduce idle times caused by evaluations of jobs with a low priority, the maximal runtime for each MBO iteration is defined by the runtime of the job with the highest priority. Lower prioritized jobs have to subordinate. At the same time we want to maximize the profit, given by the priorities, of parallel job executions for each model update. To solve this problem, we apply the $0 - 1$ multiple knapsack algorithm by interfacing the R-package `adagio` for global optimization routines [5]. Here the knapsacks are the available CPUs and their capacity is the maximally allowed computing time, defined by the runtime of the job with the highest priority. The items are the jobs J, their weights are the estimated runtimes $\hat{t}_j$ and their values are the priorities p_j. The capacity for each CPU is accordingly $\hat{t}_{j^*}$, with $j^* := \arg\max_j p_j$. To select the best subset of jobs the algorithm maximizes the profit Q:

$$Q = \sum_{j \in J} \sum_{k \in K} p_j c_{kj},$$

which is the sum of priorities of the selected jobs, under the restriction of the capacity

$$\hat{t}_{j^*} \geq \sum_{j \in J} \hat{t}_j c_{kj} \, \forall k \in K$$

per CPU. The restriction with the decision variable $c_{kj} \in \{0, 1\}$

$$1 \geq \sum_{k \in K} c_{kj} \, \forall j \in J, c_{kj} \in \{0, 1\}.$$

ensures that a job j is at most mapped to one CPU.

As the job with the highest priority defines the time bound $\hat{t}_{j*}$ it is mapped to the first CPU $k = 1$ exclusively and single jobs with higher runtimes are directly discarded. Then the knapsack algorithm is applied to assign the remaining candidates in J to the remaining $m-1$ CPUs. This leads to the best subset of J that can be run in parallel minimizing the delay of the model update. If a CPU is left without a job we query the surrogate model for a job with an estimated runtime smaller or equal to $\hat{t}_{j*}$ to fill the gaps. For a useful scheduling the set of candidates should have considerably more candidates q than available CPUs. This knapsack scheduling is a direct enhancement of the first fit scheduling strategy presented in [19].

3.4 Refinement of Job Priorities via Clustering

The refinement of job priorities has the goal to avoid parallel evaluations of very similar configurations. Approaches to specifically propose points that are promising but yet diverse are described in [4]. qLCB performed well and was chosen here because it is comparably inexpensive to create many independent candidates. However, qLCB does not include a penalty for the proximity of selected points which gets problematic if the number of parallel points is high. Therefore, we use a distance measure to reprioritize p_j to $\tilde{p}_j$, encouraging the selection sets of candidates more scattered in the domain space.

First, we oversample a set of $q > m$ candidate points from the qLCB criterion and partition them into $\tilde{q} < q$ clusters using the Euclidean distance. Next, we take the candidate with maximum priority p_j from each cluster and sort them according to their priority before pushing them to the list $\tilde{J}$ of selected jobs. Selected jobs are removed from the clusters and empty clusters are eliminated. We repeat this procedure until we have moved all q jobs into the list $\tilde{J}$. Finally, we assign new priorities $\tilde{p}_j$ based on the order of $\tilde{J}$, i.e. the first job in $\tilde{J}$ gets the highest priority q and the last job gets the lowest priority 1.

As a result, the set of candidates contains batches of jobs with similar priority that are spread in the domain space. The new priorities serve as input for scheduling which groups the q jobs to m CPUs using the runtime estimates $\hat{t}$.

4   Numerical Experiments

In our experiments, we consider two categories of synthetic functions to ensure a fair comparison in a disturbance-free environment. They are implemented in the R package smoof [6]:

1. Functions with a smooth surface: rosenbrock(d) and bohachevsky(d) with dimension $d = 2, 5$. They are likely to be fitted well by the surrogate.
2. Highly multimodal functions: ackley(d) and rastrigin(d) ($d = 2, 5$). We expect that surrogate models can have problems to achieve a good fit here.

As these are illustrative test functions, they have no significant runtime. As a resort, we also use these functions to simulate runtime behavior. First, we combine two functions: One determines the number of seconds it would take to calculate the objective value of the other function. E.g., for the combination `rastrigin(2).rosenbrock(2)` it would require `rosenbrock(2)($x$)` seconds to retrieve the objective value `rastrigin(2)($x$)` for an arbitrary proposed point x. Technically, we just sleep `rosenbrock(2)($x$)` seconds before returning the objective. We simulate the runtime with either `rosenbrock($d$)` or `rastrigin($d$)` and analyze all combinations of our four objective functions, except where the objective and the time function are identical.

A prerequisite for this approach is the unification of the input space. Thus, we simply mapped values from the input space of the objective function to the input space of the time function. The output of the time functions is scaled to return values between 5 min to 60 min.

We examine the capability of the considered optimization strategies to minimize functions with highly heterogeneous runtimes within a limited time budget. To do this, we measure the distance between the best found point at time t and a predefined target value. We call this measure accuracy. In order to make this measure comparable across different objective functions, we scale the function values to $[0, 1]$ with zero being the target value. It is defined as the best y reached by any optimization method after the complete time budget. The upper bound 1 is the best y found in the initial design (excluding the initial runs of `smac`) which is identical for all algorithms per given problem. Both values are averaged over the 10 replications.

If an algorithm needs 2 h to reach an accuracy of 0.5, this means that within 2 h half of the way to 0 has been accomplished, after starting at 1. We compare the differences between optimizers at the three accuracy levels 0.5, 0.1 and 0.01.

The optimizations are repeated 10 times and conducted on $m = 4$ and $m = 16$ CPUs. We allow each optimization to run for 4 h on 4 CPUs and for 2 h on 16 CPUs in total which includes all computational overhead and idling. All computations were performed on a Docker Swarm cluster using the R package `batchtools` [18]. The initial design is generated by Latin hypercube sampling with $n = 4 \cdot d$ points and all of the following optimizers start with the same design in the respective repetition:

> rs: Random search, serving as base-line.
> qLCB: Synchronous approach using qLCB where in each iteration $q = m$ points are proposed.
> ei.bel: Synchronous approach using Kriging believer where in each iteration m points are proposed.
> asyn.eei: Asynchronous approach using EEI (100 Monte Carlo iterations)
> asyn.ei.bel: Asynchronous Kriging believer approach.
> rambo: Synchronous approach using qLCB with our new scheduling approach where in each iteration $q = 8 \cdot m$ candidates are proposed.

qLCB and `ei.bel` are implemented in the R package `mlrMBO` [3], which builds upon the machine learning framework `mlr` [2]. `asyn.eei`, `asyn.ei.bel` and `rambo` are also based on `mlrMBO`. We use a Kriging model from the package `DiceKriging` [20] with a Matern$\frac{5}{2}$-kernel for all approaches above and add a nugget effect of $10^{-8} \cdot \mathrm{Var}(\boldsymbol{y})$, where $\boldsymbol{y}$ denotes the vector of all observed function outcomes. Additionally we compare our implementations to:

smac: Asynchronous approach that turns m independent SMAC runs into m dependent runs by sharing surrogate model data (also called shared-model-mode[1]).

SMAC was allowed the same initial budget as the other optimizers and was started with the defaults and the shared-model-mode activated. SMAC uses a random forest as surrogate and the EI criterion.

4.1 Quality of Resource Estimation

The quality of resource-aware scheduling naturally depends on the accuracy of the resource estimation. Without reliable runtime predictions, the scheduler is unable to optimize for efficient utilization. As Fig. 1 exemplary shows, the runtime prediction for the rosenbrock(5) time function works well as the residual values are getting smaller over time, while the runtime prediction for rastrigin(5) is comparably imprecise. For the 2-dimensional versions the results are similar. This encourages to consider scenarios separately where runtime prediction is possible (rosenbrock($\cdot$), Subsect. 4.2) and settings where runtime prediction is error-prone (rastrigin($\cdot$), Subsect. 4.3) for further analysis.

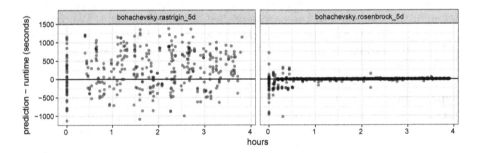

Fig. 1. Residuals of the runtime prediction in the course of time for the rosenbrock(5) and rastrigin(5) time functions on 4 CPUs and bohachevsky(5) as objective function. Positive values indicate an overestimated runtime and negative values an underestimation.

4.2 High Runtime Estimation Quality: rosenbrock

Figure 2 shows boxplots for the time required to reach the three different accuracy levels in 10 repetitions within a budget of 4 h real time on 4 CPUs (upper part) and

---

[1] Hutter, F., Ramage, S.: Manual for SMAC version v2.10.03-master. Department of Computer Science, UBC. (2015), www.cs.ubc.ca/labs/beta/Projects/SMAC/v2.10.03/manual.pdf.

Table 1. Ranking for accuracy levels 0.5, 0.1, 0.01 averaged over all problems with rosenbrock($\cdot$) time function on 4 and 16 CPUs with a time budget of 4 h and 2 h, respectively.

Algorithm	4 CPUs			16 CPUs		
	0.5	0.1	0.01	0.5	0.1	0.01
asyn.eei	3.32 (2)	3.52 (1)	4.97 (2)	3.75 (3)	4.30 (3)	5.45 (3)
asyn.ei.bel	3.55 (3)	4.10 (3)	4.97 (2)	3.48 (2)	4.08 (2)	4.53 (2)
RAMBO	3.17 (1)	3.85 (2)	4.57 (1)	3.13 (1)	3.93 (1)	4.47 (1)
ei.bel	4.38 (4)	4.98 (4)	5.90 (5)	5.00 (5)	5.48 (6)	6.28 (6)
qLCB	4.52 (5)	5.03 (5)	5.63 (4)	4.72 (4)	5.17 (4)	6.10 (4)
rs	6.02 (6)	6.67 (6)	6.83 (7)	5.50 (7)	6.48 (7)	6.87 (7)
smac	6.22 (7)	6.70 (7)	6.82 (6)	5.32 (6)	5.47 (5)	6.17 (5)

2 h on 16 CPUs (lower part). The faster an optimizer reaches the desired accuracy level, the lower the displayed box and the better the approach. If an algorithm did not reach an accuracy level within the time budget, we impute with the respective time budget (4 h or 2 h) plus a penalty of 1000 s.

Table 1 lists the aggregated ranks over all objective functions, grouped by algorithm, accuracy level, and number of CPUs. For this computation, the algorithms are ranked w.r.t. their performance for each replication and problem before they are aggregated with the mean. If there are ties (e.g. if an accuracy level was not reached), all values obtain the worst possible rank.

The benchmarks indicate an overall advantage of our proposed resource-aware MBO algorithm (rambo): On average, rambo reaches the accuracy level first in 2 of 3 setups on 4 CPUs and is always fastest on 16 CPUs. rambo is closely followed by the asynchronous variant asyn.eei on 4 CPUs but the lead becomes more clear on 16 CPUs. In comparison to the conventional synchronous algorithms (ei.bel, qLCB), rambo as well as asyn.eei and asyn.ei.bel reach the given accuracy levels in shorter time. This is especially true for objective functions that are hard to model (ackley($\cdot$), rastrigin($\cdot$)) by the surrogate as seen in Fig. 2. The simpler asyn.ei.bel performs better than asyn.eei on 16 CPUs. Except for smac, all presented MBO methods outperform base-line rs on almost all problems and accuracy levels. The bad average results for smac are partly due to its low performance on the $5d$ problems and probably because of the disadvantage of using a random forest as a surrogate on purely numerical problems. A recent benchmark in [3] was able to demonstrate the competitive performance of the Kriging based EGO approach. On 16 CPUs smac performs better than rs and comparable to ei.bel.

For a thorough analysis of the optimization, Fig. 3 exemplary visualizes the mapping of the parallel point evaluations (jobs) for all MBO algorithms on 16 CPUs for the $5d$ versions of the problems. Each gray box represents computation of a job on the respective CPU. For the synchronous MBO algorithms

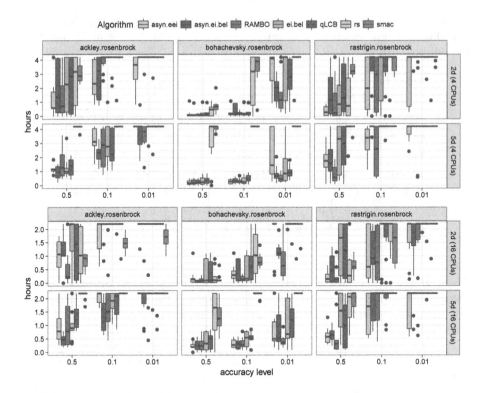

Fig. 2. Accuracy level vs. execution time for different objective functions using time function `rosenbrock(·)` (lower is better).

(`rambo`, `qLCB`, `ei.bel`) the vertical lines indicate the end of an MBO iteration. For `asyn.eei` red boxes indicate that the CPU is occupied with the point proposal. The necessity of a resource estimation for jobs with heterogeneous runtimes becomes obvious, as `qLCB` and `ei.bel` can cause long idle times by queuing jobs together with large runtime differences. The knapsack scheduling in `rambo` manages to clearly reduce this idle time. This effect of efficient resource utilization increases with the number of CPUs. `rambo` reaches nearly the same effective resource-utilization as the asynchronous `asyn.ei.bel` algorithm and `smac` (see Fig. 3) and at the same time reaches the accuracy level fastest on 16 CPUs.

The Monte Carlo approach `asyn.eei` generates a high computational overhead as indicated by the red boxes, which reduces the effective number of evaluations. Idling occurs because the calculation of the EEI is encouraged to wait for ongoing EEI calculations to include their proposals. This overhead additionally increases with the number of already evaluated points. `asyn.ei.bel` and `smac` have a comparably low overhead and thus basically no idle time. This seems to be an advantage for `asyn.ei.bel` on 16 CPUs where it performs better on average than its complex counterpart `asyn.eei`.

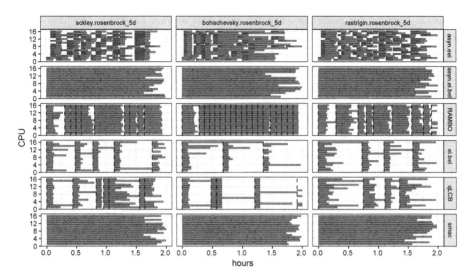

Fig. 3. Scheduling of MBO algorithms. Time on x-axis and mapping of candidates to $m = 16$ CPUs on y-axis. Each gray box represents a job. Each red box represents overhead for the asynchronous approaches. The gaps represent CPU idle time. (Color figure online)

Summed up, if the resource estimation that is used in rambo has a high quality, rambo clearly outperforms the considered synchronous MBO algorithms qLCB, ei.bel, and smac. This indicates, that the resource utilization obtained by the scheduling in rambo leads to faster and better results, especially, when the number of available CPUs increases. On average rambo performs better than all considered asynchronous approaches.

4.3   Low Runtime Estimation Quality: rastrigin

The time function rastrigin used in the following scenario is difficult to fit by surrogate models, as visualized by the residual plot in Fig. 1. For this reason, the benefit of our resource-aware knapsack strategy is expected to be minimal. For example in a possible worst case multiple supposedly short jobs are assigned to one CPU but their real runtime is considerably longer.

Similar to Subsect. 4.2, Fig. 4 shows boxplots for the benchmark results, but with rastrigin($\cdot$) as the time function. Table 2 provides the mean ranks for Fig. 4, calculated in the same way as in previous Subsect. 4.2.

Despite possible wrong scheduling decisions, rambo still manages to outperform qLCB and performs better than ei.bel on the highest accuracy level on average. asyn.eei reaches the accuracy levels fastest on 4 CPUs. Similar to the previous benchmarks on Subsect. 4.2, the simplified asyn.ei.bel seems to benefit from its reduced overhead and places first on 16 CPUs. This difference w.r.t. the scalability becomes especially visible on rosenbrock($\cdot$).

Table 2. Ranking for accuracy levels 0.5, 0.1, 0.01 averaged over all problems with rastrigin($\cdot$) time function on 4 and 16 CPUs with a time budget of 4 h and 2 h, respectively.

Algorithm	4 CPUs			16 CPUs		
	0.5	0.1	0.01	0.5	0.1	0.01
asyn.eei	3.65 (1)	3.25 (1)	4.47 (1)	4.42 (3)	4.38 (2)	5.20 (3)
asyn.ei.bel	3.88 (2)	3.50 (2)	4.52 (2)	3.90 (1)	3.75 (1)	4.33 (1)
RAMBO	4.50 (4)	4.70 (4)	4.72 (3)	4.43 (4)	4.60 (4)	5.17 (2)
ei.bel	4.22 (3)	4.42 (3)	4.87 (4)	4.33 (2)	4.55 (3)	5.27 (4)
qLCB	4.95 (5)	4.80 (5)	5.38 (5)	5.10 (5)	5.00 (5)	5.82 (5)
rs	6.30 (7)	6.42 (6)	6.63 (6)	5.78 (7)	6.23 (7)	6.43 (6)
smac	5.90 (6)	6.98 (7)	7.00 (7)	5.30 (6)	5.77 (6)	6.72 (7)

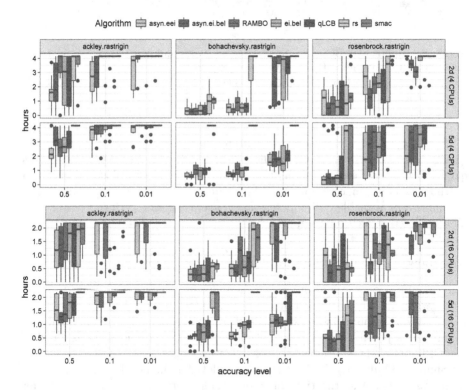

Fig. 4. Accuracy level vs. execution time for different objective functions using time function rastrigin($\cdot$) (lower is better).

smac can not compete with the Kriging based optimizers. Overall, rambo appears not to be able to outperform the asynchronous MBO methods on 4 CPUs as unreliable runtime estimates likely lead to suboptimal scheduling decisions.

However, `rambo` reaches comparable results to `asyn.eei` on 16 CPUs and compared to the default synchronous approaches it is a viable choice.

5   Conclusion

We benchmarked our knapsack based resource-aware parallel MBO algorithm `rambo` against popular synchronous and asynchronous MBO approaches on a set of illustrative test functions for global optimization methods. Our new approach was able to outperform SMAC and the default synchronous MBO approach `qLCB` on the continuous benchmark functions. On setups with high runtime estimation quality it converged faster to the optima than the competing MBO algorithms on average. This indicates, that the resource utilization obtained by our new approach improves MBO, especially, when the number of available CPUs increases. On setups with low runtime estimation quality the asynchronous Kriging based approaches performed best on 4 CPUs and only the simplified asynchronous Kriging believer kept its lead on 16 CPUs. Unreliable estimates likely lead to suboptimal scheduling decisions for `rambo`. While the asynchronous Kriging believer approach, SMAC and `rambo` benefited from increasing the number of CPUs, the overhead of the asynchronous approach based on EEI increased.

If the runtime of point proposals is predictable we suggest our new `rambo` approach for parallel MBO with high numbers of available CPUs. Even if the runtime estimation quality is obviously hard to determine in advance, for real applications like hyperparameter optimization for machine learning methods predictable runtimes can be assumed. Our results also suggest that on some setups the choice of the infill criterion determines which parallelization strategy can reach a better performance. For future work a criterion that assigns an infill value to a set of candidates that can be scheduled without causing long idle times appears promising. Furthermore we want to include the memory consumption measured by the `traceR` [16,17] tool into our scheduling decisions for experiments with high memory demands.

Acknowledgments. J. Richter and H. Kotthaus — These authors contributed equally. This work was partly supported by Deutsche Forschungsgemeinschaft (DFG) within the Collaborative Research Center SFB 876, A3 and by Competence Network for Technical, Scientific High Performance Computing in Bavaria (KONWIHR) in the project "Implementierung und Evaluation eines Verfahrens zur automatischen, massivparallelen Modellselektion im Maschinellen Lernen".

References

1. Ansótegui, C., Malitsky, Y., Samulowitz, H., Sellmann, M., Tierney, K.: Model-based genetic algorithms for algorithm configuration. In: International Joint Conference on Artificial Intelligence, pp. 733–739 (2015)
2. Bischl, B., Lang, M., Kotthoff, L., Schiffner, J., Richter, J., Studerus, E., Casalicchio, G., Jones, Z.M.: mlr: Machine learning in R. J. Mach. Learn. Res. **17**(170), 1–5 (2016)

3. Bischl, B., Richter, J., Bossek, J., Horn, D., Thomas, J., Lang, M.: mlrMBO: A modular framework for model-based optimization of expensive black-box functions. arXiv pre-print (2017). http://arxiv.org/abs/1703.03373

4. Bischl, B., Wessing, S., Bauer, N., Friedrichs, K., Weihs, C.: MOI-MBO: multi-objective infill for parallel model-based optimization. In: Pardalos, P.M., Resende, M.G.C., Vogiatzis, C., Walteros, J.L. (eds.) LION 2014. LNCS, vol. 8426, pp. 173–186. Springer, Cham (2014). doi:10.1007/978-3-319-09584-4_17

5. Borchers, H.: adagio: Discrete and Global Optimization Routines (2016). R package version 0.6.5. https://CRAN.R-project.org/package=adagio

6. Bossek, J.: smoof: Single and Multi-Objective Optimization Test Functions (2016). R package version 1.4. https://CRAN.R-project.org/package=smoof

7. Chevalier, C., Ginsbourger, D.: Fast computation of the multi-points expected improvement with applications in batch selection. In: Nicosia, G., Pardalos, P. (eds.) LION 2013. LNCS, vol. 7997, pp. 59–69. Springer, Heidelberg (2013). doi:10. 1007/978-3-642-44973-4_7

8. Ginsbourger, D., Janusevskis, J., Le Riche, R.: Dealing with asynchronicity in parallel Gaussian process based global optimization. In: 4th International Conference of the ERCIM WG on Computing & Statistics (ERCIM 2011), pp. 1–27 (2011)

9. Ginsbourger, D., Le Riche, R., Carraro, L.: Kriging is well-suited to parallelize optimization. In: Tenne, Y., Goh, C.K. (eds.) Computational Intelligence in Expensive Optimization Problems, pp. 131–162. Springer, Heidelberg (2010). doi:10.1007/978-3-642-10701-6_6

10. Hutter, F., Hoos, H.H., Leyton-Brown, K.: Sequential model-based optimization for general algorithm configuration. In: Coello, C.A.C. (ed.) LION 2011. LNCS, vol. 6683, pp. 507–523. Springer, Heidelberg (2011). doi:10.1007/978-3-642-25566-3_40

11. Hutter, F., Hoos, H.H., Leyton-Brown, K.: Parallel algorithm configuration. In: Hamadi, Y., Schoenauer, M. (eds.) LION 2012. LNCS, pp. 55–70. Springer, Heidelberg (2012). doi:10.1007/978-3-642-34413-8_5

12. Janusevskis, J., Le Riche, R., Ginsbourger, D.: Parallel expected improvements for global optimization: summary, bounds and speed-up. Technical report (2011). https://hal.archives-ouvertes.fr/hal-00613971

13. Janusevskis, J., Le Riche, R., Ginsbourger, D., Girdziusas, R.: Expected improvements for the asynchronous parallel global optimization of expensive functions: potentials and challenges. In: Hamadi, Y., Schoenauer, M. (eds.) LION 2012. LNCS, pp. 413–418. Springer, Heidelberg (2012). doi:10.1007/978-3-642-34413-8_37

14. Jones, D.R.: A taxonomy of global optimization methods based on response surfaces. J. Global Optim. 21(4), 345–383 (2001)

15. Jones, D.R., Schonlau, M., Welch, W.J.: Efficient global optimization of expensive black-box functions. J. Global Optim. 13(4), 455–492 (1998)

16. Kotthaus, H., Korb, I., Lang, M., Bischl, B., Rahnenführer, J., Marwedel, P.: Runtime and memory consumption analyses for machine learning R programs. J. Stat. Comput. Simul. 85(1), 14–29 (2015)

17. Kotthaus, H., Korb, I., Marwedel, P.: Performance analysis for parallel R programs: towards efficient resource utilization. Technical report 01/2015, Department of Computer Science 12, TU Dortmund University (2015)

18. Lang, M., Bischl, B., Surmann, D.: batchtools: Tools for R to work on batch systems. J. Open Source Softw. 2(10) (2017)

19. Richter, J., Kotthaus, H., Bischl, B., Marwedel, P., Rahnenführer, J., Lang, M.: Faster model-based optimization through resource-aware scheduling strategies. In: Festa, P., Sellmann, M., Vanschoren, J. (eds.) LION 2016. LNCS, vol. 10079, pp. 267–273. Springer, Cham (2016). doi:10.1007/978-3-319-50349-3_22

20. Roustant, O., Ginsbourger, D., Deville, Y.: DiceKriging, DiceOptim: two R packages for the analysis of computer experiments by Kriging-based metamodeling and optimization. J. Stat. Softw. **51**(1), 1–55 (2012)

21. Snoek, J., Larochelle, H., Adams, R.P.: Practical Bayesian optimization of machine learning algorithms. In: Advances in Neural Information Processing Systems, vol. 25. pp. 2951–2959. Curran Associates, Inc. (2012)

22. Strongin, R.G., Sergeyev, Y.D.: Global Optimization with Non-Convex Constraints. Kluwer Academic Publishers, Dordrecht (2000)

23. Zhigljavsky, A., Žilinskas, A.: Stochastic Global Optimization. Springer, New York (2008). doi:10.1007/978-0-387-74740-8

A New Constructive Heuristic for the No-Wait Flowshop Scheduling Problem

Lucien Mousin[(✉)], Marie-Eléonore Kessaci, and Clarisse Dhaenens

Univ. Lille, CNRS, Centrale Lille, UMR 9189 - CRIStAL - Centre de Recherche en
Informatique Signal et Automatique de Lille, 59000 Lille, France
lucien.mousin@ed.univ-lille1.fr,
{me.kessaci,clarisse.dhaenens}@univ-lille1.fr

Abstract. Constructive heuristics have a great interest as they manage
to find in a very short time, solutions of relatively good quality. Such
solutions may be used as initial solutions for metaheuristics for example.
In this work, we propose a new efficient constructive heuristic for the No-
Wait Flowshop Scheduling Problem. This proposed heuristic is based on
observations on the structure of best solutions of small instances as well
as on analyzes of efficient constructive heuristics principles of the literature. Experiments have been conducted and results show the efficiency
of the proposed heuristic compared to ones from the literature.

1 Introduction

Scheduling problems is an important class of combinatorial optimization problems, most of them being $NP - hard$. They consist in the allocation of different
operations on a set of machines over the time. The aim of scheduling problems is
to optimize different criteria such as the makespan or the flowtime. Among these
scheduling problems, some of them, for example the jobshop or the flowshop,
have been widely studied in the literature, as they represent many industrial
situations.

In this paper, we are specifically interested in the No-Wait Flowshop Scheduling Problem (NWFSP). This extension of the classical permutation Flowshop
Scheduling Problem (FSP) imposes that operations have to be processed without
any interruption between consecutive machines. This additional constraint aims
at describing real process constraints that may be found in the chemical industry for example. Regarding the solving of the problem, this additional constraint
also introduces interesting characteristics that we propose to exploit within the
design of an efficient constructive heuristic.

Indeed, while solving $NP - hard$ problems, the use of exact methods, mostly
based on enumerations, is not practicable. Therefore heuristics and metaheuristics are developed. Heuristics designed for a specific problem can use, for example, priority rules to construct the schedule. Their advantages are their speed –
most of them are greedy heuristics – and their specificity – characteristics of the
problem to be solved can be exploited. Metaheuristics, on their side, are generic

© Springer International Publishing AG 2017
R. Battiti et al. (Eds.): LION 2017, LNCS 10556, pp. 196–209, 2017.
https://doi.org/10.1007/978-3-319-69404-7_14

methods that can be applied to many optimization problems. Their efficiency is linked to their high ability of exploring the search space and the way the metaheuristic has been adapted to the problem under study. The initial solution, when one is required, may also influence the quality of the final solution provided.

Hence, as explained, the quality of solutions provided by heuristics and metaheuristics depends on the way these methods can integrate specificities of the problem under study. Several approaches exist to integrate such specificities. For example, it is possible to analyze off line characteristics of instances and to identify the best heuristic (or parameters of metaheuristic) for each kind of instances. Then, during the search, the method can identify the type of instance to solve and adapt the heuristic to use. This has been successfully used, for example, for a set partitioning problem [8]. Another approach deals with the integration of knowledge about interesting structures of solutions within the optimization method. In this context, the aim of the present work is to develop such a heuristic method for the No-Wait Flowshop Scheduling Problem. Thus, the contributions are the following: 1. analysis of interesting structure of optimal solutions, 2. integration of these observations in the design of a new constructive heuristic, 3. study the interest of the proposed approach when used alone or as the initialization of a metaheuristic.

The remainder of this paper is organized as follows. Section 2 introduces the problem formulation and provides a literature review on constructive heuristics for the NWFSP. Section 3 presents in details the proposed constructive heuristic. In Sect. 4 experiments are conducted and a comparison with efficient heuristics of the literature is provided. Some conclusions and perspectives for future works are given in Sect. 5.

2   The No-Wait Flowshop Scheduling Problem

2.1   Description of the Problem

The No-Wait Flowshop Scheduling Problem (NWFSP) is a variant of the well-known Permutation Flowshop Scheduling Problem (FSP), where no waiting time is allowed between the executions of a job on the successive machines. It may be defined as follows: Let J be a set of n jobs that have to be processed on a set of M ordered machines; $p_{i,j}$ is defined as the *processing time* of job i on machine j. A solution of the NWFSP is a sequence of jobs and so, is commonly represented by a permutation $\pi = \{\pi_1, \ldots, \pi_n\}$ where π_1 is the first job scheduled and π_n the last one. In this paper, the goal is to find a sequence that minimizes the makespan criterion (recorded as C_{max}) defined as the total completion time of the schedule.

Regarding the complexity of the problem, it has been proved by Wismer that the NWFSP can be viewed as an *Asymmetric Traveling Salesman Problem* (*ATSP*) [17]. In addition, Röck [13] proved that for m-machines ($m \geq 3$), the NWFSP is NP-hard while, the 2-machines case can be solved in $O(n * log\, n)$ [4].

However, the NWFSP possesses a characteristic not present in the classical FSP. Indeed, the start-up interval between two defined consecutive jobs of the sequence on the first machine is constant and does not depend on their position within the sequence. This interval, called the delay between two jobs, has been defined by Bertolissi [2]. Let $d_{i,i'}$ be the delay of two jobs i and i', it is computed as follows:

$$d_{i,i'} = p_{i,1} + \max_{1 \leq r \leq m} \left(\sum_{j=2}^{r} p_{i,j} - \sum_{j=1}^{r-1} p_{i',j}, 0 \right)$$

This allows to compute the completion time $C_i(\pi)$ of the job π_i of sequence π directly from the delays of the preceding jobs as follows:

$$C_i(\pi) = \sum_{k=2}^{i} d_{\pi_{k-1}, \pi_k} + \sum_{j=1}^{m} p_{\pi_i, j} \qquad (1)$$

where $i \in \{2, ..., n\}$. Note that $C_1(\pi)$ is the sum of the processing times on the m machines of the first scheduled job and then, is not concerned by the delay. Therefore, the makespan $(C_{max}(\pi))$ of a sequence π can be computed from Eq. (1) with a complexity of $O(n)$.

Experiments we will present, involve local search methods at different steps. These methods need a neighborhood operator to move from a solution to another. Such neighborhood operators are specific to the problem and more precisely to the representation of a solution. In this work, we use a permutation representation and a neighborhood structure based on the *insertion operator* [14]. For a sequence π, it consists in inserting a job π_i at position k ($i \neq k$). Hence, jobs between positions i and k are shifted. The size of this neighborhood, *i.e.*, the number of neighboring solutions, is $(n-1)^2$. Two sequences π and π' are said to be neighbors when they differ from exactly one insertion move. It is very interesting to note that exploiting the characteristics of the NWFSP, the makespan of π' can be directly computed from the makespan of π with a complexity of $O(1)$ [11].

2.2   State-of-the-Art

Many heuristics and metaheuristics have been proposed to solve scheduling problems and in particular the No-Wait Flowshop variant. The literature review proposed here, focuses on constructive heuristics as it is the scope of the article.

First, let us note that the well-known NEH (Nawaz, Enscore, Ham) heuristic, proposed for the classical permutation FSP [10], has been successfully applied on the No-Wait variant. Moreover, constructive heuristics have been designed specifically for the NWFSP. We may cite BIH (Bianco et al. [3]), BH (Bertolissi [2]) and RAJ (Rajendran [12]) among others. These heuristics define the order in which jobs are considered, regarding one criterion (e.g. decreasing order of sum of processing time of each job $i - \sum_m p_{i,m}$), but this may be improved with other jobs orders, even random. Indeed, in some contexts, it has been shown that repeating a random construction of solutions, may be more efficient than applying a constructive heuristic [1].

All these heuristics construct good quality solutions that can be further improved by a neighborhood-based metaheuristic, for example. In this paper, we focus on NEH and BIH as they provide high quality solutions and are considered as references for this problem. These two heuristics will be used later in the article for comparison.

The principle of NEH heuristic is to iteratively build a sequence π of jobs J. First, the n jobs of J are sorted by decreasing sums of processing times. Then, at each iteration, the first remaining unscheduled job is inserted in the partial sequence π in order to minimize the partial makespan. Algorithm 1 presents NEH.

Algorithm 1. NEH

Data: Set J of n jobs, π the sequence of jobs scheduled
$\pi = \emptyset$;
Sort the set J in decreasing sums of processing times;
for $k = 1$ **to** n **do**
⌊ Insert job J_k in π at the position, which minimizes the partial makespan.

This is a greedy heuristic; only n iterations are needed to build a sequence. At each iteration, a job J_k is inserted. The makespan of the new partial sequence, may be computed with a complexity of $O(n)$ thanks to the delay, when job J_k is inserted at the first position (see Sect. 2.1). Partial sequences, in which J_k is inserted at another position in the sequence, are neighbors of the previous one so, their makespan may be computed with a complexity of $O(1)$, according to the property exposed at the end of Sect. 2.1. Thus the complexity of one iteration of NEH is $O(n)$. Therefore, the complexity of the whole execution of NEH is $O(n^2)$.

BIH heuristic is another constructive heuristic based on the best insertion of a job in the partial sequence. The main difference with NEH is that all the unscheduled jobs are tested to find the best partial makespan. Indeed, while unscheduled jobs are remaining, for each one, all the possible positions to insert it into the partial sequence are evaluated; the job and the position that minimize the partial makespan are chosen. Algorithm 2 presents BIH.

If we consider the schedule of a job as one iteration, BIH builds a sequence with n iterations only. However, at each iteration, each job is a candidate to be inserted at one of all the possible positions to find the one that minimizes the partial makespan. For each job, finding the position that minimizes the partial makespan has a maximal complexity of $O(n)$ so, the complexity of one iteration of BIH is $O(n^2)$. Therefore, the complexity of the whole execution of BIH is $O(n^3)$.

BIH has been proposed in the literature after NEH in order to construct better quality solutions. However, the complexity of BIH is $O(n^3)$ whereas the one of NEH is $O(n^2)$. Thus, NEH is much faster than BIH, and for large size problems a compromise between computational time and quality is required.

Algorithm 2. BIH

Data: Set J of n jobs, π the sequence of jobs scheduled

$\pi = \emptyset$;

while $J \neq \emptyset$ **do**

> Find job $k \in J$, which can be inserted in the sequence π, such that the partial makespan is minimized. Let h be the best insertion position of job k in the sequence π;
>
> Insert job k at position h in the sequence π;
>
> Remove k from set J;

---

3   IBI: Iterated Best Insertion Heuristic

Following the construction's principles of previously exposed heuristics and after analyzing the structure of optimal solutions, we propose the Iterated Best Insertion (IBI) heuristic.

3.1   Analysis of Optimal Solutions Structure

Most of constructive heuristics are based on the best insertion principle. Indeed, the principle of the heuristic is to increase to increase, at each iteration, the problem size by one. It starts with a sub-problem of size one (only one job to schedule), then increases the size of the problem to solve (two jobs to schedule, three jobs...) until the size of the initial problem is reached.

In order to understand the dynamic of such a construction, we analyzed the structure of consecutive constructed sub-sequences and compare them to optimal sequences. Obviously this can be done only for first steps, as it is impossible to enumerate all the sequences, and find the optimal one, when the problem size is too large. Our proposition is to extend observations realized on sub-problems to the whole problem, in order to provide better solution. This approach can be compared to the Streamlined Constraint Reasoning, where the solution space is partitioned into sub-spaces whose structures are analyzed to better solve the whole problem [5].

Figure 1 gives an example starting from a sub-problem of size 8, $\mathcal{P}_8$, where 8 jobs are scheduled in the optimal order to minimize the makespan. Then, following the constructive principle, job 9 is inserted at the position that minimizes the makespan leading to the sub-problem $\mathcal{P}_9$ of size 9. This sequence corresponds to the one given by NEH strategy. When comparing the obtained sequence with the optimal solution of $\mathcal{P}_9$, we can observe that they are very close. Indeed, only two improving re-insertions (re-insertions of jobs 7 and 2) are needed to obtain this optimal solution from the NEH solution.

$\mathcal{P}_8$		2 7 8 4 3 1 5 6	$C^*_{max} = 350$	optimal solution
	insertion of job 9	2 7 **9** 8 4 3 1 5 6	$C_{max} = 447$	
$\mathcal{P}_9$	re-insertion of job 7	2 9 8 4 3 1 5 6 **7**	$C_{max} = 435$	
	re-insertion of job 2	9 8 4 3 1 5 6 7 **2**	$C^*_{max} = 427$	optimal solution

Fig. 1. Example of the evolution of the structure of the optimal solution from a sub-problem $\mathcal{P}_8$ of size 8 to a sub-problem $\mathcal{P}_9$ of size 9.

Experimental analysis on different instances have confirmed the following observations:

- at each step the structure of the obtained solution is not far from the structure of the optimal solution for the sub-problem,
- a few improving neighborhood applications may often lead to this optimal solution.

3.2 Design of IBI

Following these observations, we propose a new constructive heuristic called Iterated Best Insertion (IBI). At each iteration, two phases are successively achieved: (i) the first remaining unscheduled job is inserted in the partial sequence in order to minimize the partial makespan (same strategy than NEH) and (ii), an iterative improvement is performed on this partial solution to re-order jobs and to expect to be closer to the optimal solution of the sub-problem (see Algorithm 3). An iterative improvement is a method that moves, using a neighborhood operator, from a solution to one of its improving neighbors, which optimizes the quality. It naturally stops when a local optimum (a solution with no improving neighbors) is reached. Here, the neighborhood operator is based on the insertion and then possible move are applied while a better makespan is found.

Algorithm 3. IBI

Data: Set J of n jobs, π the sequence of jobs scheduled, σ criterion to sort J,
$\quad\quad$ *cycle* number of iterations without iterative improvement.
$\pi = \emptyset$;
Sort set J according to σ ;
for $k = 1$ **to** n **do**
$\quad$ Insert job J_k in π at the position which minimizes the partial makespan.;
$\quad$ **if** $k \equiv 0$ [*cycle*] **then**
$\quad\quad$ Perform an iterative improvement from π.

Two parameters have been introduced in the proposed method. First, σ a criterion to initially sort the set of jobs J. It will indicate in which order jobs will be considered during the construction. Second, *cycle* the number of iterations without applying the iterative improvement procedure. Indeed, even if the experimental analysis shows that the sequence built in phase (i) is not far from the optimum of the sub-problem, it is known that the exploration of the neighborhood is more and more time-consuming and the optimum more and more difficult to reach with the increase of the problem size. In order to control the time-performance of IBI, the *cycle* parameter allows the iterative improvement to be executed at a regular number of iterations only.

3.3   Experimental Analysis of Parameters

As the initial sort of IBI σ, may impact its performance as well as the *cycle* parameter, the performance of IBI is evaluated under these two parameters. This part, first presents the experimental protocol used, and then compare several versions of the IBI algorithm, using several initial sorts and finally analyses the impact of the cycle parameter.

Experimental Protocol. The benchmark used to evaluate the performance of the different variants of IBI is the Taillard's instances [15], initially provided for the flowshop problem but also widely used in the literature for the NWFSP. This benchmark proposes 120 instances, organized by 10 instances of 12 different sizes. Our experiments are conducted on the 10 available instances of size N_M where the number of jobs N is ranging from 20 to 200 and the number of machines M from 5 to 20 (total of 110 instances).

To compare the algorithms, the Relative Percentage Deviation (RPD) with the best known solution is computed. For a quality Q to minimize, the RPD is computed as follows:

$$RPD = \frac{Q(\text{Solution}) - Q(\text{Best known solution})}{Q(\text{Best known solution})} * 100 \tag{2}$$

Because of its Iterative Improvement phase, IBI is a stochastic method. Thus, 30 runs are required to allow making conclusions about results obtained. To follow this recommendation, all variants of the algorithm are executed 30 times per instance and the statistical Friedman test is performed on the RPD of all executions to compare algorithms.

Each algorithm is implemented using C++ and the experiments were executed on an Intel(R) Xeon(R) 3.5 GHz processor.

IBI: Initial Sort. Greedy heuristics, such as IBI, consider jobs in a given order. For example, in NEH, jobs are ordered by the decreasing sum of processing times. The specificity of the NWFSP gives other useful information as the GAP between two jobs on each machine that represents the idle time of the machine between the two jobs [9]. This measure does not depend on the schedule. Hence, for each machine, the GAP between each pair of jobs can be computed independently from the solving. The sum of GAPs for a pair of jobs represents the sum of their GAP on all the machines. To obtain a single value per job, we propose to define the *total GAP* of one job as the sum of the sum of GAPs. Hence, a high value of total GAP for a job indicates that the job has no good matching in the schedule. On the other hand, a low total GAP indicates that the job fits well with the others. In order to evaluate the impact of the initial sort and to define the most efficient one for the IBI heuristic, we experiment several initial sorts σ, based on (*i*) the total GAP or on (*ii*) the sum of processing times in decreasing or increasing order for both.

Table 1. Average RPD obtained on each size of Taillard's instances for different sorts: No Sort, Decreasing Mean GAP (DecrMeanGAP), Increasing Mean GAP (IncrMean-GAP), Decreasing sum of processing times (DecrPI) and the increasing sum of processing times (IncrPI). RPD values in bold stand for algorithms outperforming the other ones according to the statistical Friedman test.

Instance	NoSort	DecrMeanGAP	IncrMeanGAP	DecrPI	IncrPI
20_5	**1.79**	2.32	2.57	2.13	**1.38**
20_10	2.04	1.72	1.67	1.30	1.77
20_20	1.44	1.26	1.04	1.12	1.15
50_5	3.74	3.65	4.01	**3.23**	3.04
50_10	2.78	3.17	3.25	2.74	2.57
50_20	2.53	2.41	2.47	2.80	2.50
100_5	4.68	4.52	4.67	**4.13**	**4.12**
100_10	3.45	3.36	3.53	3.24	3.27
100_20	3.05	2.82	2.81	2.79	2.89
200_10	4.32	4.19	4.22	**3.81**	3.87
200_20	3.24	3.26	3.23	**3.03**	**2.99**

Table 1 provides a detailed comparison of the different initial sorts on the 110 Taillard's instances. This table shows that when a significant difference is observed the best initial sort is to consider the sum of processing times. Ordering jobs by increasing or decreasing order has no significant influence. Thus for the following, we choose the increasing sum of processing times as the initial sort of IBI.

IBI: Cycle. The use of an iterative improvement at each iteration can be time expensive. In order to minimize the execution time, we introduce a cycle. A cycle is a sequence of x iterations without any iterative improvement.

In Table 2, we study the impact of the size of the cycle on both the quality of solutions and the execution time.

These results show that the quality decreases with a large cycle size, but is obviously faster. As the objective of such a constructive heuristic is to provide in a very reasonable time a solution as good as possible, and as the time required here even for large instances stays reasonable, we propose not to use

Table 2. Average RPD and time (milliseconds) obtained on Taillard's instances for different sizes of cycle. RPD values in bold stand for algorithms outperforming the other ones according to the Friedman test. A cycle of size x, indicates the iterative improvement is executed every x iterations. Thus, a cycle of size n indicates the iterative improvement is executed only once, at the end of the construction.

Instance	1		2		5		10		n	
	RPD	Time	RPD	Time	RPD	Time	RPD	Time	RPD	Time
20_5	1.38	0.77	1.24	0.55	1.43	0.27	1.78	0.20	1.62	0.18
20_10	1.77	0.83	1.82	0.52	1.85	0.26	1.49	0.20	1.69	0.19
20_20	1.15	0.77	1.35	0.47	1.71	0.25	1.74	0.19	1.68	0.16
50_5	**3.04**	12.16	**3.25**	6.76	3.46	3.52	3.62	2.49	3.94	1.74
50_10	2.57	11.55	2.49	6.69	2.75	3.62	3.07	2.39	3.26	1.48
50_20	2.50	11.73	2.37	6.67	2.58	3.58	2.65	2.42	3.14	1.59
100_5	**4.12**	95.74	4.22	55.96	4.42	30.22	4.65	19.40	5.29	9.62
100_10	**3.27**	93.43	**3.27**	54.56	3.30	29.19	3.45	18.78	4.17	9.71
100_20	**2.89**	92.90	**2.93**	53.60	**3.07**	28.35	3.16	18.28	3.56	9.88
200_10	**3.87**	755.09	**3.91**	435.40	4.07	225.52	4.09	146.26	4.74	55.04
200_20	**2.99**	746.62	3.10	431.33	3.11	222.80	3.18	146.26	3.93	57.83

any cycle, that is to say to execute the iterative improvement at each step of the construction.

In conclusion of these experiments we propose to fix for the remainder of this work as IBI parameters: an initial sort based on the increasing sum of processing times and no cycle (*i.e.* cycle of size 1).

4   Experiments

The aim of this section is to analyze the efficiency of the proposed IBI method in two situations. First, IBI alone will be compared to other constructive heuristics of the literature, and in a second time these different heuristics will be used as initialization of a classical local search. Along this section, the same experimental protocol as before is used, and parameters used for IBI are those resulting from experiments of Sect. 3.3.

4.1   Efficiency of IBI

We choose to compare IBI with two other heuristics of the literature: NEH and BIH. Indeed, as exposed in Sect. 2.2 NEH and BIH are both interesting and efficient constructive heuristics for the NWFSP:

1. NEH is a classical heuristic for flow-shop problems and it has the advantage to be very fast. Moreover, the proposed method IBI can be viewed as an improvement of NEH, as an iterative improvement is added at each step.

2. BIH is a classical heuristic for the NWFSP and very efficient as, according to some preliminary experiments, it offers the best performance among several tested heuristics.

Table 3 shows the comparative study between IBI and the two other constructive heuristics. Both RPD and computation time are given to exhibit the gain in quality with regards to the time. NEH and BIH are deterministic and so, both values of an instance size are average ones computed from the ten RPD values obtained for the 10 instances respectively. On the other hand, IBI values correspond to the average ones over the 30 runs and the 10 instances of each instance size.

Table 3. RPD and time (milliseconds) obtained on Taillard's instances for NEH, BIH and IBI (average values). RPD values in bold stand for algorithms statistically outperforming the other ones according to the Friedman test.

Instance	NEH		BIH		IBI	
	RPD	Time	RPD	Time	RPD	Time
20_5	3.95	0.03	3.03	0.10	**1.38**	0.77
20_10	4.18	0.02	3.60	0.09	**1.77**	0.82
20_20	2.92	0.02	**1.74**	0.08	**1.15**	0.76
50_5	6.84	0.10	5.98	1.10	**3.04**	12.16
50_10	4.59	0.09	4.10	1.13	**2.57**	11.54
50_20	4.84	0.10	4.01	1.18	**2.50**	11.72
100_5	8.14	0.32	6.85	9.07	**4.12**	95.74
100_10	6.10	0.33	5.66	9.29	**3.27**	93.43
100_20	5.17	0.33	4.65	9.30	**2.89**	92.90
200_10	6.72	1.21	5.85	72.44	**3.87**	755.09
200_20	5.74	1.21	4.51	71.95	**2.99**	746.62

Solutions built by IBI have a better quality (lower RPD) than those built by NEH or BIH. It has statistically been verified with the Friedman Test. Beside, IBI was able to better perform than NEH or BIH for 106 instances over the 110 available instances. The counterpart of this performance is the computational time required to execute IBI regarding to the two other heuristics. However, this time remains reasonable as, less than one second is required even for larger instances. IBI is an efficient alternative to the classical heuristics used to build good quality solutions.

4.2   IBI as Initialization of a Local Search

A general use of constructive heuristics is to execute them as an initialization phase of meta-heuristics. In order to evaluate the pertinence of using IBI in such an initialization phase, it has been combined with a Tabu Search (TS). This local search is widely used to solve flowshop problems [6,7,16] and so is a good candidate to show the pertinence to build solutions with IBI rather than NEH or BIH. The three heuristics are combined with TS. In order to be fair, every combined approaches have the same running time fixed to $1000 * n$ ms (with n the number of jobs). These experiments aim at checking if the quality reached is enough to justify the use of IBI as initialization or if its *higher* execution time penalizes its use.

Table 4 presents results about the average RPD values among the 30 runs for each combined approach for each instance size. Results exposed in this table compared to those of Table 3 first indicate that the Tabu Search manages to improve solutions produced by the different heuristics. It also indicates that IBI used as an initialization phase, gets better results than the others, even if the difference is not always statistically proved. In particular for instances with a high number of jobs to schedule, IBI shows a very good performance.

Table 4. Average RPD on Taillard's instances for NEH+TS, BIH+TS and IBI+TS. RPD values in bold stand for algorithms statistically outperforming the other ones according to the Friedman test.

Instance	NEH+TS	BIH+TS	IBI+TS
20_5	1.12	1.30	0.83
20_10	1.30	1.35	1.39
20_20	1.15	0.71	0.73
50_5	3.17	3.49	2.63
50_10	2.68	2.53	2.15
50_20	2.53	2.41	2.03
100_5	4.66	4.55	**3.78**
100_10	3.70	3.48	**2.94**
100_20	2.97	2.93	2.62
200_10	4.25	4.20	**3.67**
200_20	3.56	**2.99**	**2.77**

As an illustration, Fig. 2 shows the evolution of the average makespan for a particular instance with 200 jobs and 20 machines (mean of the 30 runs). Regarding only the quality after the initialization, IBI is better than the two other heuristics, as shown in Table 3 previously. After the improvement of the Tabu Search for the two other heuristics, the final quality seems to be equivalent to the one obtained by IBI without applying the local search. That shows the efficiency of this approach. Finally, the Tabu Search does not help a lot IBI. Indeed, thanks to the successive local improvement in the second phase of IBI, the solution provided by IBI is already a local optimum. Therefore, it is difficult for the Tabu Search to improve the solution obtained with IBI.

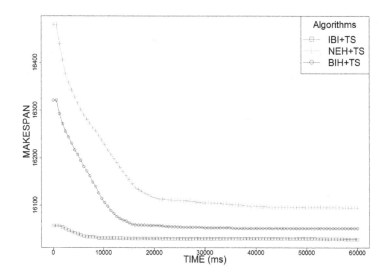

Fig. 2. Evolution of the average makespan on 30 runs for the 7th instance of Taillard (200 jobs, 20 machines) for the three combined approaches. IBI+TS is represented by squares, NEH+TS by crosses and BIH+TS by circles.

In conclusion, these experiments show the good performance of IBI for both aspects: as a constructive heuristics and as initialization of a meta-heuristic. In particular, they show the contribution of the local improvement to build the solution.

5   Conclusion and Perspectives

This work presents IBI, a new heuristic to minimize the makespan for the No-Wait Flowshop Scheduling Problem. IBI has been designed from the analysis of existing heuristics of the literature as well as of the structure of optimal solutions. Indeed, IBI is an improvement of the widely used heuristic (NEH), where an iterative improvement procedure has been added after each insertion of a job. Even if this additional procedure increases the computational time compared to other constructive heuristics of the literature (NEH and BIH), the improvement of the quality of the final solution built is noticeable and has statistically been validated with the Friedman test.

Then, we analyzed if this extra computational time is justified in term of quality obtained by evaluating the capacity of IBI to provide a good solution. IBI and the others heuristics have been used as initialization for a metaheuristic *i.e.*, the initial solution of the approach is a solution built by a heuristic. The experiments, on the Taillard instances, show that IBI helps the metaheuristic to be more efficient and the extra time needed to build an initial solution is not a drawback.

What is interesting in this work is that the addition of a simple modification of an existing constructive heuristic improves the results. This modification has been proposed from properties observed which make the application of an iterative improvement procedure at each step 1. useful – only a few steps are required to reach the optimal solutions for the small sub-problems – and 2. no time consuming–the makespan of a neighborhood solution can be computed in $O(1)$. We can now wonder whether such a modification can be performed to other constructive heuristics for other variants of the flowshop or more generally, for other problems (if they also present equivalent properties).

References

1. Balas, E., Carrera, M.C.: A dynamic subgradient-based branch-and-bound procedure for set covering. Oper. Res. **44**(6), 875–890 (1996)
2. Bertolissi, E.: Heuristic algorithm for scheduling in the no-wait flow-shop. J. Mater. Process. Technol. **107**(1–3), 459–465 (2000)
3. Bianco, L., Dell'Olmo, P., Giordani, S.: Flow shop no-wait scheduling with sequence dependent setup times and release dates. INFOR Inf. Syst. Oper. Res. **37**(1), 3–19 (1999)
4. Gilmore, P.C., Gomory, R.E.: Sequencing a one state-variable machine: a solvable case of the traveling salesman problem. Oper. Res. **12**(5), 655–679 (1964)
5. Gomes, C., Sellmann, M.: Streamlined constraint reasoning. In: Wallace, M. (ed.) CP 2004. LNCS, vol. 3258, pp. 274–289. Springer, Heidelberg (2004). doi:10.1007/978-3-540-30201-8_22
6. Grabowski, J., Pempera, J.: The permutation flow shop problem with blocking. A Tabu Search approach. Omega **35**(3), 302–311 (2007)
7. Grabowski, J., Wodecki, M.: A very fast Tabu Search algorithm for the permutation flow shop problem with makespan criterion. Comput. Oper. Res. **31**(11), 1891–1909 (2004)
8. Kadioglu, S., Malitsky, Y., Sellmann, M.: Non-model-based search guidance for set partitioning problems. In: Hoffmann, J., Selman, B. (eds.) Proceedings of the Twenty-Sixth AAAI Conference on Artificial Intelligence, Toronto, Ontario, Canada, 22–26 July 2012. AAAI Press (2012)
9. Nagano, M.S., Araújo, D.C.: New heuristics for the no-wait flowshop with sequence-dependent setup times problem. J. Braz. Soc. Mech. Sci. Eng. **36**(1), 139–151 (2013)
10. Nawaz, M., Enscore, E.E., Ham, I.: A heuristic algorithm for the m-machine, n-job flow-shop sequencing problem. Omega **11**(1), 91–95 (1983)
11. Pan, Q.-K., Wang, L., Tasgetiren, M.F., Zhao, B.-H.: A hybrid discrete particle swarm optimization algorithm for the no-wait flow shop scheduling problem with makespan criterion. Int. J. Adv. Manuf. Technol. **38**(3–4), 337–347 (2007)
12. Rajendran, C.: A no-wait flowshop scheduling heuristic to minimize makespan. J. Oper. Res. Soc. **45**(4), 472–478 (1994)
13. Röck, H.: The three-machine no-wait flow shop is NP-complete. J. ACM **31**(2), 336–345 (1984)
14. Schiavinotto, T., Stützle, T.: A review of metrics on permutations for search landscape analysis. Comput. Oper. Res. **34**, 3143–3153 (2007)
15. Taillard, E.: Benchmarks for basic scheduling problems. Eur. J. Oper. Res. **64**(2), 278–285 (1993)

16. Wang, C., Li, X., Wang, Q.: Accelerated tabu search for no-wait flowshop scheduling problem with maximum lateness criterion. Eur. J. Oper. Res. **206**(1), 64–72 (2010)
17. Wismer, D.A.: Solution of the flowshop-scheduling problem with no intermediate queues. Oper. Res. **20**(3), 689–697 (1972)

Sharp Penalty Mappings for Variational Inequality Problems

Evgeni Nurminski$^{(\boxtimes)}$

School of Natural Sciences, Far Eastern Federal University, Vladivostok, Russia
nurminski.ea@dvfu.ru

Abstract. First, this paper introduces a notion of a sharp penalty mapping which can replace more common exact penalty function for convex feasibility problems. Second, it uses it for solution of variational inequalities with monotone operators or pseudo-varitional inequalities with oriented operators. Appropriately scaled the sharp penalty mapping can be used as an exact penalty in variational inequalities to turn them into fixed point problems. Then they can be approximately solved by simple iteration method.

Keywords: Monotone variational inequalities · Oriented mappings · Sharp penalty mappings · Exact penalty mappings · Approximate solution

1 Introduction

Variational inequalities (VI) became one of the common tools for representing many problems in physics, engineering, economics, computational biology, computerized medicine, to name but a few, which extend beyond optimization, see [1] for the extensive review of the subject. Apart from the mathematical problems connected with the characterization of solutions and development of the appropriate algorithmic tools to find them, modern problems offer significant implementation challenges due to their non-linearity and large scale. It leaves just a few options for the algorithms development as it occurs in the others related fields like convex feasibility (CF) problems [2] as well. One of these options is to use fixed point iteration methods with various attraction properties toward the solutions, which have low memory requirements and simple and easily parallelized iterations. These schemes are quite popular for convex optimization and CF problems but they need certain modifications to be applied to VI problems. The idea of modification can be related to some approaches put forward for convex optimization and CF problems in [3–5] and which is becoming known as superiorization technique (see also [6] for the general description).

From the point of view of this approach the optimization problem

$$\min f(x), \quad x \in X \tag{1}$$

© Springer International Publishing AG 2017
R. Battiti et al. (Eds.): LION 2017, LNCS 10556, pp. 210–221, 2017.
https://doi.org/10.1007/978-3-319-69404-7_15

or VI problem to find $x^\star \in X$ such that

$$F(x^\star)(x - x^\star) \geq 0, \quad x \in X \qquad (2)$$

are conceptually divided into the feasibility problem $x \in X$ and the second-stage optimization or VI problems next. Then these tasks can be considered to a certain extent separately which makes it possible to use their specifics to apply the most suitable algorithms for feasibility and optimization/VI parts.

The problem is to combine these algorithms in a way which provides the solution of the original problems (1) or (2). As it turns out these two tasks can be merged together under rather reasonable conditions which basically require a feasibility algorithm to be resilient with respect to diminishing perturbations and the second-stage algorithm to be something like globally convergent over the feasible set or its small expansions.

Needless to say that this general idea meets many technical difficulties one of them is to balance in intelligent way the feasibility and optimization/VI steps. If optimization steps are essentially "smaller" than feasibility steps then it is possible to prove general convergence results [3, 4] under rather mild conditions. However it looks like that this requirements for optimization steps to be smaller (in fact even vanishing compared to feasibility) slows down the overall optimization in (1) considerably.

This can be seen in the text-book penalty function method for (1) which consists in the solution of the auxiliary problem of the kind

$$\min_{x}\{\, \Phi_X(x) + \epsilon f(x) \,\} = \Phi_X(x_\epsilon) + \epsilon f(x_\epsilon) \qquad (3)$$

where $\Phi_X(x) = 0$ for $x \in X$ and $\Phi_X(x) > 0$ otherwise. The term $\epsilon f(x)$ can be considered as the perturbation of the feasibility problem $\min_x \Phi(x)$ and for classical smooth penalty functions the penalty parameter $\epsilon > 0$ must tend to zero to guarantee convergence of x_ϵ to the solution of (1). Definitely it makes the objective function $f(x)$ less influential in solution process of (1) and hinders the optimization.

To overcome this problem the exact penalty functions $\Psi_X(\cdot)$ can be used which provide the exact solution of (1)

$$\min_{x}\{\, \Psi_X(x) + \epsilon f(x) \,\} = \epsilon f(x^\star) \qquad (4)$$

for small enough $\epsilon > 0$ under rather mild conditions. The price for the conceptual simplification of the solution of (2) is the inevitable non-differentiability of the penalty function $\Psi_X(x)$ and the corresponding worsening of convergence rates for instance for gradient-like methods (see [8, 9] for comparison). Nevertheless the idea has a certain appeal, keeping in mind successes of nondiffereniable optimization, and the similar approaches with necessary modifications were used for VI problems starting from [10] and followed by [11–14] among others. In these works the penalty functions were introduced and their gradient fields direct the iterations to feasibility.

Here we suggest a more general definition of a sharp penalty mapping P : $E \to \mathcal{C}(E)$, not necessarily potential, which is oriented toward a feasible set (for details of notations see the Sect. 2). It also admits a certain problem-dependent penalty constant $\lambda > 0$ such that the sum $F + \lambda P$ of variational operator F of (2) and P scaled by λ possesses a desirable properties to make the iteration algorithm converge at least to a given neighborhood of solution of (2). In the preliminary form this idea was suggested in [15] using a different definition of a sharp penalty mapping which resulted in rather weak convergence result. Here we show that it is possible to reach stronger convergence result with all limit points of the iteration method being an ϵ-solutions of VI problem (2).

2   Notations and Preliminaries

Let E denotes a finite-dimensional space with the inner product xy for $x, y \in E$, and the standard Euclidean norm $\|x\| = \sqrt{xx}$. The one-dimensional E is denoted as $\mathbb{R}$ and $\mathbb{R}_\infty = \mathbb{R} \cup \{\infty\}$. The unit ball in E is denoted as $B = \{x : \|x\| \leq 1\}$. The space of bounded closed convex subsets of E is denoted as $\mathcal{C}(E)$. The distance function $\rho(x, X)$ between point x and set $X \subset E$ is defined as $\rho(x, X) = \inf_{y \in X} \|x - y\|$. The norm of a set X is defined as $\|X\| = \sup_{x \in X} \|x\|$.

For any $X \subset E$ its interior is denoted as $\mathrm{int}(X)$, the closure of X is denoted as $\mathrm{cl}(X)$ and the boundary of X is denoted as $\partial X = X \setminus \mathrm{int}(X)$.

The sum of two subsets A and B of E is denoted as $A + B$ and understood as $A + B = \{a + b, a \in A, b \in B\}$. If A is a singleton $\{a\}$ we write just $a + B$.

Any open subset of E containing zero vector is called a neighborhood of zero in E. We use the standard definition of upper semi-continuity and monotonicity of set-valued mappings:

Definition 1. *A set-valued mapping $F : E \to \mathcal{C}(E)$ is called upper semi-continuous if at any point $\bar{x}$ for any neighborhood of zero U there exists a neighborhood of zero V such that $F(x) \subset F(\bar{x}) + U$ for all $x \in \bar{x} + V$.*

Definition 2. *A set-valued mapping $F : E \to \mathcal{C}(E)$ is called a monotone if $(f_x - f_y)(x - y) \geq 0$ for any $x, y \in E$ and $f_x \in F(x), f_y \in F(y)$.*

We use standard notations of convex analysis: if $h : E \to \mathbb{R}_\infty$ is a convex function, then $\mathrm{dom}(h) = \{x : h(x) < \infty\}$, $\mathrm{epi}\, h = \{(\mu, x) : \mu \geq h(x), x \in \mathrm{dom}(h)\} \subset \mathbb{R} \times E$, the sub-differential of h is defined as follows:

Definition 3. *For a convex function $h : E \to \mathbb{R}_\infty$ a sub-differential of h at point $\bar{x} \in \mathrm{dom}(h)$ is the set $\partial h(\bar{x})$ of vectors g such that $h(x) - h(\bar{x}) \geq g(x - \bar{x})$ for any $x \in \mathrm{dom}(h)$.*

This defines a convex-valued upper semi-continuous maximal monotone set-valued mapping $\partial h : \mathrm{int}(\mathrm{dom}(h)) \to \mathcal{C}(E)$. At the boundaries of $\mathrm{dom}(h)$ the sub-differential of h may or may not exist. For differentiable $h(x)$ the classical gradient of h is denoted as $h'(x)$.

We define the convex envelope of $X \subset E$ as follows.

Definition 4. *An inclusion-minimal set $Y \in \mathcal{C}(E)$ such that $X \subset Y$ is called a convex envelope of X and denoted as $\mathrm{co}(X)$.*

Our main interest consists in finding a solution $x^\star$ of a following finite-dimensional VI problem with a single-valued operator $F(x)$:

$$\text{Find } x^\star \in X \subset \mathcal{C}(E) \text{ such that } F(x^\star)(x - x^\star) \geq 0 \text{ for all } x \in X. \tag{5}$$

This problem has its roots in convex optimization and for $F(x) = f'(x)$ VI (5) is the geometrical formalization of the optimality conditions for (1).

If F is monotone, then the pseudo-variational inequality (PVI) problem

$$\text{Find } x^\star \in X \text{ such that } F(x)(x - x^\star) \geq 0 \text{ for all } x \in X. \tag{6}$$

has a solution $x^\star$ which is a solution of (5) as well. However it is not necessary for F to be monotone to have a solution of (6) which coincides with a solution of (5) as Fig. 1 demonstrates.

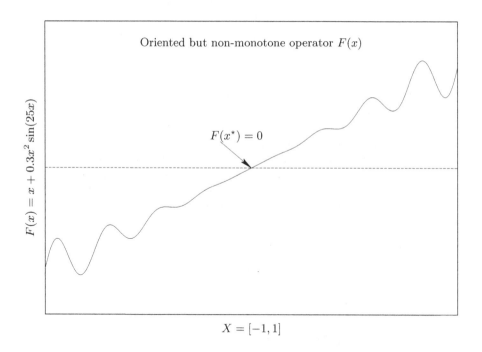

Fig. 1. Non-monotone operator $F(x)$ oriented toward $x^\star = 0$.

For simplicity we assume that both problems (5) and (6) has unique and hence coinciding solutions.

To have more freedom to develop iteration methods for the problem (6) we introduce the notions of oriented and strongly oriented mappings according to the following definitions.

Definition 5. *A set-valued mapping $G : E \to \mathcal{C}(E)$ is called oriented toward $\bar{x}$ at point x if*

$$g_x(x - \bar{x}) \geq 0 \tag{7}$$

for any $g_x \in G(x)$.

Definition 6. *A set-valued mapping $G : E \to \mathcal{C}(E)$ is called strongly oriented toward $\bar{x}$ on a set X if for any $\epsilon > 0$ there is $\gamma_\epsilon > 0$ such that*

$$g_x(x - \bar{x}) \geq \gamma_\epsilon \tag{8}$$

for any $g_x \in G(x)$ and all $x \in X \setminus \{\bar{x} + \epsilon B\}$.

If G is oriented (strongly oriented) toward $\bar{x}$ at all points $x \in X$ then we will call it oriented (strongly oriented) toward $\bar{x}$ on X.

Of course if $\bar{x} = x^\star$, a solution of PVI problem (6), then G is oriented toward $x^\star$ on X by definition and the other way around.

The notion of oriented mappings is somewhat related to attractive mappings introduced in [2], which can be defined for our purposes as follows.

Definition 7. *A mapping $F : E \to E$ is called attractive with respect to $\bar{x}$ at point x if*

$$\|F(x) - \bar{x}\| \leq \|x - \bar{x}\| \tag{9}$$

It is easy to show that if F is an attractive mapping, then $G(x) = F(x) - x$ is an oriented mapping, however $G(x) = -10x$ is the oriented mapping toward $\{0\}$ on $[-1, 1]$ but neither $G(x)$ nor $G(x) + x$ are attractive.

Despite the fact that the problem (5) depends upon the behavior of F on X only, we need to make an additional assumption about global properties of F to avoid certain problems with possible divergence of iteration method due to "run-away" effect. Such assumption is the long-range orientation of F which is frequently used to ensure the desirable global behavior of iteration methods.

Definition 8. *A mapping $F : E \to E$ is called long-range oriented toward a set X if there exists $\rho_F \geq 0$ such that for any $\bar{x} \in X$*

$$F(x)(x - \bar{x}) > 0 \text{ for all } x \text{ such that } \|x\| \geq \rho_F. \tag{10}$$

We will call ρ_F the radius of long-range orientation of F toward X.

3   Sharp Penalty Mappings

In this section we present the key construction which makes possible to reduce an approximate solution of VI problem into calculation of the limit points of iterative process, governed by strongly oriented operators.

For this purpose we modify slightly the classical definition of a polar cone of a set X.

Definition 9. *The set $K_X(x) = \{p : p(x - y) \geq 0 \text{ for all } y \in X\}$ we will call the polar cone of X at a point x.*

Of course $K_X(x) = \{0\}$ if $x \in \text{int } X$.

For our purposes we need also a stronger definition which defines a certain sub-cone of $K_X(x)$ with stronger pointing toward X.

Definition 10. Let $\epsilon \geq 0$ and $x \notin X + \epsilon B$. The set

$$K_X^\epsilon(x) = \{p : p(x - y) \geq 0 \text{ for all } y \in X + \epsilon B\} \qquad (11)$$

will be called ϵ-strong polar cone of X at x.

As it is easy to see that the alternative definition of $K_X^\epsilon(x)$ is $K_X^\epsilon(x) = \{p : p(x - y) \geq \epsilon \|p\|$ for all $y \in X.\}$

To define a sharp penalty mapping for the whole space E we introduce a composite mapping

$$\tilde{K}_X^\epsilon(x) = \begin{cases} \{0\} & \text{if } x \in X \\ K_X(x) & \text{if } x \in \text{cl}\{\{X + \epsilon B\} \setminus X\} \\ K_X^\epsilon(x) & \text{if } x \in \rho_F B \setminus \{X + \epsilon B\} \end{cases} \qquad (12)$$

Notice that $\tilde{K}_X^\epsilon(x)$ is upper semi-continuous by construction.

Now we define a sharp penalty mapping for X as

$$P_X^\epsilon(x) = \{p \in \tilde{K}_X^\epsilon(x), \|p\| = 1\}. \qquad (13)$$

Clear that $P_X^\epsilon(x)$ is not defined for $x \in \text{int}\{X\}$ but we can defined it to be equal to zero on $\text{int}\{X\}$ and take a convex envelope of $P_X^\epsilon(x)$ and $\{0\}$ at the boundary of X to preserve upper semi-continuity.

For some positive λ define $F_\lambda(x) = F(x) + \lambda P_X^\epsilon(x)$. Of course by construction $F_\lambda(x)$ is upper semi-continuous for $x \notin X$.

For the further development we establish the following result on construction of an approximate globally oriented mapping related to the VI problem (5).

Lemma 1. Let $X \subset E$ is closed and bounded, F is monotone and long-range oriented toward X with the radius of orientability ρ_F and strongly oriented toward solution $x^\star$ of (5) on X with the constants $\gamma_\epsilon > 0$ for $\epsilon > 0$, satisfying (8) and $P_X^\epsilon(\cdot)$ is a sharp penalty (13). Then for any sufficiently small $\epsilon > 0$ there exists $\lambda_\epsilon > 0$ and $\delta_\epsilon > 0$ such that for all $\lambda \geq \lambda_\epsilon$ a penalized mapping $F_\lambda(x) = F(x) + \lambda P_X^\epsilon(x)$ satisfies the inequality

$$f_x(x - x^\star) \geq \delta_\epsilon \qquad (14)$$

for all $x \in \rho_F B \setminus \{x^\star + \epsilon B\}$ and any $f_x \in F_\lambda(x)$.

Proof. For monotone F we can equivalently consider a pseudo-variational inequality (6) with the same solution $x^\star$. Define the following subsets of E:

$$\begin{aligned} X_\epsilon^{(1)} &= X \setminus \{x^\star + \epsilon B\}, \\ X_\epsilon^{(2)} &= \{\{X + \epsilon B\} \setminus X\} \setminus \{x^\star + \epsilon B\}, \\ X_\epsilon^{(3)} &= \rho_F B \setminus \{\{X + \epsilon B\} \setminus \{x^\star + \epsilon B\}\}. \end{aligned} \qquad (15)$$

Correspondingly we consider 3 cases.

Case A. $x \in X_\epsilon^{(1)}$. In this case $f_\lambda(x) = F(x)$ and therefore

$$f_\lambda(x)(x - x^\star) = F(x)(x - x^\star) \geq \gamma_\epsilon > 0. \tag{16}$$

Case B. $x \in X_\epsilon^{(2)}$. In this case $f_\lambda(x) = F(x) + \lambda p_X(x)$ where $p_X(x) \in K_X(x)$, $\|p_X(x)\| = 1$ and therefore

$$f_\lambda(x)(x - x^\star) = F(x)(x - x^\star) + \lambda p_X(x)(x - x^\star) \geq \gamma_\epsilon/2 > 0. \tag{17}$$

as $\lambda p_X(x)(x - x^\star) > 0$ by construction.

Case C. $x \in X_\epsilon^{(3)}$. In this case $f_\lambda(x) = F(x) + \lambda p_X(x)$ where $p_X(x) \in K_X^\epsilon(x)$, $\|p_X(x)\| = 1$. By continuity of F the norm of F is bounded on $\rho_F B$ by some M and as $P_X^\epsilon(\cdot)$ is ϵ-strong penalty mapping

$$\begin{aligned} f_\lambda(x)(x - x^\star) = F(x)(x - x^\star) + \lambda p_X(x)(x - x^\star) \geq \\ -M\|x - x^\star\| + \lambda\epsilon \leq -2\rho_F M + \lambda\epsilon \geq \rho_F M > 0 \end{aligned} \tag{18}$$

for $\lambda \geq \rho_F M/\epsilon$.

By combining all three bounds we obtain

$$f_\lambda(x)(x - x^\star) \geq \min\{\gamma_\epsilon/2, \rho_F M\} = \delta_\epsilon > 0 \tag{19}$$

for $\lambda \geq \Lambda_\epsilon = \rho_F M/\epsilon$ which completes the proof. ■

The elements of a polar cone for a given set X can be obtained by different means. The most common are either by projection onto set X:

$$x - \Pi_X(x) \in K_X(x) \tag{20}$$

where $\Pi_X(x) \in X$ is the orthogonal projection of x on X, or by subdifferential calculus when X is described by a convex inequality $X = \{x : h(x) \leq 0\}$. If there is a point $\bar{x}$ such that $h(\bar{x}) < 0$ (Slater condition) then $h(y) < 0$ for all $y \in \text{int}\{X\}$. Therefore $0 < h(x) - h(y) \leq g_h(x)(x - y)$ for any $y \in \text{int}\{X\}$. By continuity $0 < h(x) - h(y) \leq g_h(x)(x - y)$ for all $y \in X$ which means that $g_h \in K_X(x)$.

One more way to obtain $g_h \in K_X(x)$ relies on the ability to find some $x^c \in \text{int}\{X\}$ and use it to compute Minkowski function

$$\mu_X(x, x^c) = \inf_{\theta \geq 0}\{\theta : x^c + (x - x^c)\theta^{-1} \in X\} > 1 \text{ for } x \notin X. \tag{21}$$

Then by construction $\bar{x} = x^c + (x - x^c)\mu_X(x, x^c)^{-1} \in \partial X$, i.e. $h(\bar{x}) = 0$ and for any $g_h \in \partial h(\bar{x})$ the inequality $g_h\bar{x} \geq g_h y$ holds for any $y \in X$.

By taking $y = x^c$ obtain $g_h\bar{x} \geq g_h x^c$ and therefore

$$\begin{aligned} g_h\bar{x} = g_h x^c + g_h(x - x^c)\mu_X(x, x^c)^{-1} = \mu_X(x, x^c)^{-1}g_h x + (1 - \mu_X(x, x^c)^{-1})g_h x^c \leq \\ \mu_X(x, x^c)^{-1}g_h x + (1 - \mu_X(x, x^c)^{-1})g_h\bar{x}. \end{aligned} \tag{22}$$

Hence $g_h x \geq g_h\bar{x} \geq g_h y$ for any $y \in X$, which means that $g_h \in K_X(x)$.

As for ϵ-expansion of X it can be approximated from above (included into) by the relaxed inequality $X + \epsilon B \subset \{x : h(x) \leq L\epsilon\}$ where L is a Lipschitz constant in an appropriate neighborhood of X.

4   Iteration Algorithm

After construction of the mapping F_λ, oriented toward solution $x^\star$ of (6) at the whole space E except ϵ-neighborhood of $x^\star$ we can use it in an iterative manner like

$$x^{k+1} = x^k - \theta_k f^k, \ f^k \in F_\lambda(x^k), \ k = 0, 1, \ldots, \tag{23}$$

where $\{\theta_k\}$ is a certain prescribed sequence of step-size multipliers, to get the sequence of $\{x^k\}, k = 0, 1, \ldots$ which hopefully converges under some conditions to to at least the set $X_\epsilon = x^\star + \epsilon B$ of approximate solutions of (5).

For technical reasons, however, it would be convenient to guarantee from the very beginning the boundedness of $\{x^k\}, k = 0, 1, \ldots$. Possibly the simplest way to do so is to insert into the simple scheme (23) a safety device, which enforces restart if a current iteration x^k goes too far. This prevents the algorithm from divergence due to the "run away" effect and it can be easily shown that it keeps a sequence of iterations $\{x^k\}$ bounded.

Thus the final form of the algorithm is shown as the figure Algorithm 1, assuming that the set X, the operator F and the sharp penalty mapping P_X^ϵ satisfy conditions of the Lemma 1. We prove convergence of the Algorithm 1 under common assumptions on step sizes: $\theta_k \to +0$ when $k \to \infty$ and $\sum_{k=1}^{K} \theta_k \to \infty$ when $K \to \infty$. This is not the most efficient way to control the algorithm, but at the moment we are interested mostly in the very fact of convergence.

Data: The variational inequality operator F, sharp penalty mapping P_X, positive constant ϵ, penalty constant λ which satisfy conditions of the Lemma 1, long-range orientation radius ρ_F, a sequence of step-size multipliers $\{0 < \theta_k, k = 0, 1, 2, \ldots\}$. and an initial point $x^0 \in \rho_F B$.

Result: The sequence of approximate solutions $\{x^k\}$ where every converging sub-sequence has a limit point which belongs to a set X_ϵ of ϵ-solution of variational inequality (5).

Initialization;

Define penalized mapping

$$F_\lambda(x) = F(x) + \lambda P_X(x), \tag{24}$$

and set the iteration counter k to 0;

while *The limit is not reached* **do**

  Generate a next approximate solution x_{k+1}:

$$x^{k+1} = \begin{cases} x^k - \theta_k f^k, \ f^k \in F_\lambda(x^k), & \text{if } \|x^k\| \le 2\rho_F \\ x^0 & \text{otherwise.} \end{cases} \tag{25}$$

  Increment iteration counter $k \longrightarrow k + 1$;

end

Complete: accept $\{x^k\}, k = 0, 1, \ldots$ as an approximate solution of (5).[a]

[a]The exact meaning of this will be clarified in the convergence Theorem 1.

Algorithm 1. The generic structure of a conceptual version of the iteration algorithm with exact penalty.

Theorem 1. *Let $\epsilon > 0, \Lambda_\epsilon, F, P_X$ satisfy the assumptions of the Lemma 1, $\lambda > \Lambda_\epsilon$, and $\theta_k \to +0$ when $k \to \infty$ and $\sum_{k=1}^K \theta_k \to \infty$ when $K \to \infty$. Then all limit points of the sequence $\{x^k\}$ generated by the Algorithm 1 belong to the set of ϵ-solutions $X_\epsilon = x^\star + \epsilon B$ of the problem (5).*

Proof. We show first the boundedness of the sequence $\{x^k\}$. To do so it is sufficient to demonstrate that the sequence $\{\|x^k\|, k = 1, 2, \dots\}$ crosses the interval $[\rho_F, 2\rho_F]$ a finite number of times only (any way, from below or from above). Show first that the sequence $\{x^k\}$ leaves the set $\frac{3}{2}\rho_F B$ a finite number of times only. Define (a finite or not) set T of indices $T = \{t_k, k = 1, 2, \dots\}$ such that

$$\|x^\tau\| < \frac{3}{2}\rho_F \text{ and } \|x^{\tau+1}\| \geq \frac{3}{2}\rho_F. \tag{26}$$

If $\tau \in T$ then

$$\|x^{\tau+1}\|^2 = \|x^\tau - \theta_k F_\Lambda(x^\tau)\|^2 = \|x^\tau\|^2 - 2\theta_\tau f^\tau x^\tau + \theta_\tau^2 \|f^\tau\|^2 \leq \\ \|x^\tau\|^2 - 2\theta_\tau f^\tau x^\tau + \theta_\tau^2 C^2 \leq \|x^\tau\|^2 - 2\theta_\tau \gamma\delta + \theta_\tau^2 C^2, \tag{27}$$

where C is an upper bound for $\|F_\lambda(x)\|$ with $x \in 2\rho_F B$ and $\gamma > 0$ is a lower bound for $f^\tau x$ for $x \in 2\rho_F B$ and $f^\tau \in F_\lambda(x)$. For τ large enough $\theta_\tau < \gamma C^{-2}$ and hence

$$\|x^{\tau+1}\|^2 \leq \|x^\tau\|^2 - \theta_\tau \delta \overset{\bullet}{<} \|x^\tau\|^2 \tag{28}$$

which contradicts the definition of the set T. Therefore T is a finite set and the sequence $\{x^k\}$ leaves the set $\frac{3}{2}\rho_F B$ a finite number of times only which proves the boundedness of $\{x^k\}$.

Define now $W(x) = \|x - x^\star\|^2$ and notice that due to the boundedness of $\{x^k\}$ and semi-continuity of $F_\lambda(x)$ and etc., $W(x^{k+1}) - W(x^k) \to 0$ when $k \to \infty$. It implies that the limit set

$$W_\star = \{w_\star : \text{ the sub-sequence } \{x^{k_s}\} \text{ exists such that } \lim_{s \to \infty} W(x^{k_s}) = w_\star\} \tag{29}$$

is a certain interval $[w_\star^l, w_\star^u] \subset \mathbb{R}_+$ and the statement of the theorem means that $w_\star^u \leq \epsilon^2$.

To prove this we assume contrary, that is $w_\star^u > \epsilon^2$ and hence there exists a sub-sequence $\{x^{k_s}, s = 1, 2, \dots\}$ such that $\lim_{s \to \infty} W(x^{k_s}) = w' > \epsilon^2$. Without loss of generality we may assume that $\lim_{s \to \infty} x^{k_s} = x'$ and of course $x' \notin X_\epsilon$. Therefore $f'(x' - x^\star) > 0$ for any $f' \in F_\lambda(x')$ and by upper semi-continuity of F_λ there exists an $\upsilon > 0$ such that $F_\lambda(x)(x - x^\star) \geq \delta$ for all $x \in x' + 4\upsilon B$ and some $\delta > 0$. Again without loss of generally we may assume that $\upsilon < (\sqrt{w'} - \epsilon)/4$ so $(x' + 4\upsilon B) \cap (x^\star + \epsilon B) = \emptyset$.

For for s large enough $x^{k_s} \in x' + \upsilon B$ and let us assume that for all $t > k_s$ the sequence $\{x^t, t > k_s\} \subset x^{k_s} + \upsilon B \subset x' + 2\upsilon B$.

Then

$$W(x^{t+1}) = \|x^t - \theta_t F_\lambda(x^t) - x^\star\|^2 = W(x^t) - 2\theta_t F_\lambda(x^t)(x^t - x^\star) + \theta_t^2 \|F_\lambda(x^t)\|^2 \leq \\ W(x^t) - 2\theta_t F_\lambda(x^t)(x^t - x^\star) + \theta_t^2 C^2 \leq W(x^t) - 2\theta_t \delta + \theta_t^2 C^2 < W(x^t) - \theta_t \delta, \tag{30}$$

for all $t > k_s$ and s large enough that $\sup_{t > k_s} \theta_t < \delta/C^2$. Summing up last inequalities from $t = k_s$ to $t = T - 1$ obtain

$$W(x^T) \leq W(x^{k_s}) - \delta \sum_{t=k_s}^{T-1} \theta_t \to -\infty \tag{31}$$

when $T \to \infty$ which is of course impossible.

Hence for each k_s there exists $r_s > k_s$ such that $\|x^{k_s} - x^{r_s}\| > v > 0$ Assume that r_s is in fact a minimal such index, i.e. $\|x^{k_s} - x^t\| \leq v$ for all t such that $k_s < t < r_s$ or $x^t \in x^{t_k} + vB \subset x' + 2vB$ for all such t. Without any loss of generality we may assume that $x^{r_s} \to x''$ where by construction $\|x' - x''\| \geq v > 0$ and therefore $x' \neq x''$.

As all conditions which led to (31) hold for $T = r_s$ then by letting $T = r_s$ we obtain

$$W(x^{r_s}) \leq W(x^{k_s}) - \delta \sum_{t=k_s}^{r_s-1} \theta_t. \tag{32}$$

On the other hand

$$v < \|x^{k_s} - x^{r_s}\| \leq \sum_{t=k_s}^{r_s} \|x^{t+1} - x^t\| \leq \sum_{t=k_s}^{r_s-1} \theta_t \|F_\lambda(x^t)\| \leq K \sum_{t=k_s}^{r_s} \theta_t \tag{33}$$

where K is the upper estimate of the norm of $F_\lambda(x)$ on $2\rho_F B$.

Therefore $\sum_{t=k_s}^{r_s-1} \theta_t > v/K > 0$ and finally

$$W(x^{r_s}) \leq W(x^{k_s}) - \delta v/K. \tag{34}$$

Passing to the limit when $s \to \infty$ obtain $W(x'') \leq W(x') - \delta v/K < W(x')$ Also $W(x'') > \epsilon^2$ as $x'' \in x' + 4vB$ which does not intersect with $x^\star + \epsilon B$. To save on notations denote $W(x') = w'$ and $W(x'') = w''$.

In other words, assuming that $w' > \epsilon^2$ we constructed another limit point w'' of the sequence $\{W(x^k)\}$ such that $\epsilon^2 < w'' < w'$. It follows from this that the sequence $\{W(x^k)\}$ infinitely many times crosses any sub-interval $[\tilde{w}'', \tilde{w}'] \subset (w'', w')$ both in "up" and "down" directions and hence there exist sub-sequences $\{p_s, s = 1, 2, \dots\}$ and $\{q_s, s = 1, 2, \dots\}$ such that $p_s < q_s$ and

$$W(x^{p_s}) \leq \tilde{w}'', W(x^{q_s}) \geq \tilde{w}', W(x^t) \in (w'', w') \text{ for } p_s < t < q_s \tag{35}$$

Then

$$0 < W(x^{q_s}) - W(x^{p_s}) = \sum_{t=p_s}^{q_s-1} (W(x^{t+1}) - W(x^t)) \tag{36}$$

and hence for any s there is an index $t_s : p_s < t_s < q_s$ such that

$$0 < W(x^{t_s+1}) - W(x^{t_s}). \tag{37}$$

However as $W(x^{t_s}) > w''$, $x^{t_s} \notin x^\star + \epsilon B$ and therefore

$$
\begin{aligned}
W(x^{t_s+1}) - W(x^{t_s}) &= \|x^{t_s+1} - x^\star\|^2 - \|x^{t_s} - x^\star\|^2 = \\
&\quad \|x^{t_s} - x^\star + \theta_{t_s} f^{t_s}\|^2 - \|x^{t_s} - x^\star\|^2 = \\
2\theta_{t_s} f^{t_s}(x^{t_s} - x^\star) &+ \theta_{t_s}^2 \|f^{t_s}\|^2 = \theta_{t_s}(2f^{t_s}(x^{t_s} - x^\star) + \theta_{t_s}\|f^{t_s}\|^2),
\end{aligned}
\tag{38}
$$

where $f^{t_s} \in F_\lambda(x^{t_s})$. Notice that $f^{t_s}(x^{t_s} - x^\star) < -\delta > 0$ and $\|f^{t_s}\|^2 \leq C$. Using these estimates we obtain

$$
W(x^{t_s+1}) - W(x^{t_s}) \leq \theta_{t_s}(-2\delta - \theta_{t_s}C) \leq -\theta_{t_s}\delta < 0
\tag{39}
$$

for all s large enough. This contradicts (37) and therefore proves the theorem.   ∎

5   Conclusions

In this paper we define and use a sharp penalty mapping to construct the iteration algorithm converging to an approximate solutions of monotone variational inequalities. Sharp penalty mappings are analogues of gradient fields of exact penalty functions but do not need to be potential mappings. Three examples of sharp penalty mappings are given with one of them seems to be a new one. The algorithm consists in recursive application of a penalized variational inequality operator, but scaled by step-size multipliers which satisfy classical diverging series condition. As for practical value of these result it is generally believed that the conditions for the step-size multipliers used in this theorem result in rather slow convergence of the order $O(k^{-1})$. However the convergence rate can be improved by different means following the example of non-differentiable optimization. The promising direction is for instance the least-norm adaptive regulation, suggested probably first by A. Fiacco and McCormick [16] as early as 1968 and studied in more details in [17] for convex optimization problems. With some modification in can be easily used for VI problems as well. Experiments show that under favorable conditions it produces step multipliers decreasing as geometrical progression which gives a linear convergence for the algorithm. This may explain the success of [7] where geometrical progression for step multipliers was independently suggested and tested in practice.

Acknowledgements. This work is supported by the Ministry of Science and Education of Russian Federation, project 1.7658.2017/6.7

References

1. Facchinei, F., Pang, J.-S.: Finite-dimensional Variational Inequalities and Complementarity Problems. Springer, Berlin (2003)
2. Bauschke, H., Borwein, J.: On projection algorithms for solving convex feasibility problems. SIAM Rev. **38**, 367–426 (1996)

3. Nurminski, E.A.: Féjer processes with diminishing disturbances. Doklady Math. **78**(2), 755–758 (2008)
4. Nurminski, E.A.: Use of additional diminishing disturbances in Féjer models of iterative algorithms. Comput. Math. Math. Phys. **48**(12), 2154–2161 (2008)
5. Censor, Y., Davidi, R., Herman, G.T.: Perturbation resilience and superiorization of iterative algorithms. Inverse Probl. **26** (2010). doi:10.1088/0266-5611/26/6/065008
6. Censor, Y., Davidi, R., Herman, G.T., Schulte, R.W., Tetruashvili, L.: Projected subgradient minimization versus superiorization. J. Optim. Theory Appl. **160**, 730–747 (2014)
7. Butnariu, D., Davidi, R., Herman, G.T., Kazantsev, I.G.: Stable convergence behavior under summable perturbations of a class of projection methods for convex feasibility and optimization problems. IEEE J. Sel. Top. Sig. Process. **1**, 540–547 (2007). doi:10.1109/JSTSP.2007.910263
8. Nesterov, Y.E.: A method for solving the convex programming problem with convergence rate $O(1/k^2)$. Sov. Math. Dokl. **27**, 372–376 (1983)
9. Nesterov, Y.E.: Introductory Lectures on Convex Optimization. Kluwer, Boston (2004)
10. Browder, F.E.: On the unification of the calculus of variations and the theory of monotone nonlinear operators in banach spaces. Proc. Nat. Acad. Sci. U.S.A. **56**, 419–425 (1966)
11. Konnov, I.V., Pinyagina, O.V.: D-gap functions and descent methods for a class of monotone equilibrium problems. Lobachevskii J. Math. **13**, 57–65 (2003)
12. de André, T., Silva, P.S.: Exact penalties for variational inequalities with applications to nonlinear complementarity problems. Comput. Optim. Appl. **47**(3), 401–429 (2010). doi:10.1007/s10589-008-9232-3
13. Kokurin, M.Y.: An exact penalty method for monotone variational inequalities and order optimal algorithms for finding saddle points. Izv. Vyssh. Uchebn. Zaved. Mat. **8**, 23–33 (2011)
14. Patriksson, M.: Nonlinear Programming and Variational Inequality Problems: A Unified Approach. Springer Science & Business Media, New York (2013)
15. Nurminski, E.A.: arXiv:1611.09697 [math.OC]
16. Fiacco, A.V., McCormick, G.P.: Nonlinear Programming: Sequential Unconstrained Minimization Techniques. Wiley, New York (1968)
17. Nurminski, E.A.: Envelope step-size control for iterative algorithms based on Féjer processes with attractants. Optim. Methods Softw. **25**(1), 97–108 (2010)

A Nonconvex Optimization Approach to Quadratic Bilevel Problems

Andrei Orlov[✉]

Matrosov Institute for System Dynamics and Control Theory of Siberian
Branch of RAS, Irkutsk, Russia
anor@icc.ru
http://nonconvex.isc.irk.ru

Abstract. This paper addresses one of the classes of bilevel optimization problems in their optimistic statement. The reduction of the bilevel problem to a series of nonconvex mathematical optimization problems, together with the specialized Global Search Theory, is used for developing methods of local and global searches to find optimistic solutions. Illustrative examples show that the approach proposed is prospective and performs well.

Keywords: Bilevel optimization · Quadratic bilevel problems · Optimistic solution · KKT-approach · Global Search Theory · Computational simulation

1 Introduction

As well-known, problems with hierarchical structure arise in investigations of complex control systems [9] with the bilevel optimization being the most popular modeling tool [7]. According to Pang [17], a distinguished expert in optimization, development of methods for solving various problems with hierarchical structure is one the three challenges faced by optimization theory and methods in the 21st century. This paper investigates one of the classes of bilevel problems with a convex quadratic upper level goal function and a quadratic lower level goal function if the constraints are linear. The lower level goal function has a bilinear component. The task is to find an optimistic solution in the situation when the actions of the lower level might coordinate with the interests of the upper level [7].

During more than three decades of intensive investigation of bilevel optimization problems, many methods for finding the optimistic solutions were proposed by different authors (see the surveys [6,8]). Nevertheless, as far as we can conclude on the basis of the available literature, a few results published so far deal with numerical solutions of merely test bilevel high-dimension problems (for example, up to 100 variables at each level for linear bilevel problems [19]). Most frequently authors consider just illustrative examples with the dimension up to 10 (see [14,18]) and only the works [1,5,10,12] present some results on solving nonlinear bilevel problems of dimension up to 30 at each level.

© Springer International Publishing AG 2017
R. Battiti et al. (Eds.): LION 2017, LNCS 10556, pp. 222–234, 2017.
https://doi.org/10.1007/978-3-319-69404-7_16

At the same time, in our group we have an experience of solving linear bilevel problems with up to 500 variables at each level [11] and quadratic-linear bilevel problems of dimension up to (150×150) [23]. Here we generalize our methods for problems with quadratic goal functions at each level.

For this purpose, we use the most common approach to address a bilevel problem via its reduction to a single-level one by replacing the lower level problem with optimality conditions (the so called KKT-approach) [7,23]. Then, using the penalty method, the resulting problem with a nonconvex feasible set is reduced to a series of problems with a nonconvex function under linear constraints [7,23]. The latter ones turn out to be the d.c. optimization problems (with goal functions that can be represented as the difference of two convex functions) [20,21], which can be addressed by means of the Global Search Theory developed by Strekalovsky [20,21]. In contrast to the generally accepted global optimization methods such as branch-and-bound based techniques, approximation methods and the like, this theory employs reduction of the nonconvex problem to a family of simpler problems with a possibility of application of classic convex optimization methods.

In accordance with the Global Search Theory, this paper aims at constructing specialized local and global search methods for finding optimistic solutions to the problems under study. Illustrative examples taken from the literature were used to demonstrate that the approach proposed for numerical solution of quadratic bilevel problems performs rather well.

2 Statement of the Problem and Its Reduction

Consider the following quadratic-quadratic problem of bilevel optimization in its optimistic statement. In this case, according to the theory [7], at the upper level we perform the minimization with respect to the variables of both levels which are in cooperation:

$$
\left.
\begin{aligned}
F(x,y) &:= \frac{1}{2}\langle x, Cx \rangle + \langle c, x \rangle + \frac{1}{2}\langle y, Dy \rangle + \langle d, y \rangle \downarrow \min_{x,y}, \\
x &\in X := \{x \in \mathbb{R}^m \mid Ax \le b\}, \\
y &\in Y_*(x) := \operatorname{Arg}\min_{y}\{\frac{1}{2}\langle y, D_1 y \rangle + \langle d_1, y \rangle + \langle x, Qy \rangle \mid y \in Y(x)\}, \\
Y(x) &\stackrel{\triangle}{=} \{y \in \mathbb{R}^n \mid A_1 x + B_1 y \le b_1\},
\end{aligned}
\right\} \quad (\mathcal{QBP})
$$

where $A \in \mathbb{R}^{p \times m}$, $A_1 \in \mathbb{R}^{q \times m}$, $B_1 \in \mathbb{R}^{q \times n}$, $C \in \mathbb{R}^{m \times m}$, $D, D_1 \in \mathbb{R}^{n \times n}$, $Q \in \mathbb{R}^{m \times n}$, $c \in \mathbb{R}^m$, $d, d_1 \in \mathbb{R}^n$, $b \in \mathbb{R}^p$, $b_1 \in \mathbb{R}^q$. Additionally, $C = C^T \ge 0$, $D = D^T \ge 0$, $D_1 = D_1^T \ge 0$. The quadratic terms of the form $\frac{1}{2}\langle x, C_1 x \rangle + \langle c_1, x \rangle$, where $C_1 = C_1^T \ge 0$, are not included into the lower level goal function, because for a fixed upper level variable x they are constant and do not affect the structure of the set $Y_*(x)$.

Note that despite the bilinear component in the goal function, for a fixed $x \in X$ the lower level problem

$$\frac{1}{2}\langle y, D_1 y\rangle + \langle xQ + d_1, y\rangle \downarrow \min_y, \quad A_1 x + B_1 y \le b_1 \quad (\mathcal{FP}(x))$$

is convex with automatically fulfilled Abadie regularity conditions [2] due to affine constraints. Therefore, the Karush-Kuhn-Tacker conditions [2] are necessary and sufficient for Problem $(\mathcal{FP}(x))$. According to these conditions, the point y is a solution to $(\mathcal{FP}(x))$ $(y \in Sol(\mathcal{FP}(x)))$ if and only if there exists a vector $v \in \mathbb{R}^q$ such that

$$\left. \begin{array}{cc} D_1 y + d_1 + xQ + vB_1 = 0, \quad v \ge 0, \quad A_1 x + B_1 y \le b_1, \\ \langle v, A_1 x + B_1 y - b_1\rangle = 0. \end{array} \right\} \quad (\mathcal{KKT})$$

By replacing the lower level problem in $(\mathcal{QBP})$ with its optimality conditions $(\mathcal{KKT})$ we obtain the following mathematical optimization problem:

$$\left. \begin{array}{c} F(x, y) \downarrow \min_{x, y, v}, \quad Ax \le b, \\ D_1 y + d_1 + xQ + vB_1 = 0, \quad v \ge 0, \quad A_1 x + B_1 y \le b_1, \\ \langle v, A_1 x + B_1 y - b_1\rangle = 0. \end{array} \right\} \quad (\mathcal{DCC})$$

It is clear that $(\mathcal{DCC})$ is a problem with a nonconvex feasible set because the nonconvexity is generated by the last equality constraint (complementary constraint). The following theorem on the equivalence between $(\mathcal{QBP})$ and $(\mathcal{DCC})$ is valid.

Theorem 1 [7]. *For the pair (x^*, y^*) to be a global solution to $(\mathcal{QBP})$ it is necessary and sufficient that there exists a vector $v \in \mathbb{R}^q$ such that the triple (x^*, y^*, v^*) is a global solution to $(\mathcal{DCC})$.*

This theorem reduces the search for an optimistic solution to the bilevel problem $(\mathcal{QBP})$ to solving Problem $(\mathcal{DCC})$. Note that direct handling of the nonconvex set in this problem is quite challenging, which is why we use the penalty method to boil it down to a series of nonconvex problems with a convex feasible set.

It is easy to see that the complementary constraint can be written in the equivalent form $\langle v, b_1 - A_1 x - B_1 y\rangle = 0$. Then $\forall (x, y, v) \in D$, where

$$D := \{(x, y, v) \mid Ax \le b, \quad D_1 y + d_1 + xQ + vB_1 = 0, \quad v \ge 0, \quad A_1 x + B_1 y \le b_1\}$$

is the convex set in $\mathbb{R}^{n+m+q}$, the following inequality holds $\langle v, b_1 - A_1 x - B_1 y\rangle \ge 0$. Now the penalized problem can be written as

$$\left. \begin{array}{c} \Phi(x, y, v) := \frac{1}{2}\langle x, Cx\rangle + \langle c, x\rangle + \frac{1}{2}\langle y, Dy\rangle + \langle d, y\rangle + \\ + \mu \langle v, b_1 - A_1 x - B_1 y\rangle \downarrow \min_{x, y, v}, \quad (x, y, v) \in D, \end{array} \right\} \quad (\mathcal{DC}(\mu))$$

where $\mu > 0$ is a penalty parameter. For a fixed μ this problem belongs to the class of d.c. minimization problems [20, 21] with a convex feasible set. In what follows we show that the goal function of $(\mathcal{DC}(\mu))$ can be represented as a difference of two convex functions. Let $(x(\mu), y(\mu), v(\mu))$ be a solution to $(\mathcal{DC}(\mu))$ for some μ. Denote $r[\mu] := \langle v(\mu), b_1 - A_1 x(\mu) - B_1 y(\mu) \rangle$ and formulate the following result on the connection between the solutions to $(\mathcal{DCC})$ and $(\mathcal{DC}(\mu))$.

Proposition 1 [2,7]

(i) *Let for some $\hat{\mu} > 0$ the equality $r[\hat{\mu}] = 0$ holds for a solution $(x(\hat{\mu}), y(\hat{\mu}), v(\hat{\mu}))$ to Problem $(\mathcal{DC}(\mu))$. Then, the triple $(x(\hat{\mu}), y(\hat{\mu}), v(\hat{\mu}))$ is a solution to Problem $(\mathcal{DCC})$.*

(ii) *For all values of the parameters $\mu > \hat{\mu}$ the function $r[\mu]$ vanishes, so that the triple $(x(\mu), y(\mu), v(\mu))$ is a solution to Problem $(\mathcal{DCC})$.*

Therefore, the established connection between Problems $(\mathcal{DC}(\mu))$ and $(\mathcal{DCC})$ enables us to search for a solution to Problem $(\mathcal{DC}(\mu))$ instead of a solution to Problem $(\mathcal{DCC})$, when the parameter μ grows. It can be shown that there exists a finite value of μ with $r[\hat{\mu}] = 0$ [3]. For a fixed μ, we will address Problem $(\mathcal{DC}(\mu))$ by means of the Global Search Theory developed by Strekalovsky [20,21]. Within that theory, it is first required to construct a local search procedure that takes into consideration special features of the problem under study.

3   The Local Search

It is noteworthy that nonconvexity in Problem $(\mathcal{DC}(\mu))$ is generated by only a bilinear component $\langle v, b_1 - A_1 x - B_1 y \rangle$. On the basis of this fact, we suggest to perform the local search in Problem $(\mathcal{DC}(\mu))$ using the idea of the successive solution by different groups of variables. Earlier this idea was successfully applied in solving bimatrix games [16,22], problems of bilinear programming [15,22], and quadratic-linear bilevel problems [23]. Even though, contrarily to the problems mentioned above, the feasible set D is not split up into different groups of variables, Problem $(\mathcal{DC}(\mu))$ turns into a convex quadratic optimization problem for a fixed v; meanwhile we obtain the linear programming (LP) problem with respect to the variable v for a fixed pair (x, y).

Therefore, a specialized local search method appears. Let there be given a starting point (x_0, y_0, v_0).

V-procedure

Step 0. Set $s := 1$, $v^s := v_0$.

Step 1. Using a suitable quadratic programming method, find the $\dfrac{\rho_s}{2}$-solution (x^{s+1}, y^{s+1}) to the problem

$$\left.\begin{array}{c} \dfrac{1}{2}\langle x, Cx \rangle + \langle c, x \rangle + \dfrac{1}{2}\langle y, Dy \rangle + \langle d, y \rangle - \mu \langle v^s A_1, x \rangle - \mu \langle v^s B_1, y \rangle \downarrow \min_{x,y}, \\ Ax \le b, \quad A_1 x + B_1 y \le b_1, \quad D_1 y + d_1 + xQ + v^s B_1 = 0. \end{array}\right\} \quad (\mathcal{QP}(v^s))$$

Step 2. Find the $\frac{\rho_s}{2}$-solution v^{s+1} to the following LP-problem:

$$\left.\begin{array}{l} \langle b_1 - A_1 x^{s+1} - B_1 y^{s+1}, v \rangle \downarrow \min_v, \\ D_1 y^{s+1} + d_1 + x^{s+1} Q + v B_1 = 0, \quad v \geq 0. \end{array}\right\} \quad (\mathcal{LP}(x^{s+1}, y^{s+1}))$$

Step 3. Set $s := s + 1$ and move to **Step 1.**

Note that the name of the V-procedure was not chosen randomly. To launch the algorithm, we do not need all the components of the starting point (x_0, y_0, v_0), knowing only v_0 is sufficient. Herein, it is necessary to choose the components (x_0, y_0, v_0) so that the auxiliary problems of linear and quadratic programming were solvable at steps of the algorithm. The solvability can be guaranteed, for example, by an appropriate choice of the feasible point $(x_0, y_0, v_0) \in D$. It also should be noted that here, contrarily to the problems with disjoint constrains, local subproblems are each time solved on new sets that depend on the components of the current point. Nonetheless, it happened to be possible to prove the following convergence theorem for the V-procedure.

Theorem 2

(i) Let a sequence $\{\rho_s\}$ be such that $\rho_s > 0$, $s = 1, 2, ...$, and $\sum_{s=1}^{\infty} \rho_s < +\infty$. Then the number sequence $\{\Phi_s := \Phi(x^s, y^s, v^s)\}$ generated by the V-procedure converges.

(ii) If $(x^s, y^s, v^s) \to (\hat{x}, \hat{y}, \hat{v})$, then the accumulation point $(\hat{x}, \hat{y}, \hat{v})$ satisfies the inequalities

$$\Phi(\hat{x}, \hat{y}, \hat{v}) \leq \Phi(x, y, \hat{v}) \quad \forall (x, y) \in D(\hat{v}), \tag{1}$$

$$\Phi(\hat{x}, \hat{y}, \hat{v}) \leq \Phi(\hat{x}, \hat{y}, v) \quad \forall v \in D(\hat{x}, \hat{y}), \tag{2}$$

where $D(v) := \{(x, y) | (x, y, v) \in D\}$, $D(x, y) := \{v | (x, y, v) \in D\}$.

Definition 1. The triple $(\hat{x}, \hat{y}, \hat{v})$ satisfying (1) and (2) is said to be the critical point in Problem $(\mathcal{DC}(\mu))$. If the inequalities (1) and (2) are satisfied with a certain accuracy then we refer to this point as approximately critical.

It can be shown (see, e.g., [23]), that if we use, for example, the following inequality as a stopping criterion for the V-procedure

$$\Phi_s - \Phi_{s+1} \leq \tau, \tag{3}$$

where τ is a given accuracy, then after the finite number of iterations of the local search method we arrive at the approximately critical point. Recall that such a definition of critical points is quite advantageous when we perform a global search in problems with a bilinear structure [22, 23]. The next section describes basic elements of the global search for Problem $(\mathcal{DC}(\mu))$.

4 Global Optimality Conditions and the Global Search Procedure

As well-known, the local search does not provide, in general, a global solution in nonconvex problems of even moderate dimension [20–22]. Therefore, in what follows we discuss the procedure of escaping critical points obtained during the local search. The procedure is based on the Global Optimality Conditions (GOCs) developed by Strekalovsky for the d.c. minimization problems [20, 21]. To build the global search procedure, first of all we need an explicit d.c. representation of the goal function of $(\mathcal{DC}(\mu))$. We will employ the following representation based on the known property of scalar product:

$$\Phi(x, y, v) = g(x, y, v) - h(x, y, v), \qquad (4)$$

where $g(x, y, v) = \frac{1}{2}\langle x, Cx\rangle + \langle c, x\rangle + \frac{1}{2}\langle y, Dy\rangle + \langle d, y\rangle + \mu\langle b_1, v\rangle + \frac{\mu}{4}\|A_1 x - v\|^2 + \frac{\mu}{4}\|B_1 y - v\|^2$, $h(x, y, v) = \frac{\mu}{4}\|A_1 x + v\|^2 + \frac{\mu}{4}\|B_1 y + v\|^2$. Note that the so-called *basic nonconvexity* in Problem $(DC(\mu))$ is provided by the function h (for more detail, refer to [20]).

The necessary Global Optimality Conditions that constitute the basis of the global search procedure have the following form in terms of Problem $(\mathcal{DC}(\mu))$.

Theorem 3 [20, 21, 23]. *If the feasible point (x^*, y^*, v^*) is a (global) solution to Problem $(\mathcal{DC}(\mu))$, then $\forall(z, u, w, \gamma) \in R^{m+n+q+1}$:*

$$h(z, u, w) = \gamma - \zeta, \quad \zeta := \Phi(x^*, y^*, v^*), \qquad (5)$$

$$g(x, y, v) \le \gamma \le \sup_{x,y,v}(g, D), \qquad (6)$$

$$g(x, y, v) - \gamma \ge \langle \nabla h(z, u, w), (x, y, v) - (z, u, w)\rangle \quad \forall(x, y, v) \in D. \qquad (7)$$

The conditions (5)–(7) possess the so-called algorithmic (constructive) property: if the GOCs are violated, we can construct a feasible point that will be better than the point in question [20, 21]. Indeed, if for some $(\tilde{z}, \tilde{u}, \tilde{w}, \tilde{\gamma})$ from (5) on some level $\zeta := \zeta_k := \Phi(x^k, y^k, v^k)$ for the feasible point $(\tilde{x}, \tilde{y}, \tilde{v}) \in D$ the inequality (7) is violated:

$$g(\tilde{x}, \tilde{y}, \tilde{v}) < \tilde{\gamma} + \langle \nabla h(\tilde{z}, \tilde{u}, \tilde{w}), (\tilde{x}, \tilde{y}, \tilde{v}) - (\tilde{z}, \tilde{u}, \tilde{w})\rangle,$$

then it follows from the convexity of $h(\cdot)$ that

$$\Phi(\tilde{x}, \tilde{y}, \tilde{v}) = g(\tilde{x}, \tilde{y}, \tilde{v}) - h(\tilde{x}, \tilde{y}, \tilde{v}) < h(\tilde{z}, \tilde{u}, \tilde{w}) + \zeta - h(\tilde{z}, \tilde{u}, \tilde{w}) = \Phi(x^k, y^k, v^k),$$

or, $\Phi(\tilde{x}, \tilde{y}, \tilde{v}) < \Phi(x^k, y^k, v^k)$. Therefore, the point $(\tilde{x}, \tilde{y}, \tilde{v}) \in D$ happens to be better with respect to the goal function value than the point (x^k, y^k, v^k). Consequently, by varying parameters (z, u, w, γ) in (5) for a fixed $\zeta = \zeta_k$ and finding approximate solutions $(x(z, u, w, \gamma), y(z, u, w, \gamma), v(z, u, w, \gamma))$ of the linearized problems (see (7))

$$g(x, y, v) - \langle \nabla h(z, u, w), (x, y, v) \rangle \downarrow \min_{x,y,v} (x, y, v) \in D, \quad (\mathcal{PL}(z, u, w))$$

we obtain a family of starting points to launch the local search procedure. Additionally, we do not need to sort through all (z, u, w, γ) at each level ζ, because it is sufficient to prove that the inequality (7) is violated at the single 4-tuple $(\tilde{z}, \tilde{u}, \tilde{w}, \tilde{\gamma})$. After that we move to the new level $(x^{k+1}, y^{k+1}, v^{k+1}) := (\tilde{x}, \tilde{y}, \tilde{v})$, $\zeta_{k+1} := \Phi(x^{k+1}, y^{k+1}, v^{k+1})$ and vary parameters again.

Hence, according to the Global Search Theory developing by Strekalovsky [20, 21], Problem $(\mathcal{DC}(\mu))$ can be split into several simpler problems (linearized problems and problems with respect to other parameters from the global optimality conditions), which can be represented in the form of the following *Global Search Strategy*.

Suppose we know some approximately critical point (x^k, y^k, z^k) in Problem $(\mathcal{DC}(\mu))$ with the goal function value $\zeta_k := \Phi(x^k, y^k, v^k)$. The point has been constructed with the help of a specialized local search method (for example, we could use the V-procedure). Then we perform the following chain of operations.

(1) Choose a number $\gamma \in [\gamma_-, \gamma_+]$, where $\gamma_- := \inf(g, D)$, $\gamma_+ := \sup(g, D)$. We can take, for example, $g(x^k, y^k, z^k)$ as a starting value of the parameter γ [20, 21].

(2) Furtherfore, construct some finite approximation

$$\mathcal{A}_k = \{(z^i, u^i, w^i) \mid h(z^i, u^i, w^i) = \gamma - \zeta_k, \quad i = 1, ..., N_k\}$$

for the level surface $U(\zeta_k) = \{(x, y, v) \mid h(x, y, v) = \gamma - \zeta_k\}$ of the convex function $h(\cdot)$.

(3) For all approximation points $\mathcal{A}_k$ verify the inequality

$$g(z^i, u^i, w^i) \le \gamma, \quad i = 1, 2, ..., N_k, \tag{8}$$

that follows from the global optimality conditions for Problem $(DC(\mu))$ (see (5)). If the inequality (8) is satisfied, then the approximation point will be used in process. Otherwise, the point (z^i, u^i, w^i) is useless, because it is not able to improve the current point [20–23].

(4) For each point (z^i, u^i, w^i), $i \in \{1, 2, ..., N_k\}$ chosen at Stage (3) find approximate solutions $(\bar{z}^i, \bar{u}^i, \bar{w}^i)$ of the linearized (with respect to the basic nonconvexity) Problems $(\mathcal{PL}(z^i, u^i, w^i))$.

(5) Using the points $(\bar{z}^i, \bar{u}^i, \bar{w}^i)$, perform the local search that delivers approximately critical points $(\hat{x}^i, \hat{y}^i, \hat{v}^i)$, $i \in \{1, ..., N\}$ in Problem $(\mathcal{DC}(\mu))$.

(6) For the chosen $i \in \{1, ..., N_k\}$, solve the level problem:

$$\left. \begin{array}{l} \langle \nabla_x h(z, u, w), \hat{x}^i - z \rangle + \langle \nabla_y h(z, u, w), \hat{y}^i - u \rangle \\ + \langle \nabla_v h(z, u, w), \hat{v}^i - w \rangle \uparrow \max_{(z,u,w)}, \quad h(z, u, w) = \gamma - \zeta_k. \end{array} \right\} \quad (\mathcal{U}_i)$$

Note that definition for $h(\cdot)$ (see (4)) makes it possible to solve Problem $(\mathcal{U}_i)$ analytically. Let (z_0^i, u_0^i, w_0^i) be the approximate solution to this problem.

(7) If for some $j \in \{1, \ldots, N_k\}$ the following inequality holds

$$g(\hat{x}^j, \hat{y}^j, \hat{v}^j) - \gamma < \langle \nabla_x h(z_0^j, u_0^j, w_0^j), \hat{x}^j - z_0^j \rangle$$
$$+ \langle \nabla_y h(z_0^j, u_0^j, w_0^j), \hat{y}^j - u_0^j \rangle + \langle \nabla_v h(z_0^j, u_0^j, w_0^j), \hat{v}^j - w_0^j \rangle,$$

then, due to convexity of $h(\cdot)$, we obtain

$$\gamma - h(z_0^j, u_0^j, w_0^j) = \zeta_k = \Phi(x^k, y^k, v^k) > \Phi(\hat{x}^j, \hat{y}^j, \hat{v}^j).$$

Thus, we constructed the point $(\hat{x}^j, \hat{y}^j, \hat{v}^j) \in D$, which is better than (x^k, y^k, v^k). If we failed to improve the value of ζ_k using all approximation points $\mathcal{A}_k$, then we have to continue the one-dimensional search with respect to γ on the segment $[\gamma_-, \gamma_+]$.

Observe that the key moment of the global search strategy described above is that on Stage (2), when we construct an approximation for the level surface of the convex function $h(\cdot)$ that defines the basic nonconvexity in Problem $(\mathcal{DC}(\mu))$. Very important to note that the approximation should be representative enough to find out whether the current critical point (x^k, y^k, z^k) is a global solution or not [20,21]. To construct this approximation, we can employ, for example, special direction sets that include, in particular, the Euclidean basis vectors and the components of the current critical point (for more detail, see [20–23]).

5   Computational Simulation

To illustrate the operability of the newly constructed local and global search algorithms for finding optimistic solutions to quadratic bilevel problems, a few examples of moderate dimension were taken from the available literature. Note that here they have the form $(\mathcal{QBP})$, with constant components excluded from the goal functions on the upper and lower levels.

Example 1 ([18])

$$\left. \begin{array}{c} F(x, y) = x^2 - 10x + 4y^2 + 4y \downarrow \min_{x,y}, \\ x \in X = \{x \in \mathbb{R}^1 \mid x \geq 0\}, \\ y \in Y_*(x) = \text{Arg} \min_y \{y^2 - 2y \mid y \in Y(x)\}, \\ Y(x) = \{y \in \mathbb{R}^1 \mid -3x + y \leq -3, \ x - 0.5y \leq 4, \ x + y \leq 7, \ y \geq 0\}. \end{array} \right\}$$

It is well-known that the global optimistic solution to this problem with the goal function optimal value $F_* = -9$ is achieved at the point $(x^*, y^*) = (1, 0)$.

Example 2 ([1])

$$\left. \begin{array}{c} F(x, y) = x^2 - 10x + 4y^2 + 4y \downarrow \min_{x,y}, \\ x \in X = \{x \in \mathbb{R}^1 \mid x \geq 0\}, \\ y \in Y_*(x) = \text{Arg} \min_y \{y^2 - 2y - 1.5xy \mid y \in Y(x)\}, \\ Y(x) = \{y \in \mathbb{R}^1 \mid -3x + y \leq -3, \ x - 0.5y \leq 4, \ x + y \leq 7, \ y \geq 0\}. \end{array} \right\}$$

Interestingly, this problem differs from the previous one only by having a bilinear component in the lower level goal function. Obviously, it does not result in noncovexity of the problem at the lower level, because for a fixed x the bilinear component becomes linear. Moreover, the global optimistic solution to this problem with the goal function optimal value $F_* = -9$ is achieved at the same point $(x^*, y^*) = (1, 0)$ as above.

Example 3 ([4])

$$\left.\begin{aligned} F(x,y) &= \frac{1}{2}x^2 - x + \frac{1}{2}y^2 \downarrow \min_{x,y}, \\ x \in X &= \{x \in \mathbb{R}^1 \mid x \geq 0\}, \\ y \in Y_*(x) &= \operatorname{Arg}\min_y\{\frac{1}{2}y^2 - xy \mid y \in Y(x)\}, \\ Y(x) &= \{y \in \mathbb{R}^1 \mid x - y \leq 1, \ x + y \leq \rho, \ -x - y \leq -1. \end{aligned}\right\}$$

This problem has a form of the kernel problems used for generating quadratic bilevel problems by means of the method constructed in [4]. Depending on the value of the parameter ρ, the kernel problems are divided into following classes:

Class 1 ($\rho = 1$): the global optimistic solution $F_* = -0.5$ is achieved at the point $(x^*, y^*) = (1, 0)$.
Class 2 ($\rho = 1.5$): the global solution $F_* = -0.4375$, $(x^*, y^*) = (1.25, 0.25)$.
Class 3 ($\rho = 2$): $F_* = -0.25$, $(x^*, y^*) = (0.5, 0.5)$ or $(x^*, y^*) = (1.5, 0.5)$.
Class 4 ($\rho = 3$): $F_* = -0.25$, $(x^*, y^*) = (0.5, 0.5)$.

Even though the problems discussed above differ from each other only by a single parameter, they all have different properties and a different number of local and global solutions (for more detail, refer to [4]).

Example 4 ([14])

$$\left.\begin{aligned} F(x,y) &= -7x_1 + 4x_2 + y_1^2 + y_3^2 - y_1y_3 - 4y_2 \downarrow \min_{x,y}, \\ x \in X &= \{x \in \mathbb{R}^2 \mid x_1 + x_2 \leq 1, \ x_{1,2} \geq 0, \}, \\ y \in Y_*(x) &= \operatorname{Arg}\min_y\{y_1 - 3x_1y_1 + y_2 + x_2y_2 + y_1^2 + \frac{1}{2}y_2^2 + \frac{1}{2}y_3^2 + y_1y_2| \\ y \in Y(x)\}, \quad Y(x) &= \{y \in \mathbb{R}^3 \mid x_1 - 2x_2 + 2y_1 + y_2 - y_3 + 2 \leq 0, \ y_{1,2,3} \geq 0\}. \end{aligned}\right\}$$

Apparently, this problem is more complex than the previous ones, because the dimension of the upper level variable is 2, whereas the dimension of the lower level variable is 3. The global optimistic solution approximately equals $F_* = 0.6426$ and is achieved at the point $(x^*, y^*) = (0.609, 0.391, 0, 0, 1.828)$.

First, for each of the problems we write down its single-level equivalent. Note that the dimension of the single-level problems is bigger than that one of the original problems by exactly the number of constraints in the lower level of the bilevel problem. Further, we penalize the complementary constraint in each problem and afterwards apply the local and global search methods described above.

The software that implements the methods developed was coded in MATLAB 7.11.0.584 R2010b [13]. The auxiliary problems of linear and convex quadratic programming were solved by the standard MATLAB subroutines "linprog" and "quadprog" with the default settings [13]. To run the software, we used the computer with Intel Core i5-2400 processor (3.1 GHz) and 4 Gb RAM.

To construct feasible starting points for the local search method, we used the projection of the chosen infeasible point $(x^0, y^0, v^0) = (0, 0, 0)$ onto the feasible set D by solving the following problem:

$$\left. \begin{array}{l} \dfrac{1}{2}\|(x, y, v) - (x^0, y^0, v^0)\|^2 \downarrow \min_{x, y, v}, \\[2mm] (x, y, v) \in D. \end{array} \right\} \quad (\mathcal{PR}(x^0, y^0, v^0))$$

The solution to Problem $(\mathcal{PR}(x^0, y^0, v^0))$ was taken as a starting point (x_0, y_0, v_0). The inequality (3) with $\tau = 10^{-5}$ was the stopping criterion for the local search method.

The value of the penalty parameter was set as $\mu = 10$. Besides, similar to solving quadratic-linear bilevel problem [23], we need two extra parameters to launch the global search algorithm. The first parameter will be required at Stage (3) of the global search, when, using the inequality (8), we verify whether the point is suitable for future use. During numerical implementation we need the parameter ξ, which can be varied to change the accuracy with which the inequality is satisfied (to reduce the impact of the round-off errors, see also [15, 16, 22]). Thus, the global search algorithm will not utilize the points that fail to satisfy the inequality

$$g(z^i, u^i, w^i) \leq \gamma + \xi\gamma, \tag{9}$$

If we set, for example, $\xi = 0.0$ then the algorithm works fast but is not always efficient. When we increase ξ, the algorithm's quality improves. However, the run time increases too, because the condition (9) gets relaxed and the number of level surface approximation points to be investigated, grows.

Various values of the second parameter M are responsible for splitting the segment $[\gamma_-, \gamma_+]$ into a corresponding number of subsegments to perform the passive one-dimensional search with respect to γ. The segment lower bound γ_- was set to equal 0, meanwhile the upper bound was estimated by $\gamma_+ = (m + n + q)\mu$. When we increase the value of M, the algorithm's accuracy grows, of course, but it happens at the expense of the proportional increase of the run time. At the initial stage the values of these parameters were chosen as $\xi = 0.0$, $M = 2$.

To approximate the level surface at Stage (2) of the global search procedure, we used the following three sets which represent direction set variations from [23]:

$$Dir1 = \{((x, y) - e^l, v - e^j), \ l = 1, ..., m + n, \ j = 1, ..., q\};$$

$$Dir2 = \{(-e^l, -e^j), \ l = 1, ..., m + n, \ j = 1, ..., q\};$$

$$Dir3 = \{(e^l, -e^j), \ l = 1, ..., m + n, \ j = 1, ..., q\}.$$

Here $e^l \in \mathbb{R}^{m+n}$, $e^j \in \mathbb{R}^q$ are the Euclidean basis vectors of the corresponding dimension, (x, y, v) is a current critical point. These sets proved to be most efficient when solving test problems. Computational results are given in Table 1 with the following denotations:

No is a number of example;

Dir is the most efficient direction set which delivered its global optimistic solution for the given problem;

Loc stands for the number of start-ups of the local search procedure required to find the approximate global solution to the problem;

LP is the number of the LP problems solved during the operation of the program;

QP stands for the number of auxiliary convex quadratic problems solved;

GIt is the number of iterations of the global search method (the number of improved critical points obtained during the operation of the program);

(x^*, y^*) is an optimal solution to the bilevel problem;

v^* is an optimal value of the auxiliary variables;

$F_* = \Phi_*$ is the optimal value of the goal functions of Problems $(\mathcal{QBP})$ and $(\mathcal{DC}(\mu))$;

T is the operating time of the program (in seconds).

Table 1. Testing of the global search method

No	Dir	Loc	LP	QP	GIt	(x^*, y^*)	v^*	$F_* = \Phi_*$	T
1	Dir3	10	11	21	2	$(1; 0)$	$(2; 0; 0; 0)$	-9.0	0.21
2	Dir3	10	11	21	2	$(1; 0)$	$(3.5; 0; 0; 0)$	-9.0	0.22
3.1	Dir1	11	13	24	3	$(1; 0)$	$(0; 1; 0)$	-0.5	0.23
3.2	Dir2	16	21	37	3	$(1.25; 0.25)$	$(0; 1; 0)$	-0.4375	0.29
3.3	Dir1	8	11	19	1	$(0.5; 0.5)$	$(0; 0; 0)$	-0.25	0.22
3.4	Dir1	8	10	18	1	$(0.5; 0.5)$	$(0; 0; 0)$	-0.25	0.21
4	Dir3	60	120	180	6	$(0.602; 0.398; 0; 0; 1.807)$	$(1.81; 2.81; 3.2; 0)$	0.6396	0.94

First of all note that the values of the parameters μ, ξ, and M specified above happened to be insufficient to solve Problem 4. To find the global optimistic solution in it, we needed the values $\mu = 15$, $\xi = 0.1$, $M = 3$. On the other hand, we managed to somewhat improve the approximate solution against the results from [14]. This involved 6 global search iterations and about 1 s of operating time.

Analysis of the rest of results in the table shows that in all one-dimensional problems the known global optimistic solutions at each level were found in less than 0.3 s, which required between 1 and 3 iterations of the global search method (for $GIt = 1$, the solution was obtained already at the local search stage).

As expected, Problem 4 happened to be more complex than the others, primarily due to its dimension. This is attested by the values in columns Loc, LP,

QP, and GIt, which can be considered as some complexity measure for the problem under study.

Therefore, computational experiments demonstrated that the global search theory performs well when applied to bilevel problems of quadratic optimization, whereby varying of the parameters in the algorithm opens up great prospects for solving simple as well as more complex problems.

6   Conclusion

This paper proposes an innovative approach to solving quadratic bilevel problems based on their reduction to parametric problems of d.c. minimization with a subsequent application of the Global Search Theory [20,21]. The specialized local and global search methods have been constructed to find optimistic solutions to bilevel problems. The methods have proved to be efficient in solving test problems of moderate dimension.

Further research suggests extension of the range and dimension of problems. For this purpose, it is planned to implement a special method of generating test problems of bilevel optimization from [4].

The results of numerical testing as well as our previous computational experience [11,15,16,22,23] allow us to expect that the approach proposed will prove effective in solving quadratic bilevel problems of high dimension (probably, up to 100×100) with a supplementary possibility of exploiting modern software packages for solving auxiliary LP and convex quadratic problems (IBM CPLEX, FICO Xpress etc.).

Acknowledgments. This work has been supported by the Russian Science Foundation (Project no. 15-11-20015).

References

1. Bard, J.F.: Convex two-level optimization. Math. Prog. **40**, 15–27 (1988)
2. Bazara, M.S., Shetty, C.M.: Nonlinear Programming. Theory and Algorithms. Wiley, New York (1979)
3. Bonnans, J.-F., Gilbert, J.C., Lemarechal, C., Sagastizabal, C.A.: Numerical Optimization: Theoretical and Practical Aspects. Springer, Heidelberg (2006)
4. Calamai, P., Vicente, L.: Generating quadratic bilevel programming test problems. ACM Trans. Math. Softw. **20**, 103–119 (1994)
5. Colson, B., Marcotte, P., Savard, G.: A trust-region method for nonlinear bilevel programming: algorithm and computational experience. Comput. Optim. Appl. **30**, 211–227 (2005)
6. Colson, B., Marcotte, P., Savard, G.: An overview of bilevel optimization. Ann. Oper. Res. **153**, 235–256 (2007)
7. Dempe, S.: Foundations of Bilevel Programming. Kluwer Academic Publishers, Dordrecht (2002)
8. Dempe, S.: Bilevel programming. In: Audet, C., Hansen, P., Savard, G. (eds.) Essays and Surveys in Global Optimization, pp. 165–193. Springer, Boston (2005)

9. Dempe, S., Kalashnikov, V.V., Perez-Valdes, G.A., Kalashnykova, N.: Bilevel Programming Problems: Theory, Algorithms and Applications to Energy Networks. Springer, Heidelberg (2015)

10. Etoa, J.B.E.: Solving quadratic convex bilevel programming problems using a smoothing method. Appl. Math. Comput. **217**, 6680–6690 (2011)

11. Gruzdeva, T.V., Petrova, E.G.: Numerical solution of a linear bilevel problem. Comp. Math. Math. Phys. **50**, 1631–1641 (2010)

12. Gumus, Z.H., Floudas, C.A.: Global optimization of nonlinear bilevel programming problems. J. Glob. Optim. **20**, 1–31 (2001)

13. MATLAB—The language of technical computing. http://www.mathworks.com/products/matlab/

14. Muu, L.D., Quy, N.V.: A global optimization method for solving convex quadratic bilevel programming problems. J. Glob. Optim. **26**, 199–219 (2003)

15. Orlov, A.V.: Numerical solution of bilinear programming problems. Comput. Math. Math. Phys. **48**, 225–241 (2008)

16. Orlov, A.V., Strekalovsky, A.S.: Numerical search for equilibria in bimatrix games. Comput. Math. Math. Phys. **45**, 947–960 (2005)

17. Pang, J.-S.: Three modeling paradigms in mathematical programming. Math. Prog. Ser. B. **125**, 297–323 (2010)

18. Pistikopoulos, E.N., Dua, V., Ryu, J.-H.: Global optimization of bilevel programming problems via parametric programming. In: Floudas, C.A., Pardalos, P.M. (eds.) Frontiers in Global Optimization, pp. 457–476. Kluwer Academic Publishers, Dordrecht (2004)

19. Saboia, C.H., Campelo, M., Scheimberg, S.: A computational study of global algorithms for linear bilevel programming. Numer. Algorithms **35**, 155–173 (2004)

20. Strekalovsky, A.S.: Elements of Nonconvex Optimization. Nauka, Novosibirsk (2003). [in Russian]

21. Strekalovsky, A.S.: On solving optimization problems with hidden nonconvex structures. In: Rassias, T.M., Floudas, C.A., Butenko, S. (eds.) Optimization in Science and Engineering, pp. 465–502. Springer, New York (2014). doi:10.1007/978-1-4939-0808-0_23

22. Strekalovsky, A.S., Orlov, A.V.: Bimatrix Games and Bilinear Programming. FizMatLit, Moscow (2007). [in Russian]

23. Strekalovsky, A.S., Orlov, A.V., Malyshev, A.V.: On computational search for optimistic solution in bilevel problems. J. Glob. Optim. **48**, 159–172 (2010)

An Experimental Study of Adaptive Capping in irace

Leslie Pérez Cáceres[1(✉)], Manuel López-Ibáñez[2], Holger Hoos[3],
and Thomas Stützle[1]

[1] IRIDIA, Université Libre de Bruxelles, Brussels, Belgium
{leslie.perez.caceres,stuetzle}@ulb.ac.be
[2] Alliance Manchester Business School, University of Manchester, Manchester, UK
manuel.lopez-ibanez@manchester.ac.uk
[3] Computer Science Department, University of British Columbia, Vancouver, Canada
hoos@cs.ubc.cs

Abstract. The irace package is a widely used for automatic algorithm configuration and implements various iterated racing procedures. The original irace was designed for the optimisation of the solution quality reached within a given running time, a situation frequently arising when configuring algorithms such as stochastic local search procedures. However, when applied to configuration scenarios that involve minimising the running time of a given target algorithm, irace falls short of reaching the performance of other general-purpose configuration approaches, since it tends to spend too much time evaluating poor configurations. In this article, we improve the efficacy of irace in running time minimisation by integrating an adaptive capping mechanism into irace, inspired by the one used by ParamILS. We demonstrate that the resulting irace$_\mathsf{cap}$ reaches performance levels competitive with those of state-of-the-art algorithm configurators that have been designed to perform well on running time minimisation scenarios. We also investigate the behaviour of irace$_\mathsf{cap}$ in detail and contrast different ways of integrating adaptive capping.

1   Introduction

Algorithm configuration is the task of finding parameter settings (a configuration) of a target algorithm that achieve high performance for a given class of problem instances [6,8]. The appropriate choice of parameter settings is often crucial for obtaining good performance, particularly when dealing with computationally challenging (e.g., $\mathcal{NP}$-hard) problems. This choice usually depends on the set or distribution of problem instances to be solved as well as on the execution environment. Therefore, using appropriately chosen parameter values is not only essential for reaching peak performance, but also for conducting fair performance comparisons between different algorithms for the same problem.

Traditionally, algorithm configuration has been performed manually, relying on experience and intuition about the behaviour of a given algorithm. However, typical manual configuration processes are time-consuming and tedious; furthermore, they often leave the performance potential of a given target algorithm

© Springer International Publishing AG 2017
R. Battiti et al. (Eds.): LION 2017, LNCS 10556, pp. 235–250, 2017.
https://doi.org/10.1007/978-3-319-69404-7_17

unrealised. In light of this, several automated algorithm configuration approaches have been developed and are now used increasingly widely.Prominent examples of general-purpose algorithm configuration procedures include ParamILS [13], SMAC [12], GGA++ [1] and irace [4,17,18]. The key idea behind these and other configuration procedures is to view algorithm configuration as a stochastic optimisation problem that can be solved by effectively searching the space of configurations of a given target algorithm A. The performance metrics most commonly optimised in this context are the solution quality reached by A within a certain time budget and the running time of A for finding a solution (of a certain quality) to a given problem instance.

The irace software is an automatic configurator based on the iterated F-race procedure [4,7] and recent improvements [17,18]. It was initially developed for configuring metaheuristic algorithms that optimise solution quality. In contrast, minimisation of the running time of a given target algorithm was a major focus in the development of ParamILS and SMAC, and both of them include an *adaptive capping* [13] mechanism that is specifically designed to improve efficiency when dealing with this performance objective. The key idea behind adaptive capping is to reduce the time wasted in the evaluation of poorly performing configurations by bounding the maximum running time permitted for each such evaluation. This bound is calculated based on the best-performing configuration found so far. The use of adaptive capping allows the configurator to prune poorly performing target algorithm configurations early and to quickly focus the configuration budget on promising areas of the space of configurations being searched.

In this work, we improve the efficacy of irace on algorithm configuration scenarios involving running time minimisation. We adapt the ideas of the adaptive capping mechanism into the underlying iterated racing procedure and define an additional dominance criterion based on the performance of the elite configurations obtained by irace. As a first step, we show that by extending irace with adaptive capping, resulting in our new irace$_{cap}$ method, we can significantly increase its performance on well-known and difficult configuration scenarios. An additional analysis of various parameters of irace$_{cap}$ gives further insights into the importance of the statistical testing procedures and other aspects of irace. A final comparison with other state-of-the-art configuration procedures for running time minimisation, namely ParamILS and SMAC, shows that irace$_{cap}$ reaches highly competitive performance and, thus, broadens the range of configuration scenarios for which irace can be considered a possible method of choice.

The remainder of this article is structured as follows. First, we describe irace and the adaptive capping mechanism used by ParamILS (Sects. 2 and 3). Next, in Sect. 4, we describe how we integrated adaptive capping into irace, and we experimentally analyse the resulting irace$_{cap}$ in Sect. 5. In Sect. 6, we compare irace$_{cap}$ to state-of-the-art configurators for minimising running time, and we conclude in Sect. 7.

2   Elitist Iterated Racing in irace

irace is an iterated racing procedure [7] for automatic algorithm configuration. It explores the parameter space of a target algorithm by iteratively sampling parameter configurations and applying a racing procedure to select the best-performing configurations. The racing procedure considers a sequence of problem instances on which the candidate configurations are evaluated. At each stage of the race, all candidate configurations are run on a specific problem instance; at the end of the stage, configurations that perform statistically worse than others are eliminated from the race, while all others proceed to the next stage. Once a race is terminated, the best configurations, called *elite*, are used to update the sampling model from which new configurations are generated. The elite configurations are carried over to the next iteration to continue their evaluation within a new race together with the newly generated configurations. One iteration of irace comprises the process of (i) generation of candidate configurations, (ii) execution of the racing procedure, and (iii) update of the probabilistic model.

The irace package [17,18] is an implementation of irace that is publicly available as an R package. Recently, version 2.0 of the software was released, which implements an elitist racing procedure [17]. Differently from the non-elitist racing procedure on which the first version of irace is based, elitist irace evaluates configurations on a set of problem instances that increases in size in every iteration of irace. In particular, in elitist irace, an elite configuration carried over from the previous iteration cannot be eliminated until a better configuration is evaluated on the same instances as the elite one, including all instances on which the elite configuration was previously evaluated and at least one new instance.

In more detail, elitist irace works as follows (see Fig. 1). In the first iteration, configurations are sampled uniformly at random from the given configuration space. These configurations are evaluated on T^{first} instances, after which the

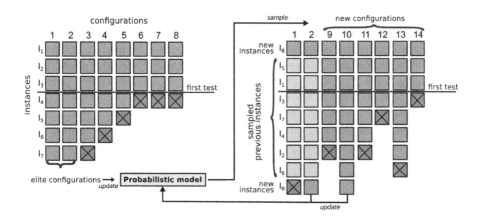

Fig. 1. Illustration of the 1st and 2nd iteration of a run of irace using $T^{\text{first}} = 3$, $T^{\text{each}} = 1$ and $T^{\text{new}} = 1$.

first statistical test is applied, and the configurations that are significantly worse performing than the best ones are eliminated. This elimination test is performed every T^{each} instances until the termination criterion of the iteration is met. The surviving configurations at the end of the iteration (elite configurations) are used to update a probabilistic model from which new configurations are sampled. The set of configurations evaluated in the next iteration is composed of the elite configurations and newly sampled configurations (non-elite ones). Algorithm 1 shows the pseudo-code of the race performed at each iteration of irace. Instances are evaluated following an execution order that is built interleaving new and old instances (procedure generateInstancesList in line 1).

More precisely, the instance list includes T^{new} previously unseen instances, followed by the list of previously evaluated instances ($\mathcal{I}^{\text{old}}$), and finally, enough new instances to complete the race. $\mathcal{I}^{\text{old}}$ is randomly shuffled to avoid a bias that could result from always using the same instance order. A race may terminate even before evaluating all instances in $\mathcal{I}^{\text{old}}$ (e.g. when a minimum number of configurations is reached), and, as a result, each elite configuration may be evaluated on some instances in $\mathcal{I}^{\text{old}}$. When the race finishes, irace therefore memorises which configuration has been evaluated on which instance. (Line 7 updates the elite status, and line 10 tests this condition.) In line 6, the configurations are evaluated on instance $\mathcal{I}[i]$ with a maximum execution time of b^{max}. If an elite configuration was already previously evaluated on $\mathcal{I}[i]$ (i.e., $\mathcal{I}[i] \in \mathcal{I}^{\text{old}}$), its result on that instance is read from memory. When the statistical elimination test is applied, only non-elite configurations (Θ^{new}) may be eliminated and elite ones are kept until they become non-elite. A configuration becomes non-elite if all instances in $\mathcal{I}^{\text{old}}$ on which it has previously been evaluated have been seen in a race. Finally, the race returns the best configurations found, which will become elite in the next iteration. For more details about irace, see [17].

3   ParamILS and Adaptive Capping

ParamILS [13] is an iterated local search [19] procedure that searches in a parameter space defined by categorical parameters only; for configuring numerical parameters with ParamILS, these need to be discretised. ParamILS uses a first-improvement local search algorithm that explores, in random order, the one-exchange neighbourhood of the current configuration.

There are two versions of ParamILS, BasicILS and FocusedILS, which differ in the number of instances evaluated when comparing two configurations [13]. BasicILS compares configurations by evaluating them on a fixed number N of instances, while FocusedILS varies the number of instances according to the quality of the configurations to be tested. The number of instances used in the comparison is adjusted based on the **dominance criterion**, by which a configuration θ_j is dominated by a configuration θ_i if (1) θ_i has been evaluated in at least as many instances as θ_j and (2) the aggregated performance of θ_i is better or equal than the one of θ_j on the N_j instances on which θ_j has been evaluated. When no dominance can be established between two configurations,

---

Algorithm 1. Racing procedure in elitist irace

Inputs are a set of newly generated configurations (Θ^{new}), a user-provided maximum execution time (b^{max}), the number of new initial instances (T^{new}), the list of unseen instances ($\mathcal{I}^{\text{new}}$), a set of elite configurations (Θ^{elite}), the list of instances on which Θ^{elite} were previously evaluated ($\mathcal{I}^{\text{old}}$), and a Boolean predicate $\text{isElite}(\theta, I)$ that returns true if configuration $\theta \in \Theta^{\text{elite}}$ was previously evaluated on instance $I \in \mathcal{I}^{\text{old}}$. In the first iteration of irace, Θ^{elite} and $\mathcal{I}^{\text{old}}$ are empty and all entries of $\text{isElite}(\cdot, \cdot)$ are set to false.

Input: $\Theta^{\text{elite}}, \Theta^{\text{new}}, b^{\text{max}}, T^{\text{new}}, \mathcal{I}^{\text{old}}, \mathcal{I}^{\text{new}}, \text{isElite}(\cdot, \cdot)$
Output: Best configurations found in the race.
begin

1    $\mathcal{I} \leftarrow \text{generateInstancesList}(T^{\text{new}}, \mathcal{I}^{\text{old}}, \mathcal{I}^{\text{new}})$

2    $i \leftarrow 1$

3    $\Theta^i \leftarrow \Theta^{\text{new}} \cup \Theta^{\text{elite}}$

4   **while** $\neg$ termination() **do**

        # execute elites only when needed; $\mathcal{I}[i]$ is the ith entry of instance list $\mathcal{I}$

5        $\Theta^{\text{exe}} \leftarrow \Theta^i \setminus \{\theta \in \Theta^{\text{elite}} \mid \text{isElite}(\theta, \mathcal{I}[i])\}$

6        $\text{execute}(\Theta^{\text{exe}}, b^{\text{max}}, \mathcal{I}[i])$

7        $\text{isElite}(\theta, \mathcal{I}[i]) \leftarrow \text{false} \quad \forall \theta \in \Theta^{\text{elite}}$

8   **if** mustTest(i) **then**

9            $\Theta^{i+1} \leftarrow \text{eliminationTest}(\Theta^i, \{\mathcal{I}[1], \ldots, \mathcal{I}[i]\})$

            # keep configurations that are still elite

10          $\Theta^{i+1} \leftarrow \Theta^{i+1} \cup \{\theta \in \Theta^{\text{elite}} \mid \bigvee_{I \in \mathcal{I}^{\text{old}}} \text{isElite}(\theta, I)\}$

11   **else**

12            $\Theta^{i+1} \leftarrow \Theta^i$

13        $i \leftarrow i + 1$

14   **return** Θ^i

---

the number of instances seen by the configuration with less instances evaluated is increased until both configurations have seen the same number of evaluations. The execution of a configuration on each instance is always bounded by a defined maximum execution time (cut-off time).

The **adaptive capping** technique further bounds the execution of a configuration by using the running time of good configurations as a bound in running time that is often less than the user-specified cut-off time. Using this technique can significantly reduce the time wasted in the evaluation of poor performing configurations. Adaptive capping adjusts the bound on running time according to the number of instances to be used in the comparison, and for this reason, it can be sensitive to the ordering of the given instances. There are two types of adaptive capping: trajectory preserving and aggressive capping [13]. The first of these bounds the running time of new configurations using the performance of the currently best configuration of each ParamILS iteration as reference, while the second additionally uses the performance of the overall best configuration multiplied by a factor, set to two by default, for bounding. This factor controls the aggressiveness of the capping strategy. Further details on adaptive capping can be found in [13].

Algorithm 2. Racing procedure in irace$_\mathsf{cap}$

For the description of the inputs, see Algorithm 1.

Input: $\Theta^{\mathrm{elite}}, \Theta^{\mathrm{new}}, b^{\mathrm{max}}, T^{\mathrm{new}}, \mathcal{I}^{\mathrm{old}}, \mathcal{I}^{\mathrm{new}}, \mathsf{isElite}(\cdot, \cdot)$
Output: Best configuration set found in the race.
begin

1     $\mathcal{I} \leftarrow \mathtt{generateInstancesList}\ (T^{\mathrm{new}}, \mathcal{I}^{\mathrm{old}}, \mathcal{I}^{\mathrm{new}})$

2     $\mathtt{execute}\ (\Theta^{\mathrm{elite}}, b^{\mathrm{max}}, \{\mathcal{I}[1], \ldots, \mathcal{I}[T^{\mathrm{new}}]\})$

3     $i \leftarrow 1$

4     $\Theta^i \leftarrow \Theta^{\mathrm{new}} \cup \Theta^{\mathrm{elite}}$

5   **while** $\neg$ termination() **do**

6         $b_i \leftarrow \mathtt{calculateEliteBound}\ (\Theta^{\mathrm{elite}}, \{\mathcal{I}[1], \ldots, \mathcal{I}[i]\})$

7         $\Theta^{\mathrm{exe}} \leftarrow \Theta^i \setminus \{\theta \in \Theta^{\mathrm{elite}} \mid \mathsf{isElite}(\theta, \mathcal{I}[i])\}$

8         $\mathtt{execute}\ (\Theta^{\mathrm{exe}}, b_i, \mathcal{I}[i])$

9         $\mathsf{isElite}(\theta, \mathcal{I}[i]) \leftarrow \mathtt{false}\quad \forall \theta \in \Theta^{\mathrm{elite}}$
        # dominancecriterion elimination

10        $\Theta^{i+1} \leftarrow \mathtt{eliminationDominance}\ (\Theta^i, \{\mathcal{I}_1, \ldots, \mathcal{I}[i]\})$
        # statistical test elimination

11   **if** $\mathtt{mustTest}\ (i)$ **then**

12           $\Theta^{i+1} \leftarrow \mathtt{eliminationTest}\ (\Theta^{i+1}, \{\mathcal{I}_1, \ldots, \mathcal{I}[i]\})$

        # keep configurations that are still elite

13        $\Theta^{i+1} \leftarrow \Theta^{i+1} \cup \{\theta \in \Theta^{\mathrm{elite}} \mid \bigvee_{I \in \mathcal{I}^{\mathrm{old}}} \mathsf{isElite}(\theta, I)\}$

14        $i \leftarrow i + 1$

15   **return** Θ^i

---

4 Adaptive Capping in irace

In this section, we describe a new version of irace that adopts the ideas underlying adaptive capping in the racing procedure. This new version, irace$_\mathsf{cap}$, introduces two new components to the algorithm: (1) the **adaptive running time bound**, used to limit the running time of new configurations on previously seen and initial instances, and (2) **dominance elimination**, a procedure that discards poorly performing configurations. Algorithm 2 shows the outline of the racing procedure implemented in irace$_\mathsf{cap}$; it follows the same structure as elitist irace, described in Sect. 2. The elite configurations are first run on the set of initial instances before the start of the race (Line 2). Line 6 calculates an initial running time bound based on the running times of the elite configurations. Let p_i^j be the average computation time of a configuration θ_j up to instance $\mathcal{I}[i]$ in the current iteration. Then, the bound b_i for running new configurations on instance $\mathcal{I}[i]$ is equal to median$_{\theta_j \in \Theta^{\mathrm{elite}}}\{p_i^j\}$. (Median is chosen to be consistent with the elimination based on dominance described next.) The bound b_i can be computed only for previously evaluated instances (including the initial instances); for any other instance, we set the bound to the cut-off time, b^{max}. This running time bound provides a reference of the minimum performance new configurations should obtain in order to compete with the current elite configurations.

The maximum running time k_i^j for each configuration θ_j on instance $I[i]$ is computed by procedure execute in line 8 using the value of b_i as follows:

$$k_i'^j = b_i \cdot i + b^{\mathrm{min}} - p_{i-1}^j \cdot (i - 1) \tag{1}$$

$$k_i^j = \begin{cases} b^{\max} & \text{if } k_i'^j > b^{\max}, \\ \min\{b_i, b^{\max}\} & \text{if } k_i'^j \leq 0, \\ k_i'^j & \text{otherwise;} \end{cases} \quad (2)$$

where $b^{\min}$ is a constant that represents a minimally measurable running time different from zero (set to a default value of 0.01). Intuitively, k_i^j is the time remaining for a configuration θ_j to improve over the median elite configuration.

We implemented a dominance-based elimination procedure inspired by the domination criterion described in Sect. 3. We compare the median performance of the elite configurations set (Θ^{elite}) on the list of instances $\{\mathcal{I}[1], \ldots, \mathcal{I}[i]\}$ considered so far with the performance of the new configurations as follows:

$$\text{Median}_{\theta_s \in \Theta^{\text{elite}}} \{p_i^s\} + b^{\min} < p_i^j \quad (3)$$

where p_i^s is the mean running time of configuration θ_s on instances $\{\mathcal{I}[1], \ldots, \mathcal{I}[i]\}$, and $b^{\min}$ is the constant defined in Eq. (2). Other choices than the median are possible and may be considered in future work. We eliminate configurations as soon as they become dominated, that is, the dominance-based elimination is applied after every instance seen within an iteration of irace.

5   Experiments

In this section, we study the impact of introducing the previously described capping procedure into irace. We compare the performance of the final configurations obtained by elitist irace and irace$_{\text{cap}}$ using different settings.

5.1   Experimental Setup

In our performance assessments of irace$_{\text{cap}}$, we use five configuration scenarios taken from previous experimental studies of other automatic algorithm configuration methods, in particular, ParamILS and SMAC. These scenarios use CPLEX [16], Lingeling [5] and Spear [3] as target algorithms, and involve parameter spaces with 74, 137 and 26 parameters, respectively. Their principal characteristics are as follows:

CPLEX - Regions100 [12,13]. 5 s cut-off time, 18 000 s total configuration budget, and a training and testing set of 1000 mixed integer programming (MIP) instances each. The instances encode a combinatorial auction winner determination problem with 100 goods and 500 bids.

CPLEX - Regions200 [11,13]. 300 s cut-off time, 172 800 s total configuration budget, and a training and testing set of 1000 MIP instances each. These instances are encodings of a combinatorial auction winner determination problem with 200 goods and 1000 bids.

CPLEX - Corlat [11]. 300 s cut-off time, 172 800 s total configuration budget, and a training and testing set of 1000 MIP instances each.

Lingeling [14]. 300 s cut-off time, 172 800 s total configuration budget, and a training and testing set of 299 and 302 SAT instances, respectively. These instances were obtained from the 2014 Configurable SAT Solver Competition (CSSC) [14].

Spear [13]. 300 s cut-off time, 172 800 s total configuration budget, and a training and testing set of 302 SAT-encoded software verification instances each.

The instance files for these scenarios are also available from the Algorithm Configuration Library (AClib) [15]. AClib specifies a cut-off time of 10 000 s for the CPLEX scenarios, which stems from their initial use in conjunction with the CPLEX auto-tuning tool. Following the experiments in [11, Sect. 5]), we use a cut-off time of 300 s.[1] Another minor difference is that we usedversion 12.4 of CPLEX, which was installed on our system, while AClib proposes to use version 12.6. However, there is no obvious reason to suspect that the particular version of CPLEX should affect our conclusions on the effect of capping inside irace, and we do not directly compare to results for the original AClib scenarios. Moreover, although both irace and SMAC are able to handle non-discrete parameter spaces, for ParamILS, all parameters have to be discretised, with all possible values specified explicitly in the scenario definition. There is some evidence that the use of non-discrete parameter spaces, where possible, leads to improved results [12], thus giving an advantage to both irace and SMAC over ParamILS, unrelated to the capping mechanism, which is the focus of our comparison presented in Sect. 6. To avoid this bias, we only consider the variants of the scenarios where all parameters are discretised and explicitly specified.

In all our experiments, we used the t-test to eliminate configurations within irace, as previously recommended for running time minimisation [21]. The comparisons presented in the following are based on 20 independent runs of all configuration procedures; multiple independent configurator runs are performed due to the inherent randomness of the configuration procedures and the configuration scenarios. The experiments were run on one core of a dual-processor 2.1 GHz AMD Opteron system with 16 cores per CPU, 16 MB cache and 64 GB RAM, running Cluster Rocks 6.2, which is based on CentOS 6.2.

In our empirical analysis of $irace_{cap}$, we use mean running time as the performance criterion to be optimised by irace. Runs that time out due to reaching the cut-off time are then counted at this maximum cut-off time. In the literature, unsuccessful runs are often more strongly penalised, computing effectively the number of timed out runs multiplied by a penalty factor p_f plus the mean computation time of the successfully terminated runs. In fact, the penalty factor p_f converts the bi-objective problem of minimising the number of timed-out runs and mean time of successful runs into a single-objective problem. In this

---

[1] A higher cut-off time, as used in AClib, would be detrimental for configuration procedures such as $irace_{cap}$, as time-outs would very strongly impact the number of configurations that can be evaluated. On the other hand, there are various techniques, such as early termination of ongoing runs or the initial use of smaller maximum cut-off times, to address this problem. In the literature, the use of smaller cut-off times has been suggested as a possible remedy [12, footnote 9].

section, runs of irace attempt to minimise mean running time (with $p_f = 1$), and we therefore assessed the performance of the resulting target algorithm configurations using this performance metric. In the supplementary material, we additionally present results for evaluating configurations using $p_f = 10$ and $p_f = 100$. In the literature, $p_f = 10$ is commonly used and referred to as PAR10; consequently, in Sect. 6, all configurator runs and target algorithm evaluations are performed using PAR10 scores.

5.2   Experimental Results

We first compare the results obtained by elitist irace and irace$_{cap}$, using their respective default settings. Table 1 presents performance statistics over the 20 runs of both irace versions. The implementation of the proposed capping procedure proves to be beneficial for the scenarios used in these experiments. For the Regions 100, Regions 200, Corlat, and Spear scenarios, the results obtained by irace$_{cap}$ are significantly better than those of elitist irace, while for the Lingeling scenario, the results are not significantly different (however, irace$_{cap}$ still achieves a better mean than irace).

Table 1. Summary statistics of the distribution of observed mean running time and percentage of timed out evaluations of 20 runs of irace$_{cap}$ and elitist irace (irace) on test sets for the various configuration scenarios. We show the first and second quartile ($q25$ and $q75$, respectively), the median, the mean, the standard deviation (sd) and the variation coefficient (sd/mean). Wilcoxon test p-values are reported in the last line. Statistically significantly better results (at $\alpha = 0.05$) are indicated in bold-face and lowest mean running times in italics.

	Regions 100		Regions 200		Corlat		Lingeling		Spear	
	irace$_{cap}$	irace	irace$_{cap}$	irace	irace$_{cap}$	irace	irace$_{cap}$	irace	irace$_{cap}$	irace
%timeout	0.08	0.085	0.01	0.015	0.695	1.205	8.377	8.659	0.397	3.328
q25	0.327	0.374	9.487	10.983	8.616	13.526	42.379	44.274	3.028	4.776
mean	**0.338**	0.395	**10.498**	13.231	**11.899**	15.935	*45.501*	46.923	**4.116**	13.068
median	0.332	0.401	10.469	12.871	9.688	14.911	44.453	47.034	3.765	14.617
q75	0.34	0.413	10.75	14.256	13.941	18.436	48.996	49.738	4.242	19.993
sd	0.018	0.033	1.335	2.908	5.645	4.325	3.799	3.658	1.848	8.092
sd/mean	0.054	0.082	0.127	0.22	0.474	0.271	0.083	0.078	0.449	0.619
p-value	5.7e-06		0.0001049		0.0055809		0.2942524		0.0002613	

The elimination criterion of the capping procedure in irace$_{cap}$ only considers aggregated running time rather than its statistical distribution, and it is not obvious whether this renders the criterion always stricter than the statistical test at the core of irace, which could, in principle, render the latter superfluous. Figure 2 shows the mean percentage of live configurations selected to be eliminated by the capping procedure and the statistical test (lines), and the mean percentage of initial configurations that become elite configurations at the end of the iteration (bars). (For results on all other scenarios, see Figure A.2.) The capping procedure selects more configurations for elimination than the statistical test in all stages of the search, while the statistical test is mainly able to eliminate configurations in the initial phases of the search. As the race progresses, capping elimination quickly becomes mainly responsible for eliminating configurations, illustrating the importance of introducing it into irace.

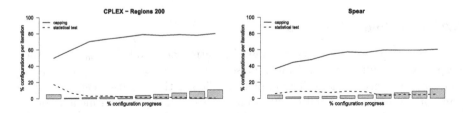

Fig. 2. Mean percentage of configurations selected for elimination by the capping procedure and the statistical test (solid and dashed lines respectively), and mean percentage of initial configurations that become elite configurations at the end of the iteration (bars). Means obtained across 20 independent runs of irace$_{cap}$ on the Regions 200 and Spear scenarios.

The capping mechanism of irace$_{cap}$ and the increased elimination of configurations induce a highly intensified search. On average, irace$_{cap}$ performs in part many more iterations than irace and shows a lower average number of elite configurations per iteration. This results in an increased number of configurations sampled overall and instances used for evaluation (Table 2).

Table 2. Statistics over 20 independent runs of irace$_{cap}$ and irace: mean number of iterations performed (iterations), mean number of instances used in the evaluation (instances), mean overall sampled configurations (candidates), mean elite configurations per iteration (elites) and mean total executions (executions).

mean	Regions 100		Regions 200		Corlat		Lingeling		Spear	
	irace$_{cap}$	irace	irace$_{cap}$	irace	irace$_{cap}$	irace	irace$_{cap}$	irace	irace$_{cap}$	irace
iterations	253.5	28.3	85.8	17.1	68.7	13	27.4	10.5	67.0	7.6
instances	258.6	47.6	91.1	36.7	75.1	29.4	35.5	26.3	83.2	16.3
candidates	27914	1136	5191	285	5318	242	2595	214	11193	718
elites	1.09	6.88	1.25	7.12	1.82	7.80	3.28	8.90	2.26	5.99
executions	30604	8362	6779	2770	8873	3147	5218	2878	28039	6109

5.3   Additional Analysis of irace$_{cap}$

In what follows, we examine in more detail the impact of some specific parameter settings of irace$_{cap}$ on its performance. For the sake of conciseness, we will only discuss overall trends;detailed results are found in supplementary material [20].

Instance order. The order of the instances may introduce a bias in irace when the configuration scenario involves a heterogeneous instance set. By default, irace shuffles the order of the training instances. Without this shuffling, irace evaluates the instances in the order provided by the user. Since the set of previously used instances is evaluated in every iteration, elitist irace randomly permutes the order of previously seen instances ($\mathcal{I}^{\mathrm{old}}$) before each iteration to further avoid any bias that the previous order may introduce. Table A.2 compares the results obtained by irace$_{cap}$ with and without this instance reshuffling. For most benchmark scenarios, disabling instance reshuffling produces better mean results and fewer timed-out runs; for Regions 200 and Lingeling, these differences are statistically significant. The main exception is the Spear scenario, where reshuffling leads to much improved results; this is probably due to the fact that this scenario contains a very heterogeneous instances set.

These results suggest that the impact of reshuffling depends on the given configuration scenario; we conjecture that for more heterogeneous instance sets, reshuffling the instance set becomes increasingly important. Investigating this conjecture in detail is an interesting direction for future work.

Confidence level of statistical test. The dominance criterion eliminates more configurations than the statistical test. Lowering the confidence level of the statistical test should lead to an even higher elimination rate of the latter and possibly improve the efficacy of the overall configuration process. We explored this possibility by lowering the confidence level in irace$_{cap}$ from its default value of 0.95 to 0.75. Table A.3 in the supplementary material shows the impact of this change. The effects on the elimination of configurations can be observed in Figure A.5 in the supplementary material. As expected, the statistical test eliminates more configurations when setting the confidence level to 0.75. This also results in a small increase in the overall number of configurations evaluated and a reduction of the mean number of elite configurations (see Table A.4 in the supplementary material). The more aggressive test slightly improved the performance for three scenarios, yielding significantly better results for Regions 200. In contrast, a confidence level of 0.75 results in slightly worse performance on the Spear scenario, indicating that the eliminations performed with lower confidence can be premature.

If we completely disable statistical testing (confidence level 1.0), the performance of irace$_{cap}$ improves on Regions 100 and Regions 200, as seen in Table A.5 in the supplementary material. This suggests that the statistical test can prematurely eliminate configurations based on an incorrect criterion. Despite this, we still recommend keeping the default confidence level of 0.95, as a safe-guard that may be useful for configuration scenarios with possibly very different properties from the ones we are testing here.

Log-transformation of running times. When used for running time minimisation, irace makes use of the t-test for elimination. However, the potentially very large variability of target algorithm running times [10] often renders the distribution of running times far from normal, a situation that may be alleviated by using a transformation of running times – in particular, a logarithmic transformation. Applying this transformation has, however, only a significant effect on the Regions 200 scenario. Increasing the difference between the performance of configurations makes the elimination more aggressive and, as seen in other experiments, the Regions 200 scenario benefits greatly of this increased intensification. For the other scenarios, the impact on performance is negligible, as seen in Table A.6 in the supplementary material, probably due to the minor impact of the statistical test on the elimination of configurations.

Number of initial new instances. Finally, we performed experiments to evaluate the impact of adding new instances at the beginning of each race. If no new instances are added at the start, then new configurations can only become elite by performing better on exactly the same instances on which the current elites performed well in previous races. Even though the new configurations may be better on instances not seen yet, they may be eliminated before seeing them, unless those new instances are evaluated at the start. On the other hand, there are no running times available for new instances; this issue is addressed by first running the elite configurations on the new instances to avoid wasting too much computation time on possibly poor newly sampled candidate configurations. Table A.7 in the supplementary material shows the results for setting the number of new initial instances (T^{new}) to 0 (new instances are never added at the start of each race), 1 (the default setting) and 5. As previously observed for elitist irace [17], a larger value of T^{new} improves the performance of irace$_{cap}$ for the Spear scenario. While for the other scenarios, the differences are minor, there appears to be a tendency for the default value of 1 (or perhaps even a slightly larger value, such as 2 or 3) to result in the most robust behaviour.[2]

6   Comparison to Other Configurators

We compare the results obtained by irace$_{cap}$ with two other automatic configurators available in the literature, ParamILS and SMAC. Both have been widely used in the literature for running time minimisation. SMAC and ParamILS, as well as irace, were run using default settings. We chose not to include instance features in the configuration process and use only fully discretised configuration spaces; this was done to isolate as much as possible the impact of the new capping mechanism in irace, and to examine whether it would become competitive

---

[2] Setting T^{new} to 0 may be beneficial for scenarios with a very large cut-off time, as used by default in AClib for the CPLEX scenarios. This should help to aggressively bound the running time at the start of each race, by using the running times of the elite configurations, thus avoiding the high cost of evaluating possibly poor configurations with a very large cut-off time.

with other configurators that already used this technique. Considering features or non-discrete parameter spaces would introduce additional factors that are likely to affect performance beyond the impact of capping. Nevertheless, SMAC can also use instance features in the configuration process, which may improve its results; therefore, the results obtained here should be considered with caution for those scenarios in cases where these features are available. Yet, identifying how much of the improvement is due to instance features or due to differences in the capping methods between SMAC and other configurators would require a more extensive analysis that is left for future research. Additionally, SMAC and irace can handle real-valued parameters and, as already shown for SMAC in [12], doing so may further improve performance.

As mentioned previously, we ran irace$_{cap}$, SMAC and ParamILS using the PAR10 evaluation on the scenarios described in Sect. 5. Table 3 shows the mean PAR10 execution times obtained from 20 runs of the configurators. In the online supplementary material, we present results with other penalty factors from $\{1, 10, 100\}$. The table shows the p-values obtained from the Wilcoxon signed-rank test comparing the performance of the two configurators with the lowest mean PAR10 score. irace$_{cap}$ obtains the statistically significantly lowest mean on the Regions 200, Corlat, and Lingeling scenarios, while SMAC obtains the statistically significantly lowest mean on the Spear scenario. On the Regions 100 scenario, irace$_{cap}$ obtains the lowest mean performance value, though its performance is not statistically different from that of ParamILS.

It is known that trajectory-based local search methods, such as ParamILS, can exhibit high performance variability over multiple independent runs due to search stagnation. A common practice for dealing with this situation, and for reducing the overall wall-clock time of the configuration process by means

Table 3. Statistics over the mean PAR10 performance and percentage of timed-out instances from 20 runs of irace$_{cap}$, SMAC and ParamILS. Wilcoxon test p-values (significance 0.05). Significantly better results in bold and best mean in cursive.

		q25	mean	median	q75	sd	sd/mean	%timeout
Regions 100 p-value: 0.5958195	ParamILS	0.318	0.38	0.37	0.416	0.066	0.173	0.130
	SMAC	0.45	0.478	0.473	0.499	0.055	0.116	0.045
	irace$_{cap}$	0.32	*0.372*	0.365	0.395	0.057	0.154	0.095
Regions 200 p-value: 0.03276825	ParamILS	9.412	11.656	10.359	13.606	3.348	0.287	0.005
	SMAC	14.205	17.917	16.452	21.925	5.419	0.302	0.045
	irace$_{cap}$	8.854	**9.926**	9.349	10.533	1.459	0.147	0.005
Corlat p-value: 0.0083084	ParamILS	30.924	193.303	48.772	74.309	360.42	1.865	5.945
	SMAC	30.866	45.847	39.855	63.805	20.903	0.456	1.005
	irace$_{cap}$	12.24	**27.974**	26.763	33.902	19.426	0.694	0.62
Lingeling p-value: 0.0362339	ParamILS	250.115	292.529	298.942	327.679	51.497	0.176	9.023
	SMAC	266.792	283.907	289.153	298.64	25.121	0.088	8.758
	irace$_{cap}$	244.313	**263.651**	259.768	271.119	31.736	0.12	8.113
Spear p-value: 9.5e-06	ParamILS	3.037	88.083	12.094	40.877	188.929	2.145	2.815
	SMAC	1.6	**3.416**	1.746	2.511	3.733	1.093	0.05
	irace$_{cap}$	5.666	23.741	22.3	25.872	21.512	0.906	0.662

Table 4. Statistics over the mean PAR10 performance for the best-out-of-ten runs sampled from the 20 original runs of irace$_{cap}$, SMAC and ParamILS. Wilcoxon test p-values ($\alpha = 0.05$). Significantly better results are shown in bold-face and best mean values in italics.

		q25	mean	median	q75	sd	sd/mean
Regions 100 p-value: 8.83e-05	ParamILS	0.303	**0.305**	0.303	0.307	0.003	0.011
	SMAC	0.386	0.392	0.386	0.391	0.012	0.031
	irace$_{cap}$	0.312	0.314	0.314	0.314	0.002	0.007
Regions 200p-value: 0.0001417	ParamILS	8.589	8.871	8.999	9.139	0.266	0.03
	SMAC	9.82	11.2	10.106	12.989	1.784	0.159
	irace$_{cap}$	8.456	**8.523**	8.518	8.581	0.069	0.008
Corlatp-value: 0.0010241	ParamILS	8.284	10.58	9.58	9.58	4.163	0.393
	SMAC	17.801	19.898	19.832	19.832	3.016	0.152
	irace$_{cap}$	7.959	**8.194**	7.959	8.256	0.627	0.076
Lingelingp-value: 0.6341078	ParamILS	210.753	*218.874*	210.753	223.379	13.256	0.061
	SMAC	230.175	241.659	236.26	257.77	12.85	0.053
	irace$_{cap}$	220.093	220.212	220.093	220.212	0.212	0.001
Spearp-value: 9e-05	ParamILS	1.911	2.075	2.012	2.073	0.27	0.13
	SMAC	1.454	**1.462**	1.454	1.463	0.018	0.013
	irace$_{cap}$	2.154	2.497	2.184	2.497	0.581	0.233

of parallelisation, is to perform multiple independent configurator runs concurrently and to return the best configuration found in any of these. This may not always be feasible when the average running times of configurations on instances are high, e.g., in the range of hours, in which case the parallelisation features of irace would be very useful. Nevertheless, we mimic this commonly applied approach and compare the performance of ParamILS, SMAC and irace$_{cap}$ based on the following resampling approach: From the 20 values of mean PAR10 performance previously obtained for each configurator on the test set of each scenario, we sample 10 values (uniformly at random and without repetition) and take the best of these samples. This is equivalent to running the configurator 10 times and determining the best of the configurations thus obtained. We repeat this process 20 times to obtain 20 replicates of the experiment. Table 4 shows the results thus obtained. ParamILS benefits most from multiple independent runs, achieving the statistically significantly best performance on the Regions 100 scenario and the best mean performance (though not statistically significantly different from that of irace$_{cap}$) on Lingeling. irace$_{cap}$ produces the statistically significantly best results on Regions 200 and Corlat, while SMAC shows the best mean performance for the Spear scenario.

7   Conclusions

In this work, we have extended irace, an automatic algorithm configuration procedure primarily designed for solution quality optimisation, with an adaptive capping mechanism. We have demonstrated that this results in substantial improvements in the efficacy of irace for running time minimisation, and our new irace$_{cap}$ configurator reaches state-of-the-art performance on prominent configuration scenarios. This considerably broadens the range of configuration scenarios on which irace should be seen as one of the methods of choice.

In future work, it would be interesting to explore which characteristics of a configuration scenario makes it particularly amenable to different variants of adaptive capping. Furthermore, we would like to investigate under which circumstances irace$_{cap}$ performs better (or worse) than other state-of-the-art configurators, notably SMAC [12], ParamILS [13] and GGA++ [1]. We see this as an important step towards automatic selection of the configurator expected to perform best on a given scenario. This could improve the state of the art in automatic algorithm configuration and further boost the appeal of the programming by optimisation (PbO) software design paradigm [9], which crucially depends on maximally effective configurators.

Acknowledgments. This research was supported through funding through COMEX project (P7/36) within the Interuniversity Attraction Poles Programme of the Belgian Science Policy Office. Thomas Stützle acknowledges support from the Belgian F.R.S.-FNRS, of which he is a Senior Research Associate. Holger Hoos acknowledges support through an NSERC Discovery Grant.

References

1. Ansótegui, C., Malitsky, Y., Samulowitz, H., Sellmann, M., Tierney, K.: Model-based genetic algorithms for algorithm configuration. In: IJCAI 2015, pp. 733–739. IJCAI/AAAI Press, Menlo Park (2015)
2. Ansótegui, C., Sellmann, M., Tierney, K.: A gender-based genetic algorithm for the automatic configuration of algorithms. In: Gent, I.P. (ed.) CP 2009. LNCS, vol. 5732, pp. 142–157. Springer, Heidelberg (2009). doi:10.1007/978-3-642-04244-7_14
3. Babić, D., Hutter, F.: Spear theorem prover. In: SAT 2008: Proceedings of the SAT 2008 Race (2008)
4. Balaprakash, P., Birattari, M., Stützle, T.: Improvement strategies for the F-Race algorithm: sampling design and iterative refinement. In: Bartz-Beielstein, T., Blesa Aguilera, M.J., Blum, C., Naujoks, B., Roli, A., Rudolph, G., Sampels, M. (eds.) HM 2007. LNCS, vol. 4771, pp. 108–122. Springer, Heidelberg (2007). doi:10.1007/978-3-540-75514-2_9
5. Biere, A.: Yet another local search solver and lingeling and friends entering the SAT competition 2014. In: Belov, A., et al. (ed.) Proceedings of SAT Competition 2014. Science Series of Publications B, vol. B-2014-2, pp. 39–40. University of Helsinki (2014)
6. Birattari, M.: The Problem of Tuning Metaheuristics as Seen from a Machine Learning Perspective. Ph.D. thesis, Université Libre de Bruxelles, Belgium (2004)
7. Birattari, M., Yuan, Z., Balaprakash, P., Stützle, T.: F-race and iterated F-race: an overview. In: Bartz-Beielstein, T., Chiarandini, M., Paquete, L., Preuss, M. (eds.) Experimental Methods for the Analysis of Optimization Algorithms, pp. 311–336. Springer, Heidelberg (2010). doi:10.1007/978-3-642-02538-9_13
8. Hoos, H.H.: Automated algorithm configuration and parameter tuning. In: Hamadi, Y., Monfroy, E., Saubion, F. (eds.) Autonomous Search, pp. 37–71. Springer, Heidelberg (2012). doi:10.1007/978-3-642-21434-9_3

9. Hoos, H.H.: Programming by optimization. Commun. ACM **55**(2), 70–80 (2012)
10. Hoos, H.H., Stützle, T.: Stochastic Local Search-Foundations and Applications. Morgan Kaufmann Publishers, San Francisco (2005)
11. Hutter, F., Hoos, H.H., Leyton-Brown, K.: Automated configuration of mixed integer programming solvers. In: Lodi, A., Milano, M., Toth, P. (eds.) CPAIOR 2010. LNCS, vol. 6140, pp. 186–202. Springer, Heidelberg (2010). doi:10.1007/978-3-642-13520-0_23
12. Hutter, F., Hoos, H.H., Leyton-Brown, K.: Sequential model-based optimization for general algorithm configuration. In: Coello, C.A.C. (ed.) LION 2011. LNCS, vol. 6683, pp. 507–523. Springer, Heidelberg (2011). doi:10.1007/978-3-642-25566-3_40
13. Hutter, F., Hoos, H.H., Leyton-Brown, K., Stützle, T.: ParamILS: an automatic algorithm configuration framework. J. Artif. Intell. Res. **36**, 267–306 (2009)
14. Hutter, F., Lindauer, M.T., Balint, A., Bayless, S., Hoos, H.H., Leyton-Brown, K.: The configurable SAT solver challenge (CSSC). Artif. Intell. **243**, 1–25 (2017)
15. Hutter, F., López-Ibáñez, M., Fawcett, C., Lindauer, M., Hoos, H.H., Leyton-Brown, K., Stützle, T.: AClib: a benchmark library for algorithm configuration. In: Pardalos, P.M., Resende, M.G.C., Vogiatzis, C., Walteros, J.L. (eds.) LION 2014. LNCS, vol. 8426, pp. 36–40. Springer, Cham (2014). doi:10.1007/978-3-319-09584-4_4
16. IBM: ILOG CPLEX optimizer. http://www.ibm.com/software/integration/optimization/cplex-optimizer/
17. López-Ibáñez, M., Dubois-Lacoste, J., Pérez Cáceres, L., Stützle, T., Birattari, M.: The irace package: iterated racing for automatic algorithm configuration. Oper. Res. Perspect. **3**, 43–58 (2016)
18. López-Ibáñez, M., Dubois-Lacoste, J., Stützle, T., Birattari, M.: The irace package, iterated race for automatic algorithm configuration. Technical report TR/IRIDIA/2011-004, IRIDIA, Université Libre de Bruxelles, Belgium (2011)
19. Lourenço, H.R., Martin, O., Stützle, T.: Iterated local search: framework and applications. In: Gendreau, M., Potvin, J.Y. (eds.) Handbook of Metaheuristics. International Series in Operations Research & Management Science, vol. 146, pp. 363–397. Springer, Boston (2010). doi:10.1007/978-1-4419-1665-5_12
20. Pérez Cáceres, L., López-Ibáñez, M., Hoos, H.H., Stützle, T.: An experimental study of adaptive capping in irace: Supplementary material (2017). http://iridia.ulb.ac.be/supp/IridiaSupp.2016-007/
21. Pérez Cáceres, L., López-Ibáñez, M., Stützle, T.: An analysis of parameters of irace. In: Blum, C., Ochoa, G. (eds.) EvoCOP 2014. LNCS, vol. 8600, pp. 37–48. Springer, Heidelberg (2014). doi:10.1007/978-3-662-44320-0_4

Duality Gap Analysis of Weak Relaxed Greedy Algorithms

Sergei P. Sidorov[1]($\boxtimes$) and Sergei V. Mironov[2]

[1] Mechanics and Mathematics Department, Saratov State University,
Saratov, Russian Federation
SidorovSP@info.sgu.ru
[2] Computer Science and Information Technologies Department,
Saratov State University, Saratov, Russian Federation
MironovSV@info.sgu.ru

Abstract. Many problems in machine learning can be presented in the form of convex optimization problems with objective function as a loss function. The paper examines two weak relaxed greedy algorithms for finding the solutions of convex optimization problems over convex hulls of atomic sets. Such problems arise as the natural convex relaxations of cardinality-type constrained problems, many of which are well-known to be NP-hard. Both algorithms utilize one atom from a dictionary per iteration, and therefore, guarantee designed sparsity of the approximate solutions. Algorithms employ the so called 'gradient greedy step' that maximizes a linear functional which uses gradient information of the element obtained in the previous iteration. Both algorithms are 'weak' in the sense that they solve the linear subproblems at the gradient greedy step only approximately. Moreover, the second algorithm employs an approximate solution at the line-search step. Following ideas of [5] we put up the notion of the duality gap, the values of which are computed at the gradient greedy step of the algorithms on each iteration, and therefore, they are inherent upper bounds for primal errors, i.e. differences between values of objective function at current and optimal points on each step. We obtain dual convergence estimates for the weak relaxed greedy algorithms.

Keywords: Greedy algorithms · Convex optimization · Sparsity · Duality gap

1 Introduction

Let X be a Banach space with norm $\| \cdot \|$. Let E be a convex function defined on X. The problem of convex optimization is to find an approximate solution to the problem

$$E(x) \to \min_{x \in X}. \tag{1}$$

Many problems in machine learning can be reduced to the problem (1) with E as a loss function [1]. In many real applications it is required that the optimal

R. Battiti et al. (Eds.): LION 2017, LNCS 10556, pp. 251–262, 2017.
https://doi.org/10.1007/978-3-319-69404-7_18

solution x^* of (1) should have a simple structure, e.g. be a *finite* linear combination of elements from a dictionary $\mathcal{D}$ in X. In another words, x^* should be a sparse element with respect to the dictionary $\mathcal{D}$ in X. Of course, one can substitute the requirement of sparsity by a constraint on cardinality (i.e. the limit on the number of elements used in linear combinations of elements from the dictionary $\mathcal{D}$ to construct a solution of the problem (1)). However, it many cases the optimization problems with cardinality-type constraint are NP-complete. By this reason, practitioners and researchers in real applications choose to use greedy methods. By its design, greedy algorithms is capable of producing sparse solutions.

A set of elements $\mathcal{D}$ from the space X is called a *dictionary* (see, e.g. [15]) if each element $g \in \mathcal{D}$ has norm bounded by one, $\|g\| \leq 1$, and the closure of span $\mathcal{D}$ is X, i.e. $\overline{\text{span}\,\mathcal{D}} = X$. A dictionary $\mathcal{D}$ is called symmetric if $-g \in \mathcal{D}$ for every $g \in \mathcal{D}$. In this paper we assume that the dictionary $\mathcal{D}$ is symmetric.

As it was pointed out, practitioners and researchers would like to find the solutions of the optimization problem (1), which are sparse with respect to the dictionary $\mathcal{D}$, i.e. they are looking for solving the following problem:

$$E(x) \to \inf_{x \in \Sigma_m(\mathcal{D})}, \tag{2}$$

where $\Sigma_m(\mathcal{D})$ is the set of all m-term polynomials with respect to $\mathcal{D}$:

$$\Sigma_m(\mathcal{D}) = \left\{ x \in X \ : \ x = \sum_{i=1}^{m} c_i g_i, \ g_i \in \mathcal{D} \right\}. \tag{3}$$

One of the apparent choices among constructive methods for finding the best m-term approximations are greedy algorithms. The design of greedy algorithms allows us to obtain sparse solutions with respect to $\mathcal{D}$. Perhaps, the Frank-Wolfe method [2], which is also known as the "conditional gradient" method [3], is one of the most prominent algorithms for finding optimal solutions of constrained convex optimization problems. Important contributions to the development of Frank-Wolfe type algorithms can be found in [4–6]. The paper [5] provides general primal-dual convergence results for Frank-Wolfe-type algorithms by extending the duality concept presented in the work [4]. Recent convergence results for greedy algorithms one can find in the works [7–14, 16–18, 20].

This paper examines two weak relaxed greedy algorithms

– Weak Relaxed Greedy Algorithm (WRGA(co)),
– Weak Relaxed Greedy Algorithm with Error δ (WRGA(δ))

for finding solutions of convex optimization problem, which are sparse with respect to some dictionary, in Banach spaces. Primal convergence results for the weak relaxed greedy algorithms were obtained in [15, 18]. In this paper, extending the ideas of [4,5] we force into application the notion of the duality gap for weak relaxed greedy algorithms to obtain dual convergence estimates for sparse-constrained convex optimization problems of type (2). In contrast to papers [15, 18, 19], in this paper we focus on obtaining dual convergence results based on duality gap analysis.

It should be noted, that the paper of Temlyakov [15] shows that the greedy algorithms (WRGA(co) and WRGA(δ)) for finding the solutions of (2) with respect to the dictionary $\mathcal{D}$ solve the problem (1) as well. In many real applications the dimension of the search space while is finite, but is too large. Therefore, our interest lies in obtaining estimates on the rate of convergence not depending on the dimension of X. Obviously, results for the infinite Banach spaces provide such estimates on the convergence rate. Following [15], we examine the problem in an infinite dimensional Banach space setting.

Note that duality for global convex minimization problems has been well-examined. The recent paper [21] shows that not all nonconvex global minimization problems are NP-hard and that the complexity of a problem depends essentially on modeling and intrinsic symmetry of the problem.

2 Weak Relaxed Greedy Algorithms

We will suppose that function E is Fréchet differentiable. We note that it follows from convexity of E that for any x, y

$$E(y) \geq E(x) + \langle E'(x), y - x \rangle,$$

where $E'(x)$ denotes Fréchet differential of E at x.

To solve the optimization problem (2), the paper [15] uses iterative search optimizer described in Algorithm 1. The algorithm for each $m \geq 1$ finds the next element G_m by means of induction with use of

– the current element G_{m-1}
– element ϕ_m obtained in the gradient greedy step.

In the gradient greedy step we maximize a functional which uses gradient information at element G_{m-1} of X obtained in the previous iteration of the algorithm. Algorithm 1 belongs to the class of Frank-Wolfe type methods, since at each current element G_{m-1} of X it drifts to a minimizer of the linearization of the objective function E taken over the feasible set, which in this case is the dictionary $\mathcal{D}$.

We would like to note that the gradient greedy step of WRGA(co) is looking for supremum over the dictionary $\mathcal{D}$ (not its convex hull $A_1(\mathcal{D})$), since points from $A_1(\mathcal{D})$ are mostly linear combinations of infinite number of the dictionary elements. Thus, the optimal solution obtained by this way is not obliged to be sparse with respect to $\mathcal{D}$.

Let $\tau := \{t_m\}_{m=1}^{\infty}$, $t_m \in [0, 1]$, be a weakness sequence. To solve the subproblem $\sup_{s \in \mathcal{D}} \langle -E'(G_{m-1}), s - G_{m-1} \rangle$ exactly may be too expensive in many real cases. Algorithm 1 uses the weakness sequence τ in the gradient greedy step to find approximate minimizer ϕ_m instead, which has approximation (multiplicative) quality at least t_m in step m. That is why the algorithm is called "weak".

Algorithm 1. WEAK RELAXED GREEDY ALGORITHM (WRGA(CO))

---

begin
   · Let $G_0 = 0$;
   for each $m = 1, 2, \ldots, M$ **do**
      · (*Gradient greedy step*) Find the element $\phi_m \in \mathcal{D}$ such that
      $\langle -E'(G_{m-1}), \phi_m - G_{m-1}\rangle \geq t_m \sup\limits_{s \in \mathcal{D}}\langle -E'(G_{m-1}), s - G_{m-1}\rangle$;
      · (*Line-search step*) Find the real number $0 \leq \lambda_m \leq 1$, such that
      $E\left((1 - \lambda_m)G_{m-1} + \lambda_m\phi_m\right) = \inf\limits_{0 \leq \lambda \leq 1} E\left((1 - \lambda)G_{m-1} + \lambda\phi_m\right)$;
      · (*Update step*) $G_m = (1 - \lambda_m)G_{m-1} + \lambda_m\phi_m$;
end

The line-search step of Algorithm 1 finds the best point lying on the line segment between the current point G_{m-1} and ϕ_m.

Let $\Omega := \{x \in X : E(x) \leq E(0)\}$ and suppose that Ω is bounded. As it turns out, the convergence analysis of greedy algorithms essentially depends on a measure of "non-linearity" of the objective function E over set Ω, which can be depicted via the modulus of smoothness of function E.

Let us remind that the modulus of smoothness of function E on the bounded set Ω can be defined as

$$\rho(E, u) = \frac{1}{2} \sup_{x \in \Omega, \|y\|=1} |E(x + uy) + E(x - uy) - 2E(x)|, \quad u > 0. \qquad (4)$$

E is called uniformly smooth function on Ω if $\lim_{u \to 0} \rho(E, u)/u = 0$.

Let $A_1(\mathcal{D})$ denote the closure (in X) of the convex hull of $\mathcal{D}$.

Exploiting the geometric properties of the objective function E, the paper [15] proves the following estimate of the convergence rate of the WRGA(co).

Theorem 1. *Let E be a uniformly smooth convex function with modulus of smoothness $\rho(E, u) \leq \gamma u^q$, $1 < q \leq 2$, $\gamma > 0$. Then, for a weakness sequence $\tau = \{t_k\}_{k=1}^{\infty}$, $0 < t_k \leq 1$, $k = 1, 2, \ldots$, we have for any $f \in A_1(\mathcal{D})$ that*

$$E(G_m) - E(f) \leq \left(1 + C_1(q, \gamma) \sum_{k=1}^{m} t_k^p\right)^{1-q}, \quad p := \frac{q}{q-1}, \quad m \geq 2, \qquad (5)$$

where $C_1(q, \gamma)$ is positive constants not depending on k.

The paper [18] notes that values of E may not be calculated exactly for many real application problems. Moreover, very often the exact optimal value of λ in the problem

$$\inf_{0 \leq \lambda \leq 1} E\left((1 - \lambda)G_{m-1} + \lambda\phi\right) \qquad (6)$$

in Step 2 of WRGA(co) can not be found. Therefore, the paper [18] examines the weak relaxed greedy algorithm with error δ (Algorithm 2), which is a slightly modified version of WRGA(co) with changing in the second step. In comparison with the WRGA(co), WGAFR(δ) uses the error δ in the line-search step with the aim of getting the optimization problem of a different kind. It may have better complexity than the original optimization problem (6).

Algorithm 2. WEAK RELAXED GREEDY ALGORITHM WITH ERROR δ
(WRGA(δ))

---

begin
> · Let $\delta > 0$ and $G_0 = 0$;
> **for each** $m = 1, 2, \ldots, M$ **do**
>> · (*Gradient greedy step*) Find the element $\phi_m \in \mathcal{D}$ such that
>> $$\langle -E'(G_{m-1}), \phi_m - G_{m-1} \rangle \geq t_m \sup_{s \in \mathcal{D}} \langle -E'(G_{m-1}), s - G_{m-1} \rangle;$$
>> · (*Line-search step*) Find the real number $0 \leq \lambda_m \leq 1$, such that
>> $$E\left((1 - \lambda_m)G_{m-1} + \lambda_m \phi_m\right) \leq \inf_{0 \leq \lambda \leq 1} E\left((1 - \lambda)G_{m-1} + \lambda \phi_m\right) + \delta \ ;$$
>> · (*Update step*) $G_m = (1 - \lambda_m)G_{m-1} + \lambda_m \phi_m;$

end

The following estimate of the convergence rate of the WRGA(δ) is proved in the paper [18].

Theorem 2. *Let E be uniformly smooth on $A_1(\mathcal{D})$ whose modulus of smoothness $\rho(E, u)$ satisfies $\rho(E, u) \leq \gamma u^q$, $1 < q \leq 2$, $\gamma > 0$. If $t_k = \theta$, $k = 1, 2, \ldots$, then WRGA(δ) satisfies*

$$E(G_m) - E^* \leq C(q, \gamma, \theta, E)m^{1-q}, \ m \leq \delta^{-1/q},$$

where $E^* := \inf_{f \in A_1(\mathcal{D})} E(x)$.

3   Dual Convergence Results

3.1   Duality Gap

Following ideas of [5], let us introduce the notion of the duality gap for optimization problem as follows.

Definition 1. *Let $G \in A_1(\mathcal{D})$. Let us define the (surrogate) duality gap $g(G)$ at element G by*

$$g(G) := \sup_{s \in \mathcal{D}} \langle E'(G), G - s \rangle. \tag{7}$$

The decisive property of the duality gap is the following one.

Proposition 1. *Let E be a uniformly smooth convex function defined on Banach space X. Let $x^* = \arg\min_{x \in A_1(\mathcal{D})} E(x)$. For any $G \in A_1(\mathcal{D})$*

$$E(G) - E(x^*) \leq g(G).$$

Proof. Since E is convex on X we have for any $y \in A_1(\mathcal{D})$

$$E(y) \geq E(x) + \langle E'(x), y - x \rangle \geq E(x) - \sup_{s \in A_1(\mathcal{D})} \langle E'(x), x - s \rangle. \qquad (8)$$

Lemma 2.2 in [15] states that

$$\sup_{s \in A_1(\mathcal{D})} \langle E'(x), x - s \rangle = \sup_{s \in \mathcal{D}} \langle E'(x), x - s \rangle. \qquad (9)$$

Then Proposition follows from (8) and (9) with $y = x^*$.

Proposition 1 shows that the duality gap $g(G)$ is a bound for the current approximation $E(G)$ to the optimal solution $E(x^*)$.

The duality gap g is calculated as a derivative product on every iteration of both Algorithms 1 and 2. If the linearized problem at the gradient greedy step for element G_{m-1} has optimal solution ϕ_m, then the element ϕ_m is a reference for the current duality gap

$$g(G_{m-1}) = \langle E'(G_{m-1}), G_{m-1} - \phi_m \rangle.$$

Such references for the approximation quality of current iteration can be used as a stopping criterion, or to verify the numerical stability of an optimizer.

Theorems 1 and 2 give upper estimates for primal errors for WRGA(co) and WRGA(δ), respectively. In the next subsections we will obtain dual estimates for the algorithms in terms of duality gap g.

3.2   Dual Convergence Result for WRGA(co)

We need some preliminary results.

Lemma 1. *Let E be a uniformly smooth convex function defined on Banach space X. Let $\rho(E, u)$ denote the modulus of smoothness of E. Then the following inequality holds:*

$$E(G_m) \leq E(G_{m-1}) + \inf_{0 \leq \lambda \leq 1} (-\lambda t_m g(G_{m-1}) + 2\rho(E, 2\lambda)), \quad m = 1, 2, \ldots.$$

Proof. The definition of G_m in Step 3 of WRGA(co) implies that

$$G_m = (1 - \lambda_m)G_{m-1} + \lambda_m \phi_m = G_{m-1} + \lambda_m(\phi_m - G_{m-1}),$$

and

$$E(G_m) = \inf_{0 \leq \lambda \leq 1} E(G_{m-1} + \lambda(\phi_m - G_{m-1})). \qquad (10)$$

It follows from Lemma 2.3 of [15] that

$$E(G_{m-1} + \lambda(\phi_m - G_{m-1}))$$
$$\leq E(G_{m-1}) - \lambda \langle -E'(G_{m-1}), \phi_m - G_{m-1} \rangle + 2\rho(E, 2\lambda). \quad (11)$$

The step 1 of WRGA(co) gives

$$\langle -E'(G_{m-1}), \phi_m - G_{m-1} \rangle$$
$$\geq t_m \sup_{s \in \mathcal{D}} \langle -E'(G_{m-1}), s - G_{m-1} \rangle = t_m g(G_{m-1}). \quad (12)$$

Then the lemma follows from (10), (11) and (12).

Lemma 2. *Let* $\tau = \{t_k\}_{k=1}^{\infty}$, $\theta < t_k \leq 1$, $k = 1, 2, \ldots$, *for a fixed real* $\theta > 0$. *Denote*

$$s_m := s_m(m, \tau) := \sum_{k=1}^{m} t_k^p, \quad p = \frac{q}{q-1}, \quad 1 < q \leq 2. \quad (13)$$

Let $0 < \mu < 1$ *be a real and* M *be an integer. Then*

$$s_{[\mu M]+1}^{1-q} \leq (\mu \theta^p)^{1-q} s_M^{1-q},$$

where square brackets denote the integer part.

Proof. Denote $m_0 := [\mu M] + 1$. We have $\frac{m_0}{M} \geq \mu$ and

$$\frac{\sum_{k=1}^{m_0} t_k^p}{\sum_{k=1}^{M} t_k^p} \geq \frac{\sum_{k=1}^{m_0} \theta^p}{\sum_{k=1}^{M} 1} = \frac{m_0 \theta^p}{M} \geq \mu \theta^p, \quad \text{or} \quad \frac{s_{[\mu M]+1}}{s_M} \geq \mu \theta^p.$$

Theorem 3. *Let* E *be a uniformly smooth convex function defined on Banach space* X. *Let* $\rho(E, u)$ *be the modulus of smoothness of* E *and suppose that* $\rho(E, u) \leq \gamma u^q$, $1 < q \leq 2$. *Let* $\tau = \{t_k\}_{k=1}^{\infty}$, $\theta < t_k \leq 1$, $k = 1, 2, \ldots$, *be a weakness sequence,* $\theta > 0$. *Assume that WRGA(co) is run for* $M > 2$ *iterations. Then there is an iterate* $1 \leq \tilde{m} \leq M$ *such that*

$$g(G_{\tilde{m}}) \leq \beta C_2 \left(\sum_{k=1}^{M} t_k^p \right)^{1-q}, \quad p := \frac{q}{q-1} \quad (14)$$

where $C_2 := C_2(q, \gamma) := (\min\{1, C_1(q, \gamma)\})^{1-q}$ *and* β *depends only on* M, q, γ, θ.

Proof. It follows from Theorem 1 that for any $f \in A_1(\mathcal{D})$

$$E(G_m) - E(f) \leq C_2 \left(\sum_{k=1}^{m} t_k^p \right)^{1-q}, \quad p := \frac{q}{q-1}, \quad m \geq 2. \quad (15)$$

Let us assume (by contradiction) that

$$g(G_m) \geq \beta C_2 s_M^{1-q}, \quad (16)$$

for all $[\mu M] + 1 \leq m \leq M$ ($0 < \mu < 1$ is fixed and will be chosen later), s_M is defined in (13).

It easy to check that $t_1^q s_m^{1-q} \leq 1$ for all $m = 1, 2, \ldots$. Then Lemma 1 with $\lambda = t_1^q s_m^{1-q}$ implies that

$$E(G_{m+1}) - E(f) \leq E(G_m) - E(f) - t_1^q s_m^{1-q} t_m g(G_m) + 2\gamma (2 t_1^q s_m^{1-q})^q. \quad (17)$$

Applying our assumption (16) to the inequality (17), we obtain

$$E(G_{m+1}) - E(f) \leq E(G_m) - E(f) - \beta C_2 t_1^q t_m s_m^{1-q} s_M^{1-q} + \gamma 2^{q+1} t_1^{q^2} s_m^{q(1-q)}. \quad (18)$$

We will need the following inequalities:

1. $\theta \leq t_k \leq 1$, $k = 1, 2, \ldots$;
2. $s_{[\mu M]+1}^{1-q} \leq (\mu \theta^p)^{1-q} s_M^{1-q}$ (Lemma 2);
3. since $[\mu M] + 1 \leq m \leq M$, we have $s_{[\mu M]+1} \leq s_m \leq s_M$, and consequently,

$$s_{[\mu M]+1}^{1-q} \geq s_m^{1-q} \geq s_M^{1-q}.$$

It follows from (18) that

$$E(G_{m+1}) - E^*$$
$$\leq E(G_m) - E^* - \beta C_2 \theta^{q+1} s_M^{2(1-q)} + \gamma 2^{q+1} (\mu \theta^p)^{q(1-q)} s_M^{q(1-q)}, \quad (19)$$

where $E^* := \inf_{f \in A_1(\mathcal{D})} E(x)$. Let us write the chain of inequalities for all m_0 from $[\mu M] + 1$ to M, then

$$E(G_M) - E^* \leq E(G_{m_0}) - E^* - (M - m_0) s_M^{1-q} \Theta_1$$
$$\leq C_2 s_M^{1-q} - (M(1-\mu) - 1) s_M^{1-q} \Theta_1 = s_M^{1-q} [C_2 - (M(1-\mu) - 1)\Theta_1], \quad (20)$$

where

$$\Theta_1 := \beta C_2 \theta^{q+1} s_M^{1-q} - \gamma 2^{q+1} \mu^{q(1-q)} \theta^{-q^2} s_M^{-(q-1)^2}.$$

Let us take any β satisfying

$$\beta > \frac{\frac{C_2}{M(1-\mu)-1} + \gamma 2^{q+1} \mu^{q(1-q)} \theta^{-q^2} s_M^{-(q-1)^2}}{C_2 \theta^{q+1} s_M^{1-q}}$$

then we obtain

$$E(G_M) - E^* < 0$$

that can not be impossible. The smallest value of β leads to a better estimate in (14). To be sure that β is smallest we can choose the parameter μ as follows:

$$\mu := \arg \min_{0 \leq \mu \leq 1} \left(\frac{C_2}{M(1-\mu)-1} + \gamma 2^{q+1} \mu^{q(1-q)} \theta^{-q^2} s_M^{-(q-1)^2} \right).$$

3.3   Dual Convergence Result for WRGA(δ)

We need some lemmas to prove the main result.

Lemma 3. *Let E be a uniformly smooth convex function defined on Banach space X. Let $\rho(E, u)$ denote the modulus of smoothness of E. Then the following inequality holds for the WRGA(δ):*

$$E(G_m) \leq E(G_{m-1}) + \inf_{\lambda \geq 0}(-\lambda t_m g(G_{m-1}) + 2\rho(E, C_0\lambda)) + \delta, \quad m = 1, 2, \ldots,$$

where C_0 does not depend on m.

Proof. From the definition of G_m in the update step of WRGA(δ) we have

$$G_m = (1 - \lambda_m)G_{m-1} + \lambda_m \phi_m.$$

The line-search step of WRGA(δ) implies

$$E(G_m) \leq \inf_{0 \leq \lambda \leq 1} E(G_{m-1} - \lambda G_{m-1} + \lambda \phi_m) + \delta. \tag{21}$$

It follows from Lemma 1.1 of [15] that

$$E(G_{m-1} - \lambda G_{m-1} + \lambda \phi_m)) \leq E(G_{m-1})$$
$$- \lambda \langle -E'(G_{m-1}), \phi_m - G_{m-1} \rangle + 2\rho(E, \lambda\|\phi_m - G_{m-1}\|). \tag{22}$$

Using the gradient greedy step of WRGA(δ) and the definition of duality gap (7), we have

$$\langle -E'(G_{m-1}), \phi_m - G_{m-1} \rangle$$
$$\geq t_m \sup_{s \in D} \langle -E'(G_{m-1}), s - G_{m-1} \rangle = t_m g(G_{m-1}). \tag{23}$$

It follows from (21), (22) and (23) that

$$E(G_m) \leq E(G_{m-1}) + \inf_{\lambda \geq 0}(-\lambda t_m g(G_{m-1}) + 2\rho(E, \lambda\|\phi_m - G_{m-1}\|)).$$

It follows from $E(G_{m-1}) \leq E(0)$ that $G_{m-1} \in \Omega$. Our assumption on boundedness of Ω implies that there exists a constant C_1 such that $\|G_{m-1}\| \leq C_1$. Since $\phi_m \in D$, we have $\|\phi_m\| \leq 1$. Thus,

$$\|G_{m-1} - \phi_m\| \leq C_1 + 1 =: C_0.$$

This completes the proof of Lemma.

Theorem 4. *Let E be a uniformly smooth convex function defined on Banach space X. Let $\rho(E, u)$ be the modulus of smoothness of E and suppose that $\rho(E, u) \leq \gamma u^q$, $1 < q \leq 2$. Let $\tau = \{t_m\}_{m=1}^{\infty}$, $t_k = \theta$, $k = 1, 2, \ldots$, be a weakness sequence. Assume that WRGA(δ) is run for $0 < M \leq \delta^{-\frac{1}{q}}$ iterations. Then there is an iterate $1 \leq \tilde{m} \leq M$ such that*

$$g(G_{\tilde{m}}) \leq \beta C(E, q, \gamma)M^{1-q}. \tag{24}$$

Proof. It follows from Theorem 2 that

$$E(G_m) - E^* \leq C(E,q,\gamma)m^{1-q}, \quad m \leq \delta^{-\frac{1}{q}}, \tag{25}$$

where $E^* := \inf_{f \in A_1(\mathcal{D})} E(x)$. Let us suppose that

$$g(G_m) > \beta C(E,q,\gamma)M^{1-q} \tag{26}$$

for all $[\mu M] + 1 \leq m \leq M$, $0 < \mu < 1$ (μ is fixed and will be chosen later).
It follows from Lemma 3 with $\lambda = m^{1-q}$,

$$E(G_{m+1}) - E^* \leq E(G_m) - E^* - m^{1-q}t_m g(G_m) + 2\gamma(C_0 m^{1-q})^q + \delta. \tag{27}$$

Using our assumption (26), the inequality (27) can be rewritten in the form

$$\begin{aligned} E(G_{m+1}) - E^* \\ \leq E(G_m) - E^* - m^{1-q}t_m \beta C(E,q,\gamma)M^{1-q} + 2\gamma(C_0 m^{1-q})^q + \delta. \end{aligned} \tag{28}$$

Since $m_0 \leq m \leq M$, where $m_0 := [\mu M] + 1$, the following inequalities hold:

1. $m_0^{1-q} \leq \mu^{1-q}M^{1-q}$;
2. $m_0^{1-q} \geq m^{1-q} \geq M^{1-q}$.

Then (28) gives

$$\begin{aligned} E(G_{m+1}) - E^* \\ \leq E(G_m) - E^* - \beta\theta C(E,q,\gamma)M^{2(1-q)} + 2\gamma(C_0)^q \mu^{q(1-q)} M^{q(1-q)} + \delta. \end{aligned} \tag{29}$$

If we write the chain of inequalities for all $m = m_0, \ldots, M$, we get

$$\begin{aligned} E(G_{m+1}) - E^* &\leq E(G_{m_0}) - E^* - (M - m_0)M^{1-q}\Theta_2 \\ &\leq C(E,q,\gamma)m_0^{1-q} - (M(1 - \mu) - 1)M^{1-q}\Theta_2 \\ &\leq C(E,q,\gamma)\mu^{1-q}M^{1-q} - (M(1 - \mu) - 1)M^{1-q}\Theta_2 \\ &= M^{1-q}\Big(C(E,q,\gamma)\mu^{1-q} - (M(1 - \mu) - 1)\Theta_2 \Big), \end{aligned}$$

where

$$\Theta_2 := \beta M^{1-q}\theta C(E,q,\gamma) - 2\gamma(C_0)^q \mu^{q(1-q)} M^{(1-q)(q-1)} + \frac{\delta}{M^{1-q}}.$$

If we take

$$\beta > \frac{\frac{C(E,q,\gamma)\mu^{1-q}}{M(1-\mu)-1} + 2\gamma(C_0)^q \mu^{q(1-q)} M^{(1-q)(q-1)} - \frac{\delta}{M^{1-q}}}{M^{1-q}\theta C(E,q,\gamma)}$$

then we get $E(G_m) - E^* < 0$ which is impossible. We are interested in obtaining a better value of the constant β in (24). The smallest β can be attained if we choose μ as follows:

$$\mu := \arg \min_{0 \leq \mu \leq 1} \left(\frac{C(E,q,\gamma)\mu^{1-q}}{M(1-\mu)-1} + 2\gamma(C_0)^q \mu^{q(1-q)} M^{(1-q)(q-1)} - \frac{\delta}{M^{1-q}} \right).$$

4   Conclusion

Theorems 1 and 2 cited in Sect. 2 show that primal errors for the weak relaxed greedy algorithms are small and heavily depend on geometric properties of the objective function E. On the other hand, the paper [5] remarks that very often both the optimal value E^* and the constant γ in the modulus of smoothness of E are unknown, and therefore, estimates for the quality of current approximation to optimal solution are considerably in demand. Following ideas of [5], we defined the notion of the duality gap by the equality (7). The values of duality gap are calculated on each iteration of both WRGA(co) and WRGA(δ) at the gradient greedy step, and therefore, they are inherent upper bounds for primal errors, i.e. differences between values of objective function at current and optimal points on each step. We obtain dual convergence estimates for the weak relaxed greedy algorithms in Theorems 3 and 4.

Acknowledgments. This work was supported by the Russian Fund for Basic Research under Grant 16-01-00507. We would like to thank the reviewers profoundly for very helpful suggestions and commentaries.

References

1. Bubeck, S.: Convex optimization: algorithms and complexity. Found. Trends Mach. Learn. **8**(3–4), 231–358 (2015)
2. Frank, M., Wolfe, P.: An algorithm for quadratic programming. Naval Res. Logis. Quart. **3**, 95–110 (1956)
3. Levitin, E.S., Polyak, B.T.: Constrained minimization methods. USSR Comp. Math. & M. Phys. **6**(5), 1–50 (1966)
4. Clarkson, K.L.: Coresets, sparse greedy approximation, and the Frank-Wolfe algorithm. ACM Trans. Algorithms **6**(4), 1–30 (2010)
5. Jaggi, M., Frank-Wolfe, R.: Projection-free sparse convex optimization. In: Proceedings of the 30th International Conference on Machine Learning (ICML 2013), pp. 427–435 (2013)
6. Freund, R.M., Grigas, P.: New analysis and results for the Frank-Wolfe method. Math. Program. **155**(1), 199–230 (2016)
7. Friedman, J.: Greedy function approximation: a gradient boosting machine. Ann. Stat. **29**(5), 1189–1232 (2001)
8. Davis, G., Mallat, S., Avellaneda, M.: Adaptive greedy approximation. Constr. Approx. **13**, 57–98 (1997)
9. Zhang, Z., Shwartz, S., Wagner, L., Miller, W.: A greedy algorithm for aligning DNA sequences. J. Comput. Biol. (1–2), 203–214 (2000)
10. Huber, P.J.: Projection pursuit. Ann. Statist. **13**, 435–525 (1985)
11. Jones, L.: On a conjecture of Huber concerning the convergence of projection pursuit regression. Ann. Statist. **15**, 880–882 (1987)
12. Barron, A.R., Cohen, A., Dahmen, W., DeVore, R.A.: Approximation and learning by Greedy algorithms. Ann. Stat. **36**(1), 64–94 (2008)
13. DeVore, R.A., Temlyakov, V.N.: Some remarks on greedy algorithms. Adv. Comput. Math. **5**, 173–187 (1996)

14. Konyagin, S.V., Temlyakov, V.N.: A remark on greedy approximation in Banach spaces. East J. Approx. **5**(3), 365–379 (1999)
15. Temlyakov, V.N.: Greedy approximation in convex optimization. Constr. Approx. **41**(2), 269–296 (2015)
16. Nguyen, H., Petrova, G.: Greedy strategies for convex optimization. Calcolo **41**(2), 1–18 (2016)
17. Temlyakov, V.N.: Dictionary descent in optimization. Anal. Math. **42**(1), 69–89 (2016)
18. DeVore, R.A., Temlyakov, V.N.: Convex optimization on Banach spaces. Found. Comput. Math. **16**(2), 369–394 (2016)
19. Temlyakov, V.N.: Convergence and rate of convergence of some greedy algorithms in convex optimization. Proc. Steklov Inst. Math. **293**(1), 325–337 (2016)
20. Sidorov, S., Mironov, S., Pleshakov, M.: Dual greedy algorithm for conic optimization problem. CEUR Workshop Proc. **1623**, 276–283 (2016)
21. Gao, D.Y.: On unified modeling, theory, and method for solving multi-scale global optimization problems. AIP Conference Proc. **1776**(1), 0200051–0200058 (2016)

Controlling Some Statistical Properties of Business Rules Programs

Olivier Wang[1,2(✉)] and Leo Liberti[2]

[1] IBM France, 9 Rue de Verdun, 94250 Gentilly, France
[2] CNRS LIX, Ecole Polytechnique, 91128 Palaiseau, France
{olivier.wang,leo.liberti}@polytechnique.edu

Abstract. Business Rules programs encode decision-making processes using "if-then" constructs in a way that is easy for non-programmers to manipulate. A common example is the process of automatic validation of a loan request for a bank. The decision process is defined by bank managers relying on the bank strategy and their own experience. Bank-side, such processes are often required to meet goals of a statistical nature, such as having at most some given percentage of rejected loans, or having the distribution of requests that are accepted, rejected, and flagged for examination by a bank manager be as uniform as possible. We propose a mathematical programming-based formulation for the cases where the goals involve constraining or comparing values from the quantized output distribution. We then examine a simulation for the specific goals of (1) a max percentage for a given output interval and (2) an almost uniform distribution of the quantized output. The proposed methodology rests on solving mathematical programs encoding a statistically supervised machine learning process where known labels are an encoding of the required distribution.

Keywords: Distribution learning · Mixed-integer programming · Statistical goals · Business Rules

1 Introduction

Business Rules (BR) are a "programming for non programmers" paradigm that is often used by large corporations to store industrial process knowledge formally. BR replaces the two most abstract concepts of programming, namely loops and function calls, by means of an implicit outer loop and meta-variables used within a set of easy-to-manage "if-then" type instructions. BR interpreters are implemented by all BR management systems, e.g. [14]. BR programs are often used by corporations to encode their policies and empirical knowledge: given some technical input, they produce a decision, often in the form of a YES/NO output. Corporations often require their internal processes to perform according to a prescribed statistical behavior, which could be imposed because of strategy or by law. This required behavior is typically independent of the BR input data. The problem is then to parametrize the BR program so it will behave as prescribed on average, while still providing meaningful YES/NO answers on given inputs.

© Springer International Publishing AG 2017
R. Battiti et al. (Eds.): LION 2017, LNCS 10556, pp. 263–276, 2017.
https://doi.org/10.1007/978-3-319-69404-7_19

In [30] we studied a simplified version of the problem where the statistical behavior was limited to a given mean. In this paper we provide a solution methodology for a more general (and difficult) case, where the statistical behavior is described by a given discrete distribution. We achieve this goal by encoding a Machine Learning (ML) procedure by means of a Mathematical Program (MP) of the Mixed-Integer Linear Programming (MILP) type. The ML procedure relies on non-input specific labels that encode the given knowledge about the distribution. Controlling the statistical behavior of a complex process such as a BR program is a very hard task, and to the best of our knowledge this work is the first of its kind in this respect. Methodologically speaking, we think our MILP formulation is also innovative in that it encodes an ML training process having labels which, instead of applying to individual inputs, apply to the entire input distribution at once. Such an ML process bypasses the usual difficulties of trying to label the training set data, thereby being more practical for industrial applications.

The motivation for this study is a real industrial need expressed by IBM (which co-funds this work) with respect to their BR package ODM. Our previous paper [30] laid some of the groundwork, limited to the most basic statistical indicator (the mean of a distribution). Though that was a necessary step to the current work, the methodology described herein is the first to actually address the need expressed by industry: we feel this is one of the main feature that sets this work apart from our previous work. We still rely on MILP-based methodology, but now the input is a whole discrete distribution, the cardinality of which largely determines the size of the new MILP formulations presented below. Our tests show that this has an acceptable impact on empirical solution complexity.

As an experimental illustration, we consider the two cases of the statistical behavior being (1) a maximum percentage of a certain output value and (2) the output values being distributed in a fashion close to the uniform distribution, for integer outputs. We provide an optimization based approach to solving the learning problem for each of those cases, then examine some test results.

1.1 Preliminaries

We formally represent a BR program as an ordered list of sentences of the form:

> **if** $\mathsf{cond}(p, x)$ **then**
> $\quad x \leftarrow \mathsf{act}(p, x)$
> **end if**

where p is a *control parameter* vector (with c components) which encodes a possible "tuning" of the program (e.g. thresholds which can be adjusted by the user), $x \in X \subseteq \mathbb{R}^d$ is a *variable* vector representing intermediate and final stages of computation, cond is a boolean function, and act a function with values in X. We call *rule* such a sentence, *condition* an expression $\mathsf{cond}(p, x)$ and *action* an instruction $x \leftarrow \mathsf{act}(p, x)$, which indicates a modification of the value of x. We write the final value of the variable x as $x^f = P(p, q)$, where P represents the BR program and q is an *input parameter* vector representing a problem instance and equal to the initial value of x. Although in general BR programs may have any type of output, we consider only integer outputs, since BR programs are mostly used to take discrete decisions. We remark that p, x are *symbolic vectors* (rather than numeric vectors) since their components are decision variables.

BR programs are executed in an external loop construct which is transparent to the user. Without getting into the details of BR semantics, the loop executes a single action from a BR whose condition is True at each iteration. Which BR is executed depends on a conflict resolution strategy with varying complexity. De Sainte-Marie et al. [23] describe typical operational semantics, including conflict resolution strategy, for industrial BR management systems. In this paper, the list of rules is ordered and the loop executes the first BR of the list with a condition evaluating to True at each iteration. The loop only terminates once every condition of the BRs is False. We proved in [29] that there is a universal BR program which can simulate any Turing Machine (TM), which makes the BR language Turing-complete.

We consider the problem where the $q \in Q$ are the past, known instances of the BR program, and the outputs $P(p, q)$ of those instances are divided into N evenly sized intervals $[H_0, H_1], \ldots, [H_{N-1}, H_N]$, forming a *quantized output distribution*. Denoting $\nu_1(p), \ldots, \nu_N(p)$ the number of outputs in these categories, we can formalize the problem as:

$$
\left.
\begin{array}{c}
\min_{p,x} \|p - p^0\|_1 \\
\mathscr{C}(\nu_1(p), \ldots, \nu_N(p))
\end{array}
\right\}
\tag{1}
$$

where $\|\cdot\|_1$ is the L1 norm and $\mathscr{C}$ is a constraint or set of constraints. While this formulation uses the number of outputs rather than the probabilities themselves, the relation between the two is simply a ratio of $1/m$, where $m = \texttt{card}(Q)$ is the number of training data points.

In this paper, we suppose that P_1 and P_2 are BR programs with a rule set $\{\mathcal{R}_r \mid r \le \rho\}$ containing rules of the form:

if $L_r \le x \le G_r$ **then**
    $x \leftarrow A_r x + B_r$
end if

with $L_r, G_r, B_r \in \mathbb{R}$ and $A_r \in \{0,1\}^{d \times d}$. We note $R = \{1, \ldots, \rho\}$ and $D = \{1, \ldots, d\}$.

We discuss the concrete example of banks using a BR program in order to decide whether to grant a loan to a customer or not. The BR program depends on a variable vector x and initializes its parameter vector (a component of which is an income level threshold) to p^0. A BR program P_1 is used to decide whether a first bank will investigate the loan request further or simply accept the automated decision taken by an expert system, and therefore has a binary output value. This bank's high-level strategy requires that no more than 50% of loans are treated automatically, but P_1 currently treats 60%. Another bank instead uses a BR program $P2$ to accept, reject, or assign a bank manager to the loan request, and therefore has a ternary return value, represented by an integer in $\{0, 1, 2\}$. That bank's strategy requires that the proportion of each output is

$\{1/3, 1/3, 1/3\}$, but it is currently $\{1/4, 1/4, 1/2\}$. Our aim is in each case to adjust p, e.g. modifying the income level, so that the BR program satisfies the bank's goal regarding automatic loan treatment. This adjustment of parameters could be required after a change of internal or external conditions, for example.

The first scenario can be formulated as:

$$\left.\begin{array}{c} \min\limits_{p,x} \|p - p^0\|_1 \\ \mathbb{E}_{q \in Q}\left[P_1(p,q)\right] \le g \end{array}\right\} \tag{2}$$

where P_1 has an output in $\{0, 1\}$, $g \in [0, 1]$ is the desired max percentage of 1 outputs, the $q \in Q$ are the past known instances of the BR program, $\|\cdot\|_1$ is the L1 norm, p, q must satisfy the semantics of the BR program $P(p, q)$ when executed within the loop of a BR interpreter and $\mathbb{E}$ is the usual notation for the expected value.

Similarly, the second scenario where P_2 has an output in $\{1, \ldots, N\}$ and the desired output is as close to a uniform distribution as possible can be formalized as:

$$\left.\begin{array}{c} \min\limits_{p,x} \|p - p^0\|_1 \\ \forall s, t \in \{1, \ldots, N\}, \ |\nu_s - \nu_t| \le 1 \end{array}\right\} \tag{3}$$

Note that the solution to this problem is not always a truly uniform distribution, simply because there is no guarantee that m is divisible by N. However, it will always be as close as possible to a uniform distribution, since the constraint imposes that all the outputs will be reached by either floor(m/N) or ceil(m/N) data points. Again, we use whole numbers (of outputs in a given interval) instead of frequencies to be able to employ integer decision variables.

Such problems could be solved heuristically by treating P_1 or P_2 as a black-box, or by replacing it by means of a simplified model, such as e.g. a low-degree polynomial. We approach this problem as in [30]: we model the algorithmic dynamics of the BR by means of MIP constraints, in view to solving those equations with an off-the-shelf solver. That this should be possible at all in full generality stems from the fact that Mathematical Programming (MP) is itself Turing-complete [16].

We make a number of simplifying assumptions in order to obtain a practically useful methodology, based on solving a Mixed-Integer Linear Programming (MILP) reformulation of these equations using a solver such as CPLEX [13]:

1. We suppose Q is small enough that solving the MILP is (relatively) computationally cheap.
2. We assume finite BR programs with a known bound $(n-1)$ on the number of iterations of the loop for any input q (industrial BR programs often have a low value of n relative to the number of rules). This in turn implies that the values taken by x during the execution of the BR program are bounded. We assume that $M \gg 1$ is an upper bound of all absolute values of all p, q, and x, as well as any other values appearing in the BR program. It serves as a "big M" for the MP described in the rest of the paper.

3. We assume that the conditions and actions of the BR program give rise to constraints for which an exact MILP reformulation is possible. In order to have a linear model, each BR must thus be "linear", i.e. have the form:

if $L \leq x \leq G$ then
$\qquad x \leftarrow Ax + B$
end if

with $L, G, B \in \mathbb{R}^d$ and $A \in \{0, 1\}^{d \times d}$. In general, $A_{h,k}$ may have values in $\mathbb{R}$ if it is not a parameter and x_h has only integer values.

1.2 Related Works

We follow the formalism used in [30] pertaining to Business Rules (BR) programs and their statistical behavior.

Business Rules (also known as *Production Rules*) are well studied as a knowledge representation system [8,10,18], originating as a psychological model of human behavior [20,21]. They have further been used to encode expert systems, such as MYCIN [6,27], EMYCIN [6,25], OPS5 [5,11], or more recently ODM [14] or OpenRules [22]. On business side of things, they have been defined broadly and narrowly in many different ways [12,15,24]. We consider Business Rules as a computational tool, which to the best of our knowledge has not been explored in depth before.

Supervised Learning is also a well studied field of Machine Learning, with many different formulations [3,17,26,28]. A popular family of algorithms for the classification problem uses Association Rules [1,19]. Such Rule Learning is not to be confused with the problem treated in this article, which is more a regression problem than a classification problem. There exist many other algorithms for Machine Learning, from simple linear regression to neural networks [2] and support vector machines [9]. When the learner does not have as many known output values as it has items in the training set, the problem is known as Semi-Supervised Learning [7]. Similarly, there has been research into machine learning when the matching of the known outputs values to the inputs is not certain [4]. A previous paper has started to explore the Learning problem when the known information does not match to a single input [30].

2   Learning Goals with Histograms

In the rest of this paper, we concatenate indices so that $(L_r)_k = L_{rk}$, $(G_r)_k = G_{rk}$, $(A_r)_{h,k} = A_{rhk}$ and $(B_r)_k = B_{rk}$. We assume that rules are feasible, i.e. $\forall r, k \in R \times D, L_k \leq G_k$. In the rest of this section, we suppose that the dimension of p is $c = 1$, making p a scalar, and that p takes the place of A_{111}. Similar sets of constraints exists for when the parameter p takes the place of a scalar in B_r, L_r or G_r. Additional parameters correspond to additional constraints that mirror the ones used for the first parameter.

This formalization is taken from [30], in which we have also proved that the set of constraints described in Fig. 1 models the execution of such a BR program. The iterations of the execution loop are indexed by $i \in I = \{1, \ldots, n\}$ where $n-1$ is the upper bound on the number of iterations, the final value of x corresponds to iteration n. We use an auxiliary binary variable y_{ir} with the property: $y_{ir} = 1$ iff the rule $\mathcal{R}_r$ is executed at iteration i. The other auxiliary binary variables y_{ir}^U and y_{ir}^L are used to enforce this property.

We note (C1), (C2), etc. the constraints related to the evolution of the execution and (IC1), (IC2), etc. the constraints related to the initial conditions of the BR program:

- (C1): represents the evolution of the value of the variable x
- (C2): represents the property that at most one rule is executed per iteration
- (C3): represents the fact that a rule whose condition is False cannot be executed
- (C4)–(C6) represent the fact that only the first rule whose condition is True can be executed
- (IC1) through (IC3) represent the initial value of a
- (IC4) represents the initial value of x.

$$\forall i \in I \setminus \{n\} \qquad x^{i+1} = \sum_{r \in R}(a_r x^i + B_r)y_{ir} + \left(1 - \sum_{r \in R} y_{ir}\right)x^i \qquad (C1)$$

$$\forall i \in I \qquad \sum_{r \in R} y_{ir} \leq 1 \qquad (C2)$$

$$\forall (i,r) \in I \times R \qquad L_r - M(1 - y_{ir})e \leq x^i \leq G_r + M(1 - y_{ir})e \qquad (C3)$$

$$\forall (i,r,k) \in I \times R \times D \qquad x_k^i \geq G_{rk} - My_{irk}^U - M\sum_{r' < r} y_{ir'} \qquad (C4)$$

$$\forall (i,r,k) \in I \times R \times D \qquad x_k^i \leq L_{rk} + My_{irk}^L + M\sum_{r' < r} y_{ir'} \qquad (C5)$$

$$\forall (i,r) \in I \times R \qquad 2d - 1 + y_{ir} \geq \sum_{k \in D}(y_{irk}^U + y_{irk}^L) \qquad (C6)$$

$$\forall r \in \{2, \ldots, \rho\} \qquad a_r = A_r \qquad (IC1)$$

$$a_{111} = p \qquad (IC2)$$

$$\forall (h,k) \in D^2 \setminus \{1,1\} \qquad a_{1hk} = A_{1hk} \qquad (IC3)$$

$$x^1 = q \qquad (IC4)$$

$$\forall i \in I \qquad x^i \in X$$

$$\forall r \in R \qquad a_r \in \{0,1\}^{d \times d}$$

$$\forall (i,r,k) \in I \times R \times D \qquad y_{ir}, y_{irk}^U, y_{irk}^L \in \{0,1\}$$

Fig. 1. Set of constraints modeling the execution of a BR program ($e \in \mathbb{R}^d$ is the all-one vector).

2.1 A MIP for Learning Quantized Distributions

The Mixed-Integer Program from Fig. 2 models the problem from Eq. 1. We index the instances in Q with $j \in J = \{1, \ldots, m\}$. We also limit ourselves to solutions which result in computations that terminate in less than $n - 1$ rule executions. As modifying the parameter means modifying the BR program, the assumptions made regarding the finiteness of the program might not be verified otherwise.

We note $O = \{1, \ldots, N\}$, such that $\forall t \in O, \nu_t = \mathtt{card}\{j \in J \mid x^1_{n,j} \in [H_{t-1}, H_t]\}$. We enforce this definition of ν_t by using an auxiliary binary variable s_{tj} with the property: $s_{tj} = 1$ iff $x^1_{n,j} \in [B_{t-1}, B_t]$. The other auxiliary binary variables s^U_{tj} and s^L_{tj} are used to enforce this property.

The constraints are mostly similar to the ones in Fig. 1. We simply add the goal of minimizing the variation of the parameter value and the constraints $\mathscr{C}(\nu_1(p), \ldots, \nu_N(p))$ from Eq. 1. The new constraints are:

- (C7) represents the need for the computation to have terminated after $n - 1$ executions
- (C8)–(C12) represents the definition of $\nu_1, \ldots, \nu_N$
- (IC4') represents (IC4) with an additional index j.

$$\underset{p,a,x,y,y^U,y^L,s,s^U,s^L,\nu}{\text{minimize}} \qquad |p^0 - p|$$

subject to

$$(C1), (C2), (C3), (C4), (C5), (C6), (IC1), (IC2), (IC3)$$

$$\mathscr{C}(\nu_1, \ldots, \nu_N)$$

$$\forall j \in J \qquad \sum_{r \in R} y_{njr} = 0 \qquad (C7)$$

$$\forall (t,j) \in O \times J \qquad H_{t-1} - M(1 - s_{tj}) \leq x^{n,j}_1 \leq H_t + M(1 - s_{tj}) \qquad (C8)$$

$$\forall (t,j) \in O \times J \qquad x^{n,j}_1 \geq H_t - M s^U_{tj} \qquad (C9)$$

$$\forall (t,j) \in O \times J \qquad x^{n,j}_1 \leq H_{t-1} + M s^L_{tj} \qquad (C10)$$

$$\forall (t,j) \in O \times J \qquad s_{tj} \geq s^U_{tj} + s^L_{tj} \qquad (C11)$$

$$\forall t \in O \qquad \nu_t = \sum_{j \in J} s_{tj} \qquad (C12)$$

$$\forall j \in J \qquad x^{1,j} = q^j \qquad (IC4')$$

$$\forall (i,j) \in I \times J \qquad x^{i,j} \in X$$

$$\forall k \in R \qquad a_k \in \{0,1\}^{d \times d}$$

$$p \in \{0,1\}$$

$$\forall (i,j,r,k) \in I \times J \times R \times D \qquad y_{ijr}, y^U_{ijrk}, y^L_{ijrk} \in \{0,1\}$$

$$\forall (t,j) \in O \times J \qquad s_{tj}, s^U_{tj}, s^L_{tj} \in \{0,1\}$$

$$\forall t \in O \qquad \nu_t \in \mathbb{N}$$

Fig. 2. Mixed-Integer Program solving Eq. 1.

That solving the MIP in Fig. 2 also solves the original Eq. 1 is a direct consequence of the fact that the constraints in Fig. 1 simulate $P(p, q)$. The proof is simple since (C8) through (C12) trivially represent the definition of $\nu_1, \ldots, \nu_N$. A similar MIP can be obtained when p has values in different part of the BRs, from which a more complex MILP is obtained for when p is non-scalar. However, this formulation is still quite abstract, as it depends heavily on the form of $\mathscr{C}$. In fact, it can almost always be simplified given a particular constraint over the quantized distribution, as we see in the rest of this section.

2.2 A MILP for the Max Percentage Problem

A constraint programming formulation of Eq. 2 is the Mixed-Integer Linear Program (MILP) described in Fig. 3. In the case of the Max Percentage problem, we can linearize the MIP in Fig. 2 as well as remove some superfluous variables, since only one of the ν_t is relevant.

We now note $e = (1, \ldots, 1) \in \mathbb{R}^d$ the vector of all ones. We use the auxiliary variables $w \in \mathbb{R}^{I \times J \times R}$ and $z \in \mathbb{R}^{I \times J \times R \times D^2}$ such that $w_{ijr} = (A_r x^{ij} + B_r - x^{i,j}) y_{ijr}$ (i.e. $w_{ijr} = A_r x^{i,j} + B_r - x^{i,j}$, the difference between the new and the old values of x^j) and $z_{ijrhk} = a_{rhk} x_k^{i,j}$.

Any constraints numbered as before fulfills the same role. The additional constraints are:

- (C1'$_1$), (C1'$_2$), (C1'$_3$), (C1'$_4$) and (C1'$_5$) represent the linearization of (C1) from Fig. 1
- (C8') represents the goal from Eq. 2, that is a constraint over the average of the final values of x. It replaces $\mathscr{C}(\nu_1, \ldots, \nu_N)$ and all the constraints used to define ν_t from the MIP in Fig. 2.

The MILP from Fig. 3 finds a value of p that satisfies Eq. 2. This is again derived from the fact that Fig. 1 simulates a BR program, and from the trivial proof that (C1'$_1$), (C1'$_2$), (C1'$_3$), (C1'$_4$) and (C1'$_5$) represent the linearization of (C1).

2.3 A MILP for the Almost Uniform Distribution Problem

As before, we exhibit in Fig. 4 a MILP that solves Eq. 3. Any constraints numbered as before fulfills the same role. The additional constraints are:

- (C8") through (C10") represent the adaptation of (C8) through (C10) to the relevant case of integer outputs
- (C13) represents the equivalent to $\mathscr{C}$ from Eq. 3.

This MILP is obviously equivalent to solving Eq. 3, since it is for the most part a straight linearization of the MIP in Fig. 2.

$$\underset{p,a,x,y,y^U,y^L,w,z}{\text{minimize}} \qquad \left| p^0 - p \right|$$

subject to

$$\text{(C2), (C3), (C4), (C5), (C6), (C7), (IC1), (IC2), (IC3)}$$

$$\forall (i,j) \in I \backslash \{n\} \times J \qquad x^{i+1,j} = \sum_{r \in R} w_{ijr} + x^{i,j} \qquad \text{(C1'}_1\text{)}$$

$$\forall (i,j) \in I \times J \times R \qquad -M y_{ijr} e \leq w_{ijr} \leq M y_{ijr} e \qquad \text{(C1'}_2\text{)}$$

$$\forall (i,j,r,h) \in I \times J \times R \times D \qquad \sum_{k \in D} z_{ijrhk} + B_{rh} - x_h^{i,j} - M(1 - y_{ijr})$$

$$\leq w_{ijrh} \leq \sum_{k \in D} z_{ijrhk} + B_{rh} \qquad \text{(C1'}_3\text{)}$$

$$- x^{i,j} + M(1 - y_{ijr}) e$$

$$\forall (i,j,r) \in I \times J \times R \qquad -M a_r \leq z_{ijr} \leq M a_r \qquad \text{(C1'}_4\text{)}$$

$$\forall (i,j,r,h,k) \in I \times J \times R \times D^2 \qquad x_k^{i,j} - M(1 - a_{rhk})$$

$$\leq z_{ijhk} \leq x_k^{ij} \qquad \text{(C1'}_5\text{)}$$

$$\forall j \in J \qquad \sum_{r \in R} y_{njr} = 0 \qquad \text{(C7)}$$

$$\sum_{j \in J} x_1^{n,j} \leq mg \qquad \text{(C8')}$$

$$\forall (i,j,r,k_1,k_2) \in I \times J \times R \times D^2 \qquad x^{i,j}, z_{ijrhk} \in X$$

$$\forall (i,j,r) \in I \times J \times R \qquad w_{ijr} \in \mathbb{R}^d$$

$$\forall r \in R \qquad a_r \in \{0,1\}^{d \times d}$$

$$p \in \{0,1\}$$

$$\forall (i,j,r,k) \in I \times J \times R \times D \qquad y_{ijr}, y_{ijrk}^U, y_{ijrk}^L \in \{0,1\}$$

Fig. 3. MILP formulation for solving Eq. 2.

$$\underset{p,b,x,y,y^U,y^L,w,s,s^L,s^g,\nu}{\text{minimize}} \qquad \left| p^0 - p \right|$$

subject to

$$\text{(C1'}_1\text{), (C1'}_2\text{), (C1'}_3\text{), (C1'}_4\text{), (C1'}_5\text{), (C2),}$$

$$\text{(C3), (C4), (C5), (C6), (C7), (C11), (C12)}$$

$$\text{(IC1), (IC2), (IC3), (IC4')}$$

$$\forall (t,j) \in O \times J \qquad t - M(1 - s_{tj}) \leq x_1^{n,j} \leq t + M(1 - s_{tj}) \qquad \text{(C8'')}$$

$$\forall (t,j) \in O \times J \qquad x_1^{n,j} \geq t - M s_{tj}^U \qquad \text{(C9'')}$$

$$\forall (t,j) \in O \times J \qquad x_1^{n,j} \leq t + M s_{tj}^L \qquad \text{(C10'')}$$

$$\forall (t,\tau) \in O^2 \qquad -1 \leq \nu_t - \nu_\tau \leq 1 \qquad \text{(C13)}$$

$$\forall (i,j) \in I \times J \qquad x^{i,j} \in X$$

$$\forall r \in R \qquad a_r \in \{0,1\}^{d \times d}$$

$$p \in \{0,1\}$$

$$\forall (i,j,r,k) \in I \times J \times R \times D \qquad y_{ijr}, y_{ijrk}^U, y_{ijrk}^L \in \{0,1\}$$

$$\forall (t,j) \in O \times J \qquad s_{tj}, s_{tj}^U, s_{tj}^L \in \{0,1\}$$

$$\forall t \in O \qquad \nu_t \in \mathbb{N}$$

Fig. 4. MILP formulation for solving Eq. 3.

3   Implementation and Experiments

We use a Python script to randomly generate samples of 100 instances of P_1 and P_2 for different numbers of control parameters c, each instance having a corresponding set of inputs with $d = 3$, $n = 10$ and $m = 100$. The number of control parameters serves as an approximation of the complexity of the BR program to optimize: a more complex program will have more buttons to adjust, thus increasing the complexity, yet be more likely to have the goal be reachable at all, i.e. have the MILP be feasible. We define the space X as $X \subseteq \mathbb{R} \times \mathbb{R} \times \mathbb{Z}$. The BR programs are sets of $\rho = 10$ rules, where L_r, G_r, B_r are vectors of scalars in an interval **range** and A_r are $d \times d$ matrices of binary variables. In P_1, we use **range** $= [0, 1]$ and in P_2, we use **range** $= [0, 3]$. All input values q are generated using a uniform distribution in **range**.

We use these BR programs to study the computational properties of the MILP. The value of M used is customized according to each constraint, and is ultimately bounded by 6 and 16 in P_1 and P_2 respectively (strictly greater than five times the range of possible values for x). We write the MILP as an AMPL model, and solve it using the CPLEX solver on a Dell PowerEdge 860 running CentOS Linux.

3.1   The Max Percentage Problem

We observe the proportion of solvable instances of P_1 for c between 5 and 10 and $c = 15$ in Table 1. We use the MILP in Fig. 3 to solve Eq. 2 with the goal set to $g = 0.5$.

An instance is considered solvable if CPLEX reports an integer optimal solution or a (non-)integer optimal solution. We separate the instances where the optimal value is 0 from the others, as those indicate that the randomly generated BR program already fulfill the goal condition. We expect around fifty of those for any value of c.

In Fig. 5, we observe both the success rate and the average solving time when considering only the non-trivial, non-timed out instances of P_1. The success rate increases steadily, as expected. The solving time seems to indicate a non-linear increase for c greater than 6, even with its values being somewhat unreliable due to the small sample. Knowing that average industrial BRs are more complex than our toy examples, regularly having thousands of rules, this approach to the Maximum Percentage problem does not seem applicable to industrial cases.

Table 1. Experimental values for the maximum percentage problem.

Number of control parameters c	5	6	7	8	9	10	15
Trivial solvable instances (objective = 0)	52	53	49	49	58	48	46
Non-trivial solvable instances (objective $\neq$ 0)	5	6	5	13	6	6	8
Infeasible instances	43	43	40	36	31	35	14
Timed out instances	0	0	7	2	5	11	32

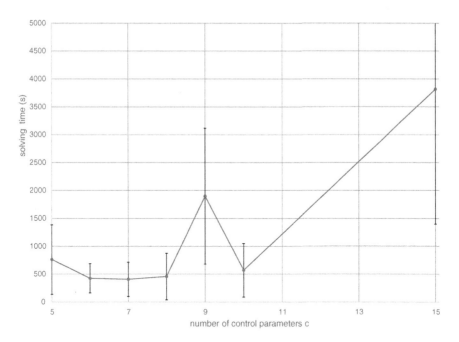

Fig. 5. Average solution time over P_1 for varying values of c in seconds.

3.2 The Almost Uniform Distribution Problem

We observe the proportion of solvable instances of P_2 for c between 5 and 10 and $c = 15$ in Table 2. We use the MILP in Fig. 4 to solve Eq. 3 with $N = 2$. Again, we separate instances where the goal is already achieved before optimization, identifiable by being solved quickly with a value of $p = p^0$, i.e. an optimal value of zero.

In Fig. 6, we display the success rate and average solving time over the non-timed out, non-presolved instances for all three values of c. We observe a sharply non-linear progression, with the average problem taking about nine minutes with 15 control parameters. Knowing that average industrial BRs are much more complex than our toy examples, regularly having thousands of rules, we conclude that this method can only be used infrequently, if at all.

Table 2. Experimental values for the almost uniform distribution problem.

Number of control parameters c	5	6	7	8	9	10	15
Trivial solvable instances (objective = 0)	8	2	1	1	4	7	4
Non-trivial solvable instances (objective $\neq$ 0)	9	2	8	5	4	15	32
Infeasible instances	83	96	91	93	92	77	63
Timed out instances	0	0	0	1	0	1	1

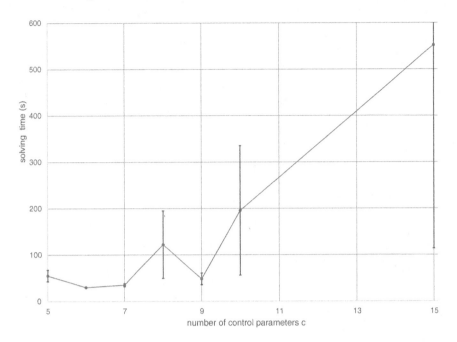

Fig. 6. Average solution time over non-trivial solvable P_2 for varying values of c.

4   Conclusion, Discussion and Future Work

We have presented a learning problem of unusual type, that of supervised learning with statistical labels. We have further explored a particular subset of those problems, those where the labels apply to a quantized output distribution. This new approach is easily applied to practical applications in industry where control parameters must be learned to satisfy a given goal. We have given a mathematical programming algorithm that solves such a learning problem given a linear BR program. Depending on the specific learning problem, the mathematical program might be easy or difficult to solve. We examined two example learning problems with practical applications for which the learning is equivalent to solving a MILP.

We observe that, though one could detect a visual similarity in the plots presented in Figs. 5 and 6, we believe that this similarity is only apparent. In fact, the error bars (which measure the standard deviation of the solution time over the instance subclass corresponding to a given size of parameters) point out that the "hard cases" (with high values of solution times) are also the cases where the error bars are longest. In other words, this "similarity" is simply a result of outliers in the corresponding peaks.

The experimental results indicate the general feasibility of this type of approach. It is clear that, due to the exponential nature of Branch-and-Bound (BB, the algorithm solving the MILPs), the performance will scale up poorly with the size of the BR program: but this can currently be said of most MILPs. This issue, which certainly requires more work, can possibly be tackled by pursuing some of the following ideas: more effective BB-based or formulation-based heuristics (also called *mat-heuristics* in the literature), cut generation based on problem structure, and decomposition. The latter, specifically, looks promising as the structure of the BR program is, up to the extent provided by automatic translation based on parsing trees, carried over to the resulting MILP.

Other avenues of research are in extending this statistical learning approach in other directions, e.g. learning other moments, or given quantiles in continuous distributions. Statistical goal learning problems are an apparently unexplored area of ML that has eminently practical applications.

Acknowledgments. The first author (OW) is supported by an IBM France/ANRT CIFRE Ph.D. thesis award.

References

1. Agrawal, R., Imieliński, T., Swami, A.: Mining association rules between sets of items in large databases. In: Buneman, P., Jajodia, S. (eds.) Proceedings of the 1993 ACM SIGMOD International Conference on Management of Data, pp. 207–216. ACM, New York (1993)
2. Atiya, A.: Learning Algorithms for Neural Networks. Ph.D. thesis, California Institute of Technology, Pasadena, CA (1991)
3. Bakir, G., Hofmann, T., Schölkopf, B., Smola, A., Taskar, B., Vishwanathan, S.: Predicting Structured Data (Neural Information Processing). The MIT Press, Cambridge (2007)
4. Brodley, C., Friedl, M.: Identifying mislabeled training data. J. Artif. Intell. Res. **11**, 131–167 (1999)
5. Brownston, L., Farrell, R., Kant, E., Martin, N.: Programming Expert Systems in OPS5: An Introduction to Rule-Based Programming. Addison-Wesley, Boston (1985)
6. Buchanan, B., Shortliffe, E. (eds.): Rule Based Expert Systems: The Mycin Experiments of the Stanford Heuristic Programming Project (The Addison-Wesley Series in Artificial Intelligence). Addison-Wesley, Boston (1984)
7. Chapelle, O., Schlkopf, B., Zien, A.: Semi-Supervised Learning. The MIT Press, Cambridge (2010)
8. Clancey, W.: The epistemology of a rule-based expert system: a framework for explanation. Artif. Intell. **20**(3), 215–251 (1983)
9. Cortes, C., Vapnik, V.: Support-vector networks. Mach. Learn. **20**(3), 273–297 (1995)
10. Davis, R., Buchanan, B., Shortliffe, E.: Production rules as a representation for a knowledge-based consultation program. Artif. Intell. **8**(1), 15–45 (1977)
11. Forgy, C.: OPS5 User's Manual. Department of Computer Science, Carnegie-Mellon University, Pittsburgh (1981)
12. Knolmayer, G., Herbst, H.: Business rules. Wirtschaftsinformatik **35**(4), 386–390 (1993)
13. IBM: ILOG CPLEX 12.2 User's Manual. IBM (2010)
14. IBM: Operational Decision Manager 8.8 (2015)

15. Kolber, A., et al.: Defining business rules - what are they really? Project Report 3, The Business Rules Group (2000)
16. Liberti, L., Marinelli, F.: Mathematical programming: turing completeness and applications to software analysis. J. Comb. Optim. **28**(1), 82–104 (2014)
17. Liu, T.Y.: Learning to rank for information retrieval. Found. Trends Inf. Retriev. **3**(3), 225–331 (2009)
18. Lucas, P., Gaag, L.V.D.: Principles of Expert Systems. Addison-Wesley, Boston (1991)
19. Malioutov, D.M., Varshney, K.R.: Exact rule learning via boolean compressed sensing. In: Dasgupta, S., McAllester, D. (eds.) Proceedings of the 30th International Conference on Machine Learning (ICML 2013). JMLR: Workshop and Conference Proceedings, vol. 28, pp. 765–773. JMLR, Brookline (2013)
20. Newell, A.: Production systems: models of control structures. In: Chase, W. (ed.) Visual Information Processing. Proceedings of the Eighth Annual Carnegie Symposium on Cognition, pp. 463–526. Academic Press, New York (1973)
21. Newell, A., Simon, H.: Human Problem Solving. Prentice-Hall, Upper Saddle River (1972)
22. OpenRules Inc.: OpenRules User Manual, Monroe (2015)
23. Paschke, A., Hallmark, G., De Sainte Marie, C.: RIF production rule dialect, 2nd edn. W3C recommendation, W3C (2013). http://www.w3.org/TR/2013/REC-rif-prd-20130205/
24. Ross, R.: Principles of the Business Rule Approach. Addison-Wesley, Boston (2003)
25. Scott, A., Bennett, J., Peairs, M.: The EMYCIN Manual. Department of Computer Science, Stanford University, Stanford (1981)
26. Settles, B.: Active learning literature survey. Computer Sciences Technical Report 1648, University of Wisconsin-Madison (2009)
27. Shortcliffe, E.: Computer-Based Medical Consultations: MYCIN. Elsevier, New York (1976)
28. Vapnik, V.: The Nature of Statistical Learning Theory. Springer, New York (1995)
29. Wang, O., Ke, C., Liberti, L., de Sainte Marie, C.: The learnability of business rules. In: International Workshop on Machine Learning, Optimization, and Big Data (MOD 2016) (2016)
30. Wang, O., Liberti, L., D'Ambrosio, C., de Sainte Marie, C., Ke, C.: Controlling the average behavior of business rules programs. In: Alferes, J.J.J., Bertossi, L., Governatori, G., Fodor, P., Roman, D. (eds.) RuleML 2016. LNCS, vol. 9718, pp. 83–96. Springer, Cham (2016). doi:10.1007/978-3-319-42019-6_6

GENOPT Paper

Hybridization and Discretization Techniques to Speed Up Genetic Algorithm and Solve GENOPT Problems

Francesco Romito[(⊠)]

ACT Operations Research, DIAG, "La Sapienza" University of Rome, Rome, Italy
francesco.romito@act-OperationsResearch.com

Abstract. One of the challenges in global optimization is to use heuristic techniques to improve the behaviour of the algorithms on a wide spectrum of problems. With the aim of reducing the probabilistic component and performing a broader and orderly search in the feasible domain, this paper presents how discretization techniques can enhance significantly the behaviour of a genetic algorithm (GA). Moreover, hybridizing GA with local searches has shown how the convergence toward better values of the objective function can be improved. The resulting algorithm performance has been evaluated during the Generalization-based Contest in Global Optimization (GENOPT 2017), on a test suite of 1800 multidimensional problems.

Keywords: Global optimization · Discretization techniques · Mixed global local search · Genetic algorithm · GENOPT

1 Introduction

One of the fundamental principles in our world is the search for an optimal state. Technological progress and the expansion of knowledge constantly bring to light the real issues that need to be explained in quantitative terms, solving global optimization problems as in [1–3]. Very often the analytical expressions of the model are missed or are not easily represented. These problems are known in the literature as black box.

The GENOPT [4] challenge allows contestants to evaluate the algorithms on randomized functions, created through suitable generators [5], provided as a binary library, in order to be treated just as black box problems. In accord with No Free Lunch Theorems for Optimization [6], an algorithm could reveal a positive result for a specific benchmark function, whereas the use of randomized functions provides a major and faithful overview of its robustness.

In literature there are now countless derivative free approaches for global optimization. For instance, recently in [7] has been proposed a *Deterministic Particle Swarm Optimization*, in order to better explore the search space. Another example is given in [8] where a *DIRECT*-type algorithm is hybridized with a derivative-free local minimization to globally solving optimization problems. Other well-known meta-heuristics like Simulated Annealing [9], Tabu Search [10], Random Optimization [11],

R. Battiti et al. (Eds.): LION 2017, LNCS 10556, pp. 279–292, 2017.
https://doi.org/10.1007/978-3-319-69404-7_20

Ant Colony Optimization [12] are widely available in literature, and used successfully on many real-world applications. Moreover, thorough overviews of global optimization, with topics such as stochastic global optimization, partitioning methods, bounding procedures, convergence studies and complexity can be found in [13–16].

This paper focuses on one of the most popular classes of algorithms belonging to evolutionary computation, namely Genetic Algorithms (GAs). Moving away from the classical scheme [17, 18], the GA has been used as an internal procedure of a larger scheme of global search and has been also successfully modified to reduce the probabilistic component that typically characterizes it.

In particular, after the preliminary Sect. 2 about useful concepts and a look on the complexity of the problem involved, in Sect. 3 a novel algorithmic scheme for global optimization is presented. Ad hoc discretization techniques have been successfully interlaced with the GA classical search operations, performing a better and wider search on a feasible domain. Moreover, a high efficient scheme of a hybridized GA with local searches is described. Section 4 details the tuning process and the results obtained during the GENOPT contest, a special session of the Learning and Intelligent Optimization Conference (LION 11: June 19–21, 2017, Nizhny Novgorod, Russia). Finally, in Sect. 5 a conclusive overview of the work is drawn, providing several lines for further research.

2   Preliminary Concepts

The focus is on the global minimization of an N-dimensional function $f(x)$ within a hyperinterval D. The problem can be stated as follows:

$$f(x^*) = \min\{f(x) : x \in D\},$$
$$D = \{x \in \mathbb{R}^N : lb_i \le x \le ub_i, 1 \le i \le N\}. \tag{1}$$

No information is available on the exact analytical expression of $f(x)$, although the belonging class is known, i.e. continuously differentiable, two time continuously differentiable, non-differentiable or high-level features such as separable function, unimodal, multimodal and so on.

In reference to GENOPT rules a problem is considered solved if the solution is identified by the algorithm within 1 million of function evaluations and so that the best found function value is within 10^{-5} from the global minimum value:

$$f(x) \le f(x^*) + \varepsilon, \qquad x \in D, \ \varepsilon = 10^{-5}. \tag{2}$$

The complexity to solve the problem (1) with an ε-approximated solution is at least exponential for any algorithm that works in the black-box model with a generic non-convex function. On this subject, a useful result due to Vavasis [19] as a special case of a theorem established by Nemirovsky and Yudin [20] is explanatory:

Theorem 1. *Let $F(k, p)$ be the class of k-times differentiable functions on D whose k^{th} derivative is bounded by p as follows: at any point $x \in D$ and for any unit vector u,*

$$\left| \frac{d^k}{dt^k} f(x + tu) \right| \le p. \tag{3}$$

Let A be any minimization algorithm that works in the black-box model (evaluating f and its derivatives). Assume that for any function $\in F(k,p)$, A is guaranteed to output a point that satisfies inequality (2). Then, there exists a function $f \in F(k,p)$ such that algorithm A will run on f for at least a number of steps given

$$c \cdot \left(\frac{p}{\varepsilon} \right)^{n/k}, \tag{4}$$

with c being a suitable positive constant.

As mentioned before, what emerges is that, even assuming bounds on the derivatives, the complexity of solving a global minimization problem increases exponentially with the problems dimension.

Usually the problem (1) cannot be solved by an exhaustive search algorithm in an efficient time. Next section will introduce an approach based on the idea of space search reduction to lead the search towards the most promising area.

3   The GABRLS Algorithm

Subsection 3.1 is aimed to the description of the modified Genetic Algorithm (GA), while in Subsect. 3.2 a novel Bounding Restart (BR) technique is described. Additionally, addressing the need to improve the convergence speed, an overall scheme with derivative free Local Searches (LS) is presented in Subsect. 3.3.

3.1   The Modified GA

Starting from a classical pattern of GA (see Algorithm 1), to take advantage of the effective geometry and to prevent premature convergence, two major changes were introduced:

Algorithm 1. Classical GA scheme

---

1. Initial Population (random points uniformly distributed in D)
2. **for** $k = 1 \rightarrow generations$ (max iterations)
3.   Evaluate Population
4.   Selection Criterion (Tournament, Elitism, Roulette wheel, etc.)
5.   Genetic Operators (Crossover – Mutation)
6.   New Population
7. **end for**

– Discrete initial population
  • Diagonal Initial Population (DiagPI).
  • Axial Initial Population (AxialPI).
– Premature convergence preserving procedure
  • Diversify.

The first change concerns the implementations of two discretization and positioning techniques to place the initial points (initial population) in the feasible domain, allowing the sampling of the objective function in such a way to get more useful information than a classical random sampling.

DiagPI routine allows the discrete positioning of points along a main diagonal of the feasible hyperrectangle (Fig. 1 provides an example in a 3D box with a vertex in the origin of the coordinate axes). Denoting with $Space_j$ the amplitude of the interval along the jth dimension, with $Increase_j$ the step along the jth dimension between two consecutive points and with Pop the number of initial points where the objective function f is evaluated, the result is the Algorithm 2.

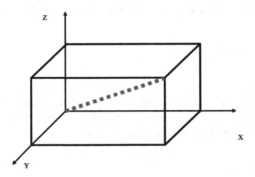

Fig. 1. Graphic view of the points (blue squared) generated with DiagPI in a 3D box. (Color figure online)

Algorithm 2. DiagPI

---

1. **for** $j = 1 \rightarrow N$
2. $Space_j = (ub_j - lb_j)$
3. $Increase_j = Space_j/(Pop - 1)$
4. **end for**
5. **for** $i = 1 \rightarrow Pop$
6.   **for** $j = 1 \rightarrow N$
7. $Point_{i,j} = lb_j + (i - 1) \cdot Increase_j$
8.   **end for**
9. **end for**

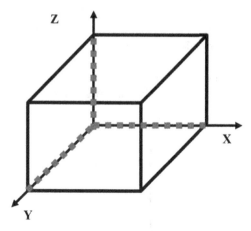

Fig. 2. Graphic view of the points (red squared) generated with AxialPI in a 3D box. (Color figure online)

Algorithm 3. AxialPI

1. $AxialPop = Pop/N$ with $Pop = k \cdot N$, k integer
2. **for** $j = 1 \rightarrow N$
3. $\quad Space_j = (ub_j - lb_j)$
4. $\quad Increase_j = Space_j/(AxialPop - 1)$
5. **end for**
6. $z = 0$
7. $t = 1$
8. **for** $i = 1 \rightarrow Pop$
9. $\quad$ **if** $(z = AxialPop)$ **then**
10. $\quad\quad z = 0$
11. $\quad\quad t = t + 1$
12. $\quad$ **end if**
13. $\quad z = z + 1$
14. $\quad$ **for** $j = 1 \rightarrow N$
15. $\quad\quad$ **if** $(j = t)$ **then**
16. $\quad\quad\quad Point_{i,j} = lb_j + (z - 1) \cdot Increase_j$
17. $\quad\quad$ **else**
18. $\quad\quad\quad Point_{i,j} = lb_j$
19. $\quad\quad$ **end if**
20. $\quad$ **end for**
21. **end for**

DiagPI and AxialPI routines allow to create, through the crossover operator during the iterations of GA (main generations loop), a multi-dimensional mesh with the points generated as nodes (Fig. 3 provides an example in a 3D box with a vertex in the origin of the coordinate axes).

AxialPI routine distributes the points of population along the coordinate axes of the feasible hyperinterval (Fig. 2), the result is the Algorithm 3.

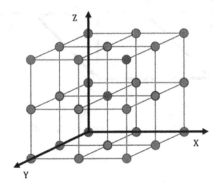

Fig. 3. Mesh of all possible points generated by the crossover (grey) through the recombination of the initial ones (AxialPI in red, DiagPI in blue). (Color figure online)

This mesh will become more dense according to the most promising areas of the feasible domain D. In particular, after every selection phase in a GA iteration, a single-point crossover operator has been adopted, without recombination of blocks to generate and place new points $\left(\mathrm{P}_i^{Son}(j) : i = 1 \rightarrow Pop, j = 1 \rightarrow N\right)$ in the mesh. Equations (5) and (6) and Fig. 4 show how it works.

$$\mathrm{P}_1^{Son} = \left\{\mathrm{P}_1^{Father}(1), \ldots, \mathrm{P}_1^{Father}(k-1), \mathrm{P}_2^{Father}(k), \ldots, \mathrm{P}_2^{Father}(N)\right\}, \qquad (5)$$

$$\mathrm{P}_2^{Son} = \left\{\mathrm{P}_2^{Father}(1), \ldots, \mathrm{P}_2^{Father}(k-1), \mathrm{P}_1^{Father}(k), \ldots, \mathrm{P}_1^{Father}(N)\right\}. \qquad (6)$$

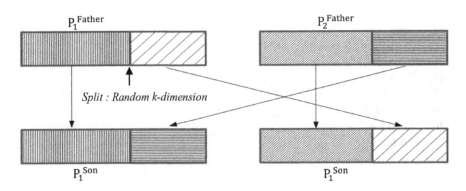

Fig. 4. Single-point crossover operator without recombination of blocks (exact exchange).

With respect to a random sampling, the discrete positioning techniques described above, together with the single-point crossover operator without recombination of blocks, allow the exploration of the search space D in a more orderly manner, and also controlled, the feasible region. In fact, all possible positions in which the crossover will be able to move the points are known a priori and are all nodes of a multi-dimensional mesh.

The second change of GA classical scheme concerns the implementation of a simple technique that allows one to avoid a premature convergence.

Inside the main loop of Algorithm 1 a check on homogeneity of current population has been inserted (Algorithm 4). With the aim of pursuing an aggressive search, a routine (hereinafter called Diversify) makes a more extensive search by replacing all the duplicate points of the current population with new points $\left(P^{New}(j), j = 1 \rightarrow N\right)$. Uniformly distributed random numbers in the range $[lb, ub]$ have been used to mutate all components $(j = 1 \rightarrow N)$ of the duplicate points, so that new random points are placed in the feasible domain. As drawback, the worst case computational cost of this routine is generally high, i.e. a simple implementation can take $(Pop - 1)^2 \cdot (2 \cdot N)$ steps, so it is reasonable to carry out a check on homogeneity of population after at least half of the iterations of the GA main loop.

Algorithm 4. Modified GA scheme

1. Initial Population (Random **or** DiagPI **or** AxialPI)
2. **for** $k = 1 \rightarrow generations$ (max iterations)
3.   Evaluate Population
4.   Selection Criterion
5.   Genetic Operators (CrossOver – Mutation)
6.   **if** $(k > generations/2)$ **then**
7.   Check premature convergence (Diversify)
8.   **end if**
9.   New Population
10. **end for**

3.2 Bounding Restart (BR) Technique

Generally, increasing the dimension of problems and the search space, GA takes more function evaluations to locate a good solution. To avoid this slowness an iterative space reduction technique has been implemented.

BR technique is a two-step scheme in which GA can be successfully integrated. The first step is the bounding one. If it is assumed that the genetic algorithm has the ability to quickly identify a promising area, however large, it's reasonable to focus the search in the given area temporarily. On that basis, by means of hyperintervals that are dynamically sized, according to the need to lead the search towards the most promising area, the bounding step at the generic iteration k is carried out through the following equations:

$$LB^k = \frac{ub + lb}{2} - \frac{ub - lb}{2 \cdot CF^{expCF}}, \quad CF \in \mathbb{R} : CF = const > 1, expCF \in \mathbb{N}. \quad (7)$$

$$UB^k = \frac{ub + lb}{2} + \frac{ub - lb}{2 \cdot CF^{expCF}}, \quad CF \in \mathbb{R} : CF = const > 1, expCF \in \mathbb{N}. \quad (8)$$

CF is a convergence factor that has a high impact on reducing the bounds. The reduction is managed by increasing $expCF$, the exponent of CF, of a unit per iteration. After updating lower and upper bounds, the reduced set is centred in the best point x^k currently known. Let $Ctrasl^k$ be the difference between x^k and the centre of the kth reduced hyperinterval, then, the set can be put centrally as follows:

$$Ctrasl^k = x^k - \frac{UB^k + LB^k}{2}, \quad (9)$$

$$LB^{k+1} = \max\{lb, LB^k + Ctrasl^k\}, \quad (10)$$

$$UB^{k+1} = \min\{ub, UB^k + Ctrasl^k\}. \quad (11)$$

At the kth BR reduction cycle, assuming $expCF = k$ the feasible space reduced is:

$$Space^k = \frac{ub - lb}{CF^k} > \delta, \quad \delta \geq machine\ precision. \quad (12)$$

In Fig. 5 the result of a bounding and positioning operation in a 3D box is drawn.

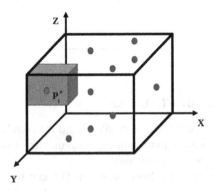

Fig. 5. Overview of a bounding step of BR. In red P_i^*, the best solution currently known. (Color figure online)

The second step of BR concerns the restart of GA inside the reduced space. Denoting with D^k the kth reduced hyperinterval, the overall algorithmic scheme is the following:

Algorithm 5. GABR scheme

---

1. Run **Algorithm 4** in D^k, $k = 0$
2. **while** $(Space_k > \delta)$ **or** (other criterion)
3.   Bounding step, $k = k + 1$
4.   Restart **Algorithm 4** in D^k
5. **end**

3.3 Hybridizing GABR with Local Searches

The global search represented by Algorithm 5 can be expensive in terms of function evaluations if the goal is to identify an optimal solution with a high precision. The reason is that the higher is the precision required, the more BR iterations must be performed. The refining of the solution can be delegated to a local search algorithm.

In order to make the whole algorithm more efficient, derivative free local searches (DFLS) have been introduced. In literature there are many ideas on the hybridization of the global search. Examples of automatic balancing techniques can be found in [21] where the estimates of local Lipschitz constants allow to accelerate significantly the global search. A suitable strategy has been implemented in the GABR algorithm, similar to that one described in [22].

Since the number of LSs performed affects the efficiency, it is necessary to locate when a LS should be started. A reasonable choice is to perform LS at every BR iteration, only after the end of GA main loop and only if an improved solution is found with respect to previous BR iteration.

Let us consider as stopping criterion the GENOPT condition of 10^6 max function evaluations. The final algorithmic scheme is the following:

Algorithm 6. GABRLS scheme

---

1. Run **Algorithm 4** in D^k, $k = 0$
2. Start LS from current best point
3. **while** (GenOpt criterion)
4.   Bounding step, $k = k + 1$
5.   Restart **Algorithm 4** in D^k
6.   **If** (new solution has been found) **then**
7.   Start LS from current best point
8.   **end if**
9. **End**

---

4   Tuning and Results of GABRLS on GENOPT Challenge

GENOPT organizers provide 1800 problems, distinguishable in 18 classes by their dimension and other high-level characteristics.

This section will report the most important settings of GABRLS algorithm that have impacted on the progressive improvement in results during the challenge.

Let us consider that DiagPI, AxialPI and Diversify routines are always active hereinafter in the GABRLS algorithm.

4.1   High Level Setting

The first strategy was the selection of a suitable local search to evaluate the improvement on convergence toward a better approximation of the global optimum with respect to the GABR scheme without LSs. In the preliminary phase of challenge, a non-monotone local search has been employed. In particular, a version of the globally convergent coordinate search algorithm called SDBOX [23] has been used, freely available in different programming languages in the software library for derivative free optimization at http://www.diag.uniroma1.it/~lucidi/DFL.

Table 1 details the results of the comparison, reports the number of global optima found over the hundred problems of each class of function. Clearly, it came out as expected that using local searches the behaviour of the algorithm is improved.

Subsequently another local search, integrated into Matlab toolbox (FMINCON [24]), has been tested. In the latest configuration of the GABRLS only FMINCON was

Table 1. Number of the problems solved by the algorithms with and without LS

Id	Type	Type-details	Dim	GABRLS	GABR
				Tasks solved	Tasks solved
0	GKLS	Non-differentiable	10	92	92
1			30	59	40
2		Differentiable	10	73	74
3			30	45	27
4		Twice differentiable	10	67	72
5			30	43	34
6	High condition	Rosenbrock	10	100	3
7			30	71	0
8		Rastrigin	10	100	99
9			30	100	0
10		Zakharov	10	100	100
11			30	2	0
12	Composite		10	100	7
13			30	100	0
14			10	100	3
15			30	100	0
16			10	100	69
17			30	100	0
Total				**1452**	**620**

integrated as LS with default parameters because of the slightly better result on several classes of function. In particular, FMINCON and NM-SDBOX showed equal performance to identify a local optimum on GKLS (classes 0, ..., 5), whereas FMINCON outperformed NM-SDBOX on all remaining classes.

The second high level strategy was to tune the amount of population and generations of GA. These parameters give an impact on the search performance and are crucial to balance efficiency (CPU time and convergence speed) and effectiveness (fast identification of the global optimum neighbourhood). Smaller values lead quickly to less quality solutions, while larger values allow to identify a more promising areas but slowing down the search.

The main aim was to find out the smallest value that allows to improve efficiency and solve the maximum number of tasks. The setting of these parameters on GKLS classes appeared more sensitive than other classes, so a mixed strategy has been implemented. In particular, two settings have been adopted as starting point of tuning, to solve GKLS functions and all other classes of functions, respectively.

Table 2 reports the selection phase of the two starting settings of population and generations indexed by scenarios.

Table 2. Selection of the starting setting of GA. The best ones are highlighted.

Scenario	Total Tasks Solved	GKLS solved	Other Classes Solved	Population	Generations
1	1580	381	1199	350	50
2	1580	380	1200	250	50
3	1577	378	1199	225	25
4	1548	349	1199	125	15
5	1276	335	941	60	10

The final amount of Population and Generations are refined for each class of GENOPT problems, so small changes have been made from the two guideline of scenarios selected.

The other high level parameters of GABRLS algorithm are self tuned or constants. After every selection phase of GA, carried out through the efficient well known *Tournament Selection* [25], a *Random Mutation* [26] with fixed rate has been integrated, as usual, inside the crossover operator to insure that the probability of reaching any point in the search space is never equal to zero. Table 3 reports the starting value and the updating rule of the most important parameters.

Table 3. Setting of other high level GA and BR parameters.

Parameter	Starting	Updating formula	Frequency
CF	1.2	$CF = CF + 0.01$	BR iterations
expCF	1	$expCF = iBR, \ iBR = i^{th} \ BR \ cycle$	BR iterations
Crossover Rate	0.8	Constant	GA iterations
Mutation Rate	0.01	Constant	GA iterations
Tournament	3	Constant	GA iterations

4.2 Results and Prizes

In the final phase of the GENOPT competition, GABRLS algorithm was able to reach the 1st prize in both partial categories. Figure 6 shows the score on the speed of convergence (High Jump), the task solved (Target Shooting) and the overall ranking (Biathlon Score).

GenOpt	HOME	UPLOAD	LEADERBOARD (1ST PHASE)	LEADERBOARD (FINAL)	LION11 CONFERENCE	CONTACT

FINAL LEADERBOARD

Position	Submission Name	High Jump ⓘ	Target Shooting ⓘ	Biathlon Score ⓘ	Submission Date	
1	F. Romito, GABRLS	1.13889	1.2222	1.18056	Apr, 7th (12 day(s) passed)	ⓘ
2	E. Segredo, E. Lalla-Ruiz, E. Hart, B. Paechter, S. Voß, HOCO	1.41667	1.7778	1.59722	Apr, 6th (13 day(s) passed)	ⓘ
3	A. Mariello, RRASH	3.41667	3.1667	3.29167	Apr, 6th (14 day(s) passed)	ⓘ
4	A. Marrero, E. Segredo, C. Segura, HCO-CMA-G	3.29167	3.3889	3.34028	Apr, 7th (12 day(s) passed)	ⓘ

Fig. 6. Final leaderboard.

The total number of solved problem was 1605 over 1800, almost 90%. Table 4 presents the number of solved problem for each class of function.

Table 4. Number of the solved tasks for each class.

Id	Type	Type-details	Dim	GABRLS Tasks solved
0	GKLS	Non-differentiable	10	86
1			30	74
2		Differentiable	10	77
3			30	56
4		Twice differentiable	10	66
5			30	47
6	High condition	Rosenbrock	10	100
7			30	100
8		Rastrigin	10	99
9			30	100
10		Zakharov	10	100
11			30	100
12	Composite		10	100
13			30	100
14			10	100
15			30	100
16			10	100
17			30	100
Total				**1605**

5   Conclusion

In this work, a novel algorithmic scheme for global optimization is presented. Several discretization techniques to place initial points and to bound the search space are described. The numerical results show that the overall scheme with local searches step up the effectiveness on locating optimal solutions within a high precision range.

Further research will involve on improving the exploratory geometry through the use of different bounding procedure. Moreover, algorithmic schemes for multi-objective optimization and constrained optimization will be investigated.

References

1. Serani, A., Fasano, G., Liuzzi, G., Lucidi, S., Iemma, U., Campana, E.F., Stern, F., Diez, M.: Ship hydrodynamic optimization by local hybridization of deterministic derivative-free global algorithms. Appl. Ocean Res. **59**, 115–128 (2016)
2. Liuzzi, G., Lucidi, S., Piccialli, V., Sotgiu, A.: A magnetic resonance device designed via global optimization techniques. Math. Program. **101**(2), 339–364 (2004)
3. Kvasov, D.E., Sergeyev, Y.D.: Deterministic approaches for solving practical black-box global optimization problems. Adv. Eng. Softw. **80**, 58–66 (2015)
4. Battiti, R., Sergeyev, Y.D., Brunato, M., Kvasov, D.E.: GENOPT 2016: design of a generalization-based challenge in global optimization. In: Sergeyev, Y.D., Kvasov, D.E., Dell'Accio, F., Mukhametzhanov, M.S. (eds.) AIP Conference Proceedings, vol. 1776, no. 060005. AIP Publishing (2016)
5. Gaviano, M., Kvasov, D.E., Lera, D., Sergeyev, Y.D.: Algorithm 829: software for generation of classes of test functions with known local and global minima for global optimization. ACM Trans. Math. Softw. **29**(4), 469–480 (2003)
6. Wolpert, D.H., Macready, W.G.: No free lunch theorems for optimization. IEEE Trans. Evol. Comput. **1**(1), 67–82 (1997)
7. Diez, M., Serani, A., Leotardi, C., Campana, E.F., Fasano, G., Gusso, R.: Dense orthogonal initialization for deterministic PSO: ORTHOinit+. In: Tan, Y., Shi, Y., Niu, B. (eds.) ICSI 2016. LNCS, vol. 9712, pp. 322–330. Springer, Cham (2016). doi:10.1007/978-3-319-41000-5_32
8. Di Pillo, G., Liuzzi, G., Lucidi, S., Piccialli, V., Rinaldi, F.: A DIRECT-type approach for derivative-free constrained global optimization. Comput. Optim. Appl. **65**(2), 361–397 (2016)
9. Kirkpatrick, S., Gelatt, C.D., Vecchi, M.P.: Optimization by simulated annealing. Science **220**(4598), 671–680 (1983)
10. Pierre, S., Houéto, F.: A tabu search approach for assigning cells to switches in cellular mobile networks. Comput. Commun. **25**(5), 464–477 (2002)
11. Baba, N.: Convergence of a random optimization method for constrained optimization problems. J. Optim. Theor. Appl. **33**(4), 451–461 (1981)
12. Dorigo, M., Blum, C.: Ant colony optimization theory: a survey. Theoret. Comput. Sci. **344**(2–3), 243–278 (2005)
13. Strongin, R.G., Sergeyev, Y.D.: Global Optimization with Non-Convex Constraints: Sequential and Parallel Algorithms. Nonconvex Optimization and Its Applications, vol. 45. Springer, New York (2000). doi:10.1007/978-1-4615-4677-1

14. Zhigljavsky, A., Žilinskas, A.: Stochastic Global Optimization. Springer Optimization and Its Applications. Springer, New York (2008). doi:10.1007/978-0-387-74740-8
15. Paulavičius, R., Žilinskas, J.: Simplicial Global Optimization. SpringerBriefs in Optimization. Springer, New York (2014). doi:10.1007/978-1-4614-9093-7
16. Locatelli, M., Schoen, F.: Global Optimization: Theory, Algorithms, and Applications. Society for Industrial and Applied Mathematics, Philadelphia (2013)
17. Whitley, D.: A genetic algorithm tutorial. Stat. Comput. 4(2), 65–85 (1994)
18. Goldberg, D.E.: Genetic Algorithms in Search, Optimization and Machine Learning. Addison-Wesley Publishing Company, Reading (1989)
19. Vavasis, S.A.: Complexity issues in global optimization: a survey. In: Horst, R., Pardalos, P. M. (eds.) Handbook of Global Optimization. Springer, Boston (1995). doi:10.1007/978-1-4615-2025-2_2
20. Nemirovskii, A., Yudin, D.B.: Problem Complexity and Method Efficiency in Optimization. A Wiley-Interscience. Wiley, New York (1983). Translated from the Russian and with a preface by Dawson E.R., Wiley-Interscience Series in Discrete Mathematics. Chichester: John Wiley and Sons (1983)
21. Sergeyev, Y.D., Kvasov, D.E.: Global search based on efficient diagonal partitions and a set of Lipschitz constants. SIAM J. Optim. 16(3), 910–937 (2006)
22. Liuzzi, G., Lucidi, S., Piccialli, V.: A DIRECT-based approach exploiting local minimizations for the solution of large-scale global optimization problems. Comput. Optim. Appl. 45 (2), 353–375 (2010)
23. Lucidi, S., Sciandrone, M.: A derivative-free algorithm for bound constrained optimization. Comput. Optim. Appl. 21(2), 119–142 (2002)
24. https://it.mathworks.com/help/optim/ug/fminunc.html#References
25. Goldberg, D.E., Kalyanmoy, D.: A comparative analysis of selection schemes used in genetic algorithms. Found. Genetic Algorithms 1, 69–93 (1991)
26. Herrera, F., Lozano, M., Verdegay, J.L.: Tackling real-coded genetic algorithms: operators and tools for behavioural analysis. Artif. Intell. Rev. 12(4), 265–319 (1998)

Short Papers

Identification of Discontinuous Thermal Conductivity Coefficient Using Fast Automatic Differentiation

Alla F. Albu[1,2], Yury G. Evtushenko[1,2], and Vladimir I. Zubov[1,2(✉)]

[1] Dorodnicyn Computing Centre, Federal Research Center
"Computer Science, and Control" of Russian Academy of Sciences,
Moscow, Russia
alla.albu@yandex.ru, evt@ccas.ru,
vladimir.zubov@mail.ru
[2] Moscow Institute of Physics and Technology,
Dolgoprudny, Moscow Region, Russia

Abstract. The problem of determining the thermal conductivity coefficient that depends on the temperature is studied. The consideration is based on the Dirichlet boundary value problem for the one-dimensional unsteady-state heat equation. The mean-root-square deviation of the temperature distribution field and the heat flux from the empirical data on the left boundary of the domain is used as the objective functional. An algorithm for the numerical solution of the problem based on the modern approach of Fast Automatic Differentiation is proposed. The case of discontinuous thermal conductivity coefficient is considered. Examples of solving the problem are discussed.

Keywords: Inverse coefficient problems · Variation problem · Fast Automatic Differentiation · Discontinuous thermal conductivity coefficient · Heat equation

1   Introduction

The classical heat equation is often used in the description and mathematical modeling of many thermal processes. The density of the substance, its specific thermal capacity, and the thermal conductivity coefficient appearing in this equation are assumed to be known functions of the coordinates and temperature. Additional boundary value conditions makes it possible to determine the dynamics of the temperature field in the substance under examination.

However, the substance properties are not always known. It often happens that the thermal conductivity coefficient depends only on the temperature, and this dependence is not known. In this case, the problem of determining the dependence of the thermal conductivity coefficient on the temperature based on experimental measurements of the temperature field arises. This problem also arises when a complex thermal process should be described by a simplified mathematical model. For example, in studying and modeling of heat propagation in complex porous composite materials, where the radiation heat transfer plays a considerable role, both the convective and radiative heat

© Springer International Publishing AG 2017
R. Battiti et al. (Eds.): LION 2017, LNCS 10556, pp. 295–300, 2017.
https://doi.org/10.1007/978-3-319-69404-7_21

transfer must be taken into account. The thermal conductivity coefficients in this case typically depend on the temperature. To estimate these coefficients, various models of the medium are used. As a result, one has to deal with a complex nonlinear model that describes the heat propagation in the composite material (see [1]). However, another approach is possible: a simplified model is constructed in which the radiative heat transfer is not taken into account, but its effect is modeled by an effective thermal conductivity coefficient that is determined based on empirical data.

Determining the thermal conductivity of a substance is an important problem, and it has been studied for a long time. This is confirmed by a large number of publications (e.g., see [2–5]).

In [6] the inverse coefficient problems are considered in new formulation. They are studied theoretically, and new numerical methods for their solution are developed. In that work the case of continuous thermal conductivity coefficient is considered.

In this paper we consider the problem studied in [6] for the case of a discontinuous thermal conductivity coefficient. The consideration is based on the Dirichlet problem for the one-dimensional unsteady-state heat equation. The inverse coefficient problem is reduced to a variation problem. A linear combination of the mean-root-square deviations of the temperature distribution field and the heat flux from the empirical data on the left boundary of the domain is used as the objective functional. An algorithm for the numerical solution of the inverse coefficient problem is proposed. It is based on the modern approach of Fast Automatic Differentiation, which made it possible to solve a number of difficult optimal control problems for dynamic systems.

2   Formulation of the Problem

A layer of material of width L is considered. The temperature of this layer at the initial time is given. It is also known how the temperature on the boundary of this layer changes with time. The distribution of the temperature field at each moment of time is described by the following initial boundary value (mixed) problem:

$$\rho C \frac{\partial T(x,t)}{\partial t} - \frac{\partial}{\partial x}\left(K(T)\frac{\partial T(x,t)}{\partial x}\right) = 0, \quad (x,t) \in Q \tag{1}$$

$$T(x,0) = w_0(x), \quad (0 \leq x \leq L), \tag{2}$$

$$T(0,t) = w_1(t), \quad T(L,t) = w_2(t), \quad (0 \leq t \leq \Theta). \tag{3}$$

Here x is the Cartesian coordinate of the point in the layer; t is time, $Q = \{(0 < x < L) \times (0 < t \leq \Theta)\}$; $T(x,t)$ is the temperature of the material at the point with the coordinate x and time t; ρ and C are the density and the heat capacity of the material, respectively; $K(T)$ is the coefficient of the convective thermal conductivity; $w_0(x)$ is the given temperature of the layer at the initial time; $w_1(t)$ is the given temperature on the left boundary of the layer; and $w_2(t)$ is the given temperature on the right boundary of the layer.

The density of the material ρ and its heat capacity C are known functions of coordinate and/or temperature.

If the dependence of the coefficient of the convective thermal conductivity $K(T)$ on the temperature T is known, then we can solve the mixed problem (1)–(3) to find the distribution of the temperature $T(x,t)$ in $\bar{Q}$. Problem (1)–(3) will be further called the direct problem.

If the dependence of the coefficient of the convective thermal conductivity of the material on the temperature is not known, it is of interest to determine this dependence. A possible statement of this problem (it is classified as an identification problem of the model parameters) is as follows: find the dependence $K(T)$ on T under which the temperature field $T(x,t)$ obtained by solving problem (1)–(3) is close to the field $Y(x,t)$, which itself is obtained empirically. The quantity

$$
\Phi(K(T)) = \int\limits_0^\Theta \int\limits_0^L [T(x,t) - Y(x,t)]^2 \cdot \mu(x,t)\, dx dt
$$
$$
+ \beta \cdot \int\limits_0^\Theta \left[K(T(0,t)) \cdot \frac{\partial T}{\partial x}(0,t) - P(t) \right]^2 dt,
$$
(4)

can be used as the measure of difference between these functions. Here $\beta \geq 0$ is a given number, $\mu(x,t) \geq 0$ is a given weighting function, and $P(t)$ is the known heat flux on the left boundary of the domain. Thus, the optimal control problem is to find the optimal control $K(T)$ and the corresponding optimal solution $T(x,t)$ of problem (1)–(3) that minimizes functional (4).

3   Numerical Solution of the Problem

The optimal control problems similar to this one are typically solved numerically using a descent method, which requires the gradient of functional (4) to be known. The unknown function $K(T)$ was approximated by a continuous piecewise linear function.

If the input data of the problem is such that the desired coefficient of thermal conductivity represents a fairly smooth function, then the problem of identification could be solved by a method proposed in [6].

This work presents the algorithm for solving the problem of identifying a discontinuous thermal conductivity coefficient and some numerical results. The proposed algorithm, as well as in [6], is based on the numerical solution of the problem of minimizing the cost functional (4). The gradient descent method was used. It is well known that it is very important for the gradient methods to determine accurate values of the gradients. For this reason, we used the efficient approach of Fast Automatic Differentiation that enables us to determine with machine precision the gradient of cost function, subject to equality constraints (see [7]).

To numerically solve the mixed problem (1)–(3) the domain $Q = \{(0 < x < L) \times (0 < t \leq \Theta)\}$ is decomposed by the grid lines $\{\tilde{x}_i\}_{i=0}^I$ and $\{\tilde{t}^j\}_{j=0}^J$ into rectangles.

At each node $(\tilde{x}_i, \tilde{t}^j)$ of Q characterized by the pair of indices (i, j), all the functions are determined by their values at the point $(\tilde{x}_i, \tilde{t}^j)$ (e.g., $T(\tilde{x}_i, \tilde{t}^j) = T_i^j$). In each rectangle, the thermal balance must be preserved.

The temperature interval $[a, b]$ (the interval of interest) is partitioned by the points, $T_0 = a, T_1, T_2, \ldots, T_N = b$ into $N = 2m + 1$ parts (they can be of equal or of different lengths). Each point T_n $(n = 0, \ldots, N)$ is assigned a number $k_n = K(T_n)$. The function $K(T) = T$, which needs to be found, is approximated by a continuous piecewise linear functions with the nodes at the points $\{(\tilde{T}_n, k_n)\}_{n=0}^{N}$ so that

$$K(T) = k_{n-1} + \frac{k_n - k_{n-1}}{T_n - T_{n-1}}(T - T_{n-1}) \text{ for } T_{n-1} \le T \le T_n, \quad (n = 1, \ldots, N).$$

The objective functional (4) was approximated by a function $F(k_0, k_1, \ldots, k_N)$ of the finite number of variables as

$$\Phi(K(T)) \approx F = \sum_{j=1}^{J} \sum_{i=1}^{I-1} \left((T_i^j - Y_i^j)^2 \cdot \mu_i^j h_i \tau^j \right)$$

$$+ \beta \cdot \sum_{j=1}^{J} \left(\left[\frac{\sigma}{2h_0} (K(T_0^j) + K(T_1^j)) \cdot (T_1^j - T_0^j) \right. \right.$$

$$\left. \left. + \frac{1 - \sigma}{2h_0} \left(K(T_0^{j-1}) + K(T_1^{j-1}) \right) \cdot \left(T_1^{j-1} - T_0^{j-1} \right) - \frac{\rho_0 C_0 h_0}{2\tau^j} \left(T_0^j - T_0^{j-1} \right) - P^j \right]^2 \cdot \tau^j \right).$$

To illustrate the efficiency of the proposed algorithm the variation problem (1)–(4) was considered with the following parameters:

$$L = 1, \qquad \Theta = 1, \quad \rho(x) \equiv C(x) \equiv 1,$$
$$w_0(x) = 2x, \quad w_1 = 0, \quad w_2 = 2,$$
$$\mu(x, t) \equiv 1, \quad \beta = 0, \quad a = 0, \quad b = 2.$$

The empirical temperature field was determined as a solution of the "direct" problem (1)–(3) with $K(T) = \begin{cases} 1, & T < 1, \\ 3, & T \ge 1. \end{cases}$

On the first m subintervals $K(T) = 1$, on the $(m + 2), (m + 3), \ldots, N$ subintervals $K(T) = 3$, and on the $(m + 1)$ subinterval $K(T)$ is a linear function varying in from 1 to 3.

For the solution of the direct and inverse problems, we used the uniform grid with the parameters $I = 300$ (the number of intervals along the axis x) and $J = 6000$ (the number of intervals along the axis t), which ensures the sufficient accuracy of computation of the temperature field.

Figure 1 shows the results (the initial approximation and optimal solution) of calculations made for $N = 19$. As an initial approximation to the thermal conductivity coefficient the function $K(T) = T$ was selected.

The value of the cost function (4) for initial approximation was equal to $\Phi_{ini} = 4.241570 \cdot 10^0$. The value of the cost function (4) for optimal control was equal

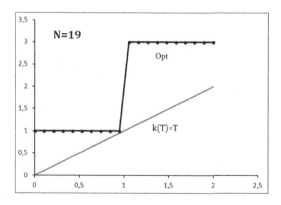

Fig. 1. Initial and optimal control ($N = 19$)

to $\Phi_{opt} = 1.013323 \cdot 10^{-21}$. In this example, we managed to identify the thermal conductivity coefficient with precision 10^{-9} in norm C.

Figure 2 presents the same results of calculations as above but made for $N = 79$ (the initial approximation and optimal solution). The value of the cost function (4) for initial approximation was equal to $\Phi_{ini} = 9.532799 \cdot 10^{-3}$. The value of the cost function (4) for optimal control was equal to $\Phi_{opt} = 2.288689 \cdot 10^{-22}$. In this case, we managed to identify the thermal conductivity coefficient with precision 10^{-10} in norm C.

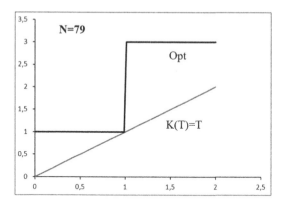

Fig. 2. Initial and optimal control ($N = 79$)

It should be noted that in this example, the thermal conductivity coefficient being identified was a continuous function, although contained narrow domain of "smoothing" jump (its width is $\approx 2/N$).

The numerous numerical results showed that the proposed algorithm is efficient and stable and allows us to restore thermal conductivity with high accuracy.

Acknowledgments. This work was supported by the Russian Foundation for Basic Research (project no. 17-07-00493a).

References

1. Alifanov, O.M., Cherepanov, V.V.: Mathematical simulation of high-porosity fibrous materials and determination of their physical properties. High Temp. **47**, 438–447 (2009)
2. Kozdoba, L.A., Krukovskii, P.G.: Methods for Solving Inverse Thermal Transfer Problems. Naukova Dumka, Kiev (1982). [in Russian]
3. Alifanov, O.M.: Inverse Problems of Heat Transfer. Mashinostroenie, Moscow (1988). [in Russian]
4. Marchuk, G.I.: Adjoint Equation and the Analysis of Complex Systems. Nauka, Moscow (1992). [in Russian]
5. Samarskii, A.A., Vabishchevich, P.N.: Computational Heat Transfer. Editorial URSS, Moscow (2003). [in Russian]
6. Zubov, V.I.: Application of fast automatic differentiation for solving the inverse coefficient problem for the heat equation. Comput. Math. Math. Phys. **56**(10), 1743–1757 (2016)
7. Evtushenko, Y.G.: Computation of exact gradients in distributed dynamic systems. Optim. Methods Softw. **9**, 45–75 (1998)

Comparing Two Approaches for Solving Constrained Global Optimization Problems

Konstantin Barkalov$^{(\boxtimes)}$ and Ilya Lebedev

Lobachevsky State University of Nizhny Novgorod, Nizhny Novgorod, Russia
{konstantin.barkalov,ilya.lebedev}@itmm.unn.ru

Abstract. In the present study, a method for solving the multiextremal problems with non-convex constrains without the use of the penalty or barrier functions is considered. This *index method* is featured by a separate accounting for each constraint. The check of the constraint fulfillment sequentially performed in every iteration point is terminated upon the first constraint violation occurs. The results of numerical comparing of the index method with the penalty function one are presented. The comparing has been carried out by means of the numerical solving of several hundred multidimensional multiextremal problems with non-convex constrains generated randomly.

Keywords: Global optimization · Multiextremal functions · Non-convex constraints

1 Introduction

In the present paper, the methods for solving the global optimization problems with non-convex constraints

$$\varphi(y^*) = \min\left\{\varphi(y) : y \in D,\ g_i(y) \leq 0,\ 1 \leq i \leq m\right\}, \tag{1}$$

$$D = \left\{y \in R^N : a_j \leq y_j \leq b_j, 1 \leq j \leq N\right\}. \tag{2}$$

are considered. The objective function as well as the constraint ones are supposed to satisfy Lipschitz condition with Lipschitz constants unknown a priori. The analytical formulae of the problem functions may be unknown, i.e. these ones may be defined by an algorithm for computing the function values in the search domain (so called "black-box"-functions). Moreover, it is suggested that even a single computing of a problem function value may be a time-consuming operation since in the applied problems it is related to the necessity of numerical modeling (see, for example, [1–4]).

The method of penalty functions is one of the most popular numerical methods for solving the problems of this kind. The idea of the method is simple and very universal; therefore, the method has found wide application to solving various practical problems. A detailed description of the method can be found, for

© Springer International Publishing AG 2017
R. Battiti et al. (Eds.): LION 2017, LNCS 10556, pp. 301–306, 2017.
https://doi.org/10.1007/978-3-319-69404-7_22

example, in [5]. Nevertheless, the method of penalty functions has a number of disadvantages.

First, when using this method, a number of unconstrained problems with different penalty coefficients should be solved. The issue of choosing the penalty coefficients should be addressed for each problem separately.

Second, the use of large values of the penalty coefficients results in the minimization of the functions with a ravine-type minimum that essentially complicates the computations.

Third, this method is not applicable in the problems with the partially computable functions, for example, in the case when the objective function is undefined if the constraints are violated.

In the present work, the use of the penalty function method is compared to the application of the index method of accounting for the constraints developed in Nizhny Novgorod State University. The method features are (i) accounting for the information on each constraint separately and (ii) solving the problems, in which the function values may be undefined out of the feasible domain.

2   Index Method

A novel approach to the minimization of the multiextremal functions with non-convex constraints (called the index method of accounting for the constraints) has been developed in [6–9]. The approach is based on a separate accounting for each constraint of the problem and is not related to the use of the penalty functions. At that, employing the continuous single-valued Peano curve $y(x)$ (*evolvent*) mapping the unit interval $[0, 1]$ on the x-axis onto the N-dimensional domain (2) it is possible to find the minimum in (1) by solving a one-dimensional problem

$$\varphi(y(x^*)) = \min \{\varphi(y(x)) : x \in [0,1],\ g_i(y(x)) \leq 0,\ 1 \leq i \leq m\}.$$

The considered dimensionality reduction scheme juxtaposes to a multidimensional problem with Lipschitzian functions a one-dimensional problem, where the corresponding functions satisfy uniform Hölder condition (see [6]).

According to the rules of index method, every iteration called trial at corresponding point of the search domain includes a sequential checking of fulfillment of the problem constraints at this point. The first occurrence of violation of any constraint terminates the trial and initiates the transition to the next iteration point, the values of the rest problem functions are not computed in this point. This allows: (i) accounting for the information on each constraint separately and (ii) solving the problems, in which the function values may be undefined out of the feasible domain.

The index algorithm can be applied to solving the unconstrained problems as well, i.e. in the case when $m = 0$ in the problem statement (1). In this case, the algorithm works analogously to its prototype – global search algorithm (GSA) – developed for solving the unconstrained optimization problems. Various modifications of the index algorithm and the corresponding theory of convergence are

presented in [6]. The algorithm is very flexible and allows an efficient parallelization as for shared memory, as for distributed memory, as for accelerators [10–14].

3 Results of Experiments

Let us compare the solving of the constrained global optimization problems using
penalty functions method (PM) and index method (IM). The penalty function
was taken in the form

$$G(y) = \sigma \sum_{j=1}^{m} \max \left\{0, g_j(y)\right\}^2,$$

where σ is the penalty coefficient. Thus, a constrained problem is reduced to an
unconstrained problem

$$F(y) = \varphi(y) + G(y)$$

which was solved using global search algorithm.

The experiments was carried out with the use of GKLS generator. This
generator for the functions of arbitrary dimensionality with known properties
(the number of local minima, the size of their domains of attraction, the global
minimizer, etc.) has been proposed in [16]. Eight GKLS classes of differentiable
test functions of the dimensions $N = 2, 3, 4$, and 5, have been used. For each
dimension, both *Hard* and *Simple* classes have been considered. The difficulty
of a class was increased either by decreasing the radius of the attraction region
of the global minimizer, or by decreasing the distance from the global minimizer
y^* to the domain boundaries. Application of this generator for studying some
unconstrained optimization algorithms has been described in [15,17–19].

According to the scheme described in [20], eight series of 100 problems each
based on the functions of *Simple* and *Hard* classes with the dimensions $N = 2, 3, 4, 5$ have been generated. There were two constraints and objective functions
in each problem. The volume fraction of the feasible domain was varied from 0.2
to 0.8. The penalty parameter σ was selected to be equal to 100.

The global minimum was considered to be found if the algorithm generates a
trial point y^k in the δ-vicinity of the global minimizer, i.e. $\left\|y^k - y^*\right\|_\infty \leq \delta$. The
size of the vicinity was selected as $\delta = 0.01 \left\|b - a\right\|$. When using IM for *Simple*
class, parameter $r = 4.5$ was selected, for *Hard* class $r = 5.6$ was taken. The
maximum allowed number of iterations was $K_{max} = 10^6$.

The average number of iterations k_{av} performed by PM for solving a series
of problems from both classes is shown in Table 1. The same values for the
index method are presented in Table 2. The number of unsolved problems is
specified in brackets. It reflects a situation when not all problems of the class
have been solved by the method. This means that the algorithm was stopped as
the maximum allowed number of iterations K_{max} was achieved. In this case, the
value $K_{max} = 10^6$ was used for calculating the average value of the number of
iterations k_{av} that corresponds to the lower estimate of the average value.

Table 1. The values of k_{av} when solving the GKLS problems by PM

Δ	$N = 2$		$N = 3$		$N = 4$		$N = 5$	
	Simple	Hard	Simple	Hard	Simple	Hard	Simple	Hard
0.2	1639	1507	56149	60651	160198(5)	119764(2)	305957(1)	488587(14)
0.4	1534	1967	67145	74860	158547	127832	396480(4)	434328(2)
0.6	1201	1514	92854	101240	138550	143818	373617(8)	561447(9)
0.8	1277	1287	108121	148479	130372	145592	488791(12)	646538(24)

Table 2. The values of k_{av} when solving the GKLS problems by IM

Δ	$N = 2$		$N = 3$		$N = 4$		$N = 5$	
	Simple	Hard	Simple	Hard	Simple	Hard	Simple	Hard
0.2	447	911	14719	20120	59680	66551(1)	391325(2)	188797(12)
0.4	465	1800	11951	17427	71248	86899	339327(1)	151998(3)
0.6	403	1988	7366	12853	58451	92007	316648	179013(4)
0.8	371	4292	4646	8702	33621	54405	309844	124952

Table 3. Solving of GKLS problems by the index method, N = 4

Δ	Simple			Hard		
	k_1	k_3	k_3	k_1	k_2	k_3
0.2	59680	20445	4401	66551	24210	6316
0.4	71248	28527	6784	86899	39682	12615
0.6	58451	31508	9505	92007	52560	19853
0.8	33621	21411	10446	54405	36838	22202

The average values presented demonstrate that the solving of the specified problems using the index method requires less number of iterations than with the use of the penalty function one. At the same time, separate accounting for the constraints in the index method provides less number of computations of the values of the problem functions as well. The numbers of computations of the values of the constraint functions $g_1(y)$, $g_2(y)$ and those for the objective function $\varphi(y)$ (k_1, k_2, and k_3, respectively) are presented in Table 3 for four-dimensional problems.

4   Conclusion

Concluding, let us note that the index method for solving constrained global optimization problems considered in the present work:

- is based on the global search algorithm, which is not inferior in the speed of work than other well-known algorithms;

- allows solving the initial problem directly, without the use of the penalty functions (thus, the issues of selection the penalty coefficient and of solving a series of unconstrained problems with different penalty coefficients are eliminated);
- allows solving the problems, which the values of the problem function are not defined everywhere (for example, the objective function values are undefined out of the problem feasible domain);
- speeds up the process of solving the constrained optimization problems (due to an essential reduction of the total number of computations of the problem function values).

The last statement has been supported by the numerical solving several hundred test problems.

Acknowledgments. This study was supported by the Russian Science Foundation, project No 16-11-10150.

References

1. Famularo, D., Pugliese, P., Sergeyev, Y.D.: A global optimization technique for checking parametric robustness. Automatica **35**, 1605–1611 (1999)
2. Kvasov, D.E., Menniti, D., Pinnarelli, A., Sergeyev, Y.D., Sorrentino, N.: Tuning fuzzy power-system stabilizers in multi-machine systems by global optimization algorithms based on efficient domain partitions. Electr. Power Syst. Res. **78**(7), 1217–1229 (2008)
3. Kvasov, D.E., Sergeyev, Y.D.: Deterministic approaches for solving practical black-box global optimization problems. Adv. Eng. Softw. **80**, 58–66 (2015)
4. Modorskii, V.Y., Gaynutdinova, D.F., Gergel, V.P., Barkalov, K.A.: Optimization in design of scientific products for purposes of cavitation problems. In: Simos, T.E. (ed.) ICNAAM 2015. AIP Conference Proceedings, vol. 1738 (2016). Article No. 400013
5. Bazaraa, M.S., Sherali, H.D., Shetty, C.M.: Nonlinear Programming: Theory and Algorithms, 2nd edn. Wiley, New York (1993)
6. Strongin, R.G., Sergeyev, Y.D.: Global Optimization with Non-convex Constraints: Sequential and Parallel Algorithms. Kluwer Academic Publishers, Dordrecht (2000)
7. Sergeyev, Y.D., Famularo, D., Pugliese, P.: Index branch-and-bound algorithm for Lipschitz univariate global optimization with multiextremal constraints. J. Glob. Optim. **21**(3), 317–341 (2001)
8. Barkalov, K.A., Strongin, R.G.: A global optimization technique with an adaptive order of checking for constraints. Comput. Math. Math. Phys. **42**(9), 1289–1300 (2002)
9. Strongin, R.G., Sergeyev, Y.D.: Global optimization: fractal approach and non-redundant parallelism. J. Glob. Optim. **27**(1), 25–50 (2003)
10. Barkalov, K., Ryabov, V., Sidorov, S.: Parallel scalable algorithms with mixed local-global strategy for global optimization problems. In: Hsu, C.-H., Malyshkin, V. (eds.) MTPP 2010. LNCS, vol. 6083, pp. 232–240. Springer, Heidelberg (2010). doi:10.1007/978-3-642-14822-4_26

11. Barkalov, K.A., Gergel, V.P.: Multilevel scheme of dimensionality reduction for parallel global search algorithms. In: Proceedings of the 1st International Conference on Engineering and Applied Sciences Optimization - OPT-i 2014, pp. 2111–2124 (2014)

12. Barkalov, K., Gergel, V., Lebedev, I.: Use of Xeon Phi coprocessor for solving global optimization problems. In: Malyshkin, V. (ed.) PaCT 2015. LNCS, vol. 9251, pp. 307–318. Springer, Cham (2015). doi:10.1007/978-3-319-21909-7_31

13. Barkalov, K., Gergel, V.: Parallel global optimization on GPU. J. Glob. Optim. **66**(1), 3–20 (2016)

14. Barkalov, K., Gergel, V., Lebedev, I.: Solving global optimization problems on GPU cluster. In: Simos, T.E. (ed.) ICNAAM 2015. AIP Conference Proceedings, vol. 1738 (2016). Article No. 400006

15. Gergel, V., Grishagin, V., Gergel, A.: Adaptive nested optimization scheme for multidimensional global search. J. Glob. Optim. **66**(1), 35–51 (2016)

16. Gaviano, M., Kvasov, D.E., Lera, D., Sergeyev, Y.D.: Software for generation of classes of test functions with known local and global minima for global optimization. ACM Trans. Math. Softw. **29**(4), 469–480 (2003)

17. Sergeyev, Y.D., Kvasov, D.E.: Global search based on efficient diagonal partitions and a set of Lipschitz constants. SIAM J. Optim. **16**(3), 910–937 (2006)

18. Paulavicius, R., Sergeyev, Y., Kvasov, D., Zilinskas, J.: Globally-biased DISIMPL algorithm for expensive global optimization. J. Glob. Optim. **59**(2–3), 545–567 (2014)

19. Sergeyev, Y.D., Kvasov, D.E.: A deterministic global optimization using smooth diagonal auxiliary functions. Commun. Nonlinear Sci. Numer. Simul. **21**(1–3), 99–111 (2015)

20. Gergel, V.: An approach for generating test problems of constrained global optimization. In: Battiti, R., Kvasov, D., Sergeyev, Y. (eds.) LION 2017. LNCS, vol. 10556, pp. 314–319. Springer, Cham (2017). doi:10.1007/978-3-319-69404-7_24

Towards a Universal Modeller
of Chaotic Systems

Erik Berglund[⊠]

Jeppesen, Gothenburg, Sweden
erik.berglund@jeppesen.com

Abstract. This paper proposes a Machine Learning (ML) algorithm
and examines its dynamic properties when trained on chaotic time series.
It will be demonstrated that the output of the ML system is itself more
chaotic if it is trained on a chaotic input than if it is trained on non-
chaotic input. The proposed ML system builds on to the Parameter-Less
Self-Organising Map 2 (PLSOM2) and introduces new developments.

Keywords: Machine learning · Self-organization · Chaotic · Recurrence · Recursion · Universal model

1 Introduction

The class of dynamical systems that are non-linear and highly sensitive to initial
conditions - commonly called *chaotic* - appear in many places in nature, and it
has even been suggested that chaos plays a part in biological intelligence [6] and
perception [7]. Since deterministic systems can be chaotic (indeed, this is where
the phenomenon was first discovered), chaos is of interest to computer scientists.

Chaotic systems have been subject of study in the Machine Learning community. Much, if not most, of this effort has been aimed at predicting the time
evolution of chaotic systems based on starting conditions and a manually constructed model of the chaotic system. This prediction is, by the very nature of
chaotic systems, extremely difficult and the predictive value of the model quickly
diminishes as the prediction time period increases.

Instead this paper presents a machine learning system that, by being trained
on the output of a dynamical system, can replicate some of the fundamental
properties of the dynamical system without necessarily attempting to predict
the evolution of the dynamical system itself, thus creating a tool that can model
chaotic behaviour without knowing the underlying rules of the chaotic system.

The algorithm is trained on four different deterministic dynamical processes,
of which one is chaotic and the others are not, and the chaotic properties of the
system are measured.

The rest of this document is laid out as follows: Sect. 2 discusses previous
works and gives background information, Sect. 3 gives details for the machine
learning system, Sect. 4 describes the experimental setup and Sect. 5 lists the
results. Section 6 contains concluding remarks and discussion.

© Springer International Publishing AG 2017
R. Battiti et al. (Eds.): LION 2017, LNCS 10556, pp. 307–313, 2017.
https://doi.org/10.1007/978-3-319-69404-7_23

2   Previous Work

Chaotic time-series and chaos in neural networks has been studied by several researchers. One prominent example is CSA [4], which has been shown to be able to solve combinatorial tasks like the Travelling Salesperson Problem (TSP) efficiently. Unlike the present work the chaos is inherent in the network model, not learned. CSA has inspired other approaches, for example [13], which also contains a good review of similar methods.

Time-series processing with SOM-type algorithms have been studied before, with the aim of predicting time series [3,8].

Another avenue of research is the role of chaotic neural networks in memory formation and retrieval [5].

The nomenclature of SOM algorithms with recurrent connections is confusing. In other ML algorithms recurrence usually means that the output of the algorithm is fed back as part of the input. The Recurrent SOM [11], on the other hand, is in essence a SOM with leaky integrators on the inputs. When recurrent connections in SOM algorithms were first investigated [12], the term "recursive" was used instead.

Self-Organising Maps are a form of unsupervised learning, where a set of nodes indexed by $i \in [1, n]$ are each associated with a weight vector w_i. Training is iterative, and proceeds as a discrete set of steps where training data is presented to the network, then the weight vectors are updated until some predetermined condition is met.

In the standard SOM the magnitude of the weight update is dependent on the iteration number or time. For the PLSOM [2] and PLSOM2 [1] algorithms, on the other hand, the weight update is a function of what can be inferred about the state of the network.

3   Learning Algorithm

Recursion in SOM type neural networks works the same way as recurrence in other Neural Network algorithms: The output at time $t - 1$ forms part of the input at time t. The particular method used here is adopted from [12].

The output is defined as a vector of length n, where n is the number of nodes, that corresponds to the excitation of a given node.

$$k_i = e^{-[\alpha||x(t-1)-w_x|| + (1-\alpha)||y(t-1)-w_y||]} \tag{1}$$

In (1), α is a tuning parameter that determines how much influence the new input has relative to the recurrent connections, k is a vector of length n, e is Euler's constant and y is the output vector, w_y is the part of the weight vector that corresponds to the output (that is, the recurrent weights) and w_x is the part of the weight vector that corresponds to the input. Following the calculation of k, the updated value of y is found by scaling and translating k so that each element lies in the range $[0, 1]$ according to (2).

$$y_i(t) = \frac{k_i - \min(k)}{\max(k) - \min(k)} \tag{2}$$

In other words, y is a vector where each entry y_i corresponds to the excitation of a node i. The entry that corresponds to the winning node, y_c, is equal to 1, all other entries are in the interval $[0, 1)$. The vector y is the excitation vector.

To get the Recursive PLSOM2 algorithm (RPLSOM2), ϵ is changed to (3).

$$\epsilon = \frac{\alpha ||x - w_{x,c}|| + (1 - \alpha)||y - w_{y,c}||}{\alpha S_x + (1 - \alpha)S_y} \tag{3}$$

The part of the weight vector of the winning node that relates to the input is $w_{x,c}$, and the part that relates to the recurrent connections is $w_{y,c}$. The scalar S_y is the diameter of the set of all recurrent inputs, and S_x is the diameter of the set of all direct inputs. The parameter α is in the range $[0, 1]$. For $\alpha = 1$ the RPLSOM2 is equal to the PLSOM2, since all recurrent connections are ignored.

3.1 Idle Mode

As is clear from the above algorithm the input is only partially responsible for selecting the winning node, a large part is also due to the excitation vector given by (2). The map can thus be iterated without taking the input into account by calculating the winning node and the next value of the excitation vector based on the current excitation vector. This will henceforth be referred to as "idle mode".

4 Experimental Setup

As training input for time series are used:

1. A sine wave with wavelength 200.
2. The sum of two sine waves, with wavelengths 200 and 61.5.
3. A simulated Mackey-Glass [9] sequence, consisting of 511 sine waves with different wavelength and amplitude added together to resemble the frequency response of the real Mackey-Glass sequence.
4. A chaotic Mackey-Glass series.

All time series were scaled and translated to span the interval $[0, 1]$. The Fourier coefficients of of the simulated Mackey-Glass series is indistinguishable from the Fourier coefficients of the real Mackey-Glass series, as the simulated series was created through inverse Fourier transform of the real series.

The Mackey-Glass time series is given by (4).

$$\frac{dx}{dt} = a\frac{x_\tau}{1 + x_\tau^n} - bx \tag{4}$$

here x_τ represents the value of the variable at time index $(t - \tau)$. In the present work the following values are used: $\tau = 17$, $a = 0.2$, $b = 0.1$, and $n = 10$.

Before each test the RPLSOM2 weights were initialised to a random state and trained with 50000 samples from one of the time series. The map has 100 nodes arranged in a 10×10 grid. The generalisation factor was set to 17, and α set to 0.6. Each experiment was repeated 1000 times for each time series.

4.1   Repetition

One characteristic of chaotic systems is that they have very long repetition periods. Therefore the repetition period for the input sequences was measured. The repetition period is defined as the number of samples one must draw from the series before the last 100 samples are repeated anywhere in any of the previous samples.

After training the map was put into idle mode for 200000 iterations. The weight vector of the winning node will describe a one-dimensional time series, which was checked for repetition.

4.2   Fractal Dimension

To calculate the fractal dimension of the series attractor, it was embedded in 3 dimensions, using (5),

$$X_t^T = \{x(t), x(t-1), x(t-2)\} \tag{5}$$

where $x(t)$ is the sample drawn from the time series at time t. This results in a trajectory of vectors in 3-space.

The fractal dimension was estimated using the information dimension based on 15000 samples drawn from the trained RPLSOM2 in idle mode. Any output sequence with fewer than 100 unique points were discarded, since this gives the information dimension computation too little to work with to give a meaningful result. Few unique points indicate a stable orbit or stable point, which would indicate a non-fractal dimension. The number of discarded output sequences for each input time series are given by Table 1.

Table 1. Number of output sequences discarded because of too few unique points.

Sequence	Number discarded
Sine	920
Double sine	764
Simulated Mackey-Glass	718
Mackey-Glass	160

4.3   Lyapunov Exponent

The Lyapunov exponent was calculated using the excitation vector p from (2) of the RPLSOM2 and the numerical algorithm described in [10]. The perturbation value used was $d_0 = 10^{-12}$, and the algorithm was in idle mode for 300 iterations before 15000 iterations were sampled and averaged to compute the Lyapunov exponent estimate.

5   Results

As can be seen from Table 2, the mean Lyapunov exponent for the map trained with a chaotic time series is clearly less negative. This becomes even clearer from Table 3, which shows the percentage of maps with positive Lyapunov exponents.

Table 2. Estimated Lyapunov exponent.

Sequence	Mean Lyapunov exponent	std err
Sine	−0.2451	0.0054
Double sine	−0.2326	0.0066
Simulated Mackey-Glass	−0.1641	0.0035
Mackey-Glass	−0.0636	0.0041

Table 3. Percentage of trained maps with a positive Lyapunov exponent

Sequence	% positive exponents	std err
Sine	2.3	0.5
Double sine	4.1	0.6
Simulated Mackey-Glass	2.1	0.5
Mackey-Glass	20.4	1.3

The connection between the chaos of the input sequence and the behaviour of the map is also evident in the number of iterations before repeat, see Table 4.

Table 4. Mean number of iterations before map starts repeating

Sequence	iterations	std err
Sine	446	35
Double sine	997	57
Simulated Mackey-Glass	832	37
Mackey-Glass	2809	81

As shown Table 5 there is a clear difference in the dimension of the different maps, and the Mackey-Glass, as expected, has the largest dimension value. The sine-trained sequence has a surprisingly high dimension, even though it is the simplest sequence. It is possibly an artefact of the high number of tests that had to be discarded due to too few unique points, see Table 1.

Table 5. Mean information dimension of map output

Sequence	dimension	std err
Sine	1.0181	0.0024
Double sine	1.0145	0.0010
Simulated Mackey-Glass	1.0115	0.0007
Mackey-Glass	1.0206	0.0006

6 Conclusion

It was observed that RPLSOM2 networks trained with a chaotic time series to a significant degree exhibit the following characteristics:

- Longer repetition periods.
- Higher Lyapunov exponent.
- Higher probability of having a positive Lyapunov exponent.
- Higher fractal dimension.

when compared to networks that have been trained on non-chaotic but otherwise similar time sequences. This is consistent with chaotic behaviour.

This is the first instance of the chaotic behaviour of a network output after training depends on its training input.

References

1. Berglund, E.: Improved PLSOM algorithm. Appl. Intell. **32**(1), 122–130 (2010)
2. Berglund, E., Sitte, J.: The parameterless self-organizing map algorithm. IEEE Trans. Neural Netw. **17**(2), 305–316 (2006)
3. Chappell, G.J., Taylor, J.G.: The temporal kohonen map. Neural Netw. **6**(3), 441–445 (1993)
4. Chen, L., Aihara, K.: Chaotic simulated annealing by a neural-network model with transient chaos. Neural Netw. **8**(6), 915–930 (1995)
5. Crook, N., Scheper, T.O.:. A novel chaotic neural network architecture. In: ESANN 2001 Proceedings, pp. 295–300, April 2001
6. Freeman, W.J.: Chaos in the brain: possible roles in biological intelligence. Int. J. Intell. Syst. **10**(1), 71–88 (1995)
7. Freeman, W.J., Barrie, J.M.: Chaotic oscillations and the genesis of meaning in cerebral cortex. In: Buzsáki, G., Llinás, R., Singer, W., Berthoz, A., Christen, Y. (eds.) Temporal Coding in the Brain. NEUROSCIENCE. Springer, Heidelberg (1994). doi:10.1007/978-3-642-85148-3_2
8. Koskela, T., Varsta, M., Heikkonen, J., Kaski, K.:. Recurrent SOM with local linear models in time series prediction. In: 6th European Symposium on Artificial Neural Networks, pp. 167–172. D-facto Publications (1998)
9. Mackey, M.C., Glass, L.: Oscillation and chaos in physiological control systems. Science **197**, 287 (1977)
10. Sprott, J.C.: Chaos and Time-Series Analysis. Oxford University Press, Oxford (2003)

11. Varstal, M., Millán, J.R., Heikkonen, J.: A recurrent self-organizing map for temporal sequence processing. In: Gerstner, W., Germond, A., Hasler, M., Nicoud, J.-D. (eds.) ICANN 1997. LNCS, vol. 1327, pp. 421–426. Springer, Heidelberg (1997). doi:10.1007/BFb0020191
12. Voegtlin, T.: Recursive self-organizing maps. Neural Netw. **15**(8–9), 979–991 (2002)
13. Wang, L., Li, S., Tian, F., Fu, X.: A noisy chaotic neural network for solving combinatorial optimization problems: stochastic chaotic simulated annealing. IEEE Trans. Syst. Man Cybern. Part B **34**(5), 2119–2125 (2004)

An Approach for Generating Test Problems of Constrained Global Optimization

Victor Gergel$^{(\boxtimes)}$

Lobachevsky State University of Nizhny Novgorod, Nizhny Novgorod, Russia
gergel@unn.ru

Abstract. In the present paper, a novel approach to the constructing of the test global optimization problems with non-convex constraints is considered. The proposed approach is featured by a capability to construct the sets of such problems for carrying out multiple computational experiments in order to obtain a reliable evaluation of the efficiency of the optimization algorithms. When generating the test problems, the necessary number of constraints and desired fraction of the feasible domain relative to the whole global search domain can be specified. The locations of the global minimizers in the generated problems are known a priori that simplifies the evaluation of the results of the computational experiments essentially. A demonstration of the developed approach in the application to well-known index method for solving complex multiextremal optimization problems with non-convex constraints is presented.

Keywords: Global optimization · Multiextremal functions · Non-convex constraints · Test optimization problems · Numerical experiments

1 Introduction

In the present paper, the methods for generating the global optimization test problems with non-convex constraints

$$\varphi(y^*) = \min\{\varphi(y) : y \in D,\ g_i(y) \leq 0,\ 1 \leq i \leq m\}, \tag{1}$$

$$D = \{y \in R^N : a_i \leq y_i \leq b_i, 1 \leq i \leq N\} \tag{2}$$

are considered. The objective function $\varphi(y)$ (henceforth denoted by $g_{m+1}(y)$) and the left-hand sides $g_i(y)$, $1 \leq i \leq m$, of the constraints are supposed to satisfy the Lipschitz condition

$$|g_i(y') - g_i(y'')| \leq L_i \|y' - y''\|,\ y', y'' \in D,\ 1 \leq i \leq m+1.$$

with the Lipschitz constants unknown a priori. The analytical formulae of the problem functions may be unknown, i.e. these ones may be defined by an algorithm for computing the function values in the search domain (so called "black-box"-functions). It is supposed that even a single computing of a problem function value may be a time-consuming operation since it is related to the necessity of numerical modeling in the applied problems (see, for example, [1–4]).

© Springer International Publishing AG 2017
R. Battiti et al. (Eds.): LION 2017, LNCS 10556, pp. 314–319, 2017.
https://doi.org/10.1007/978-3-319-69404-7_24

The evaluation of efficiency of the developed methods is one of the key problems in the optimization theory and applications. Unfortunately, it is difficult to obtain any theoretical estimates in many cases. As a result, the comparison of the methods is performed by carrying out the computational experiments on solving some test optimization problems in most cases. In order to obtain a reliable evaluation of the efficiency of the methods, the sets of test problems should be diverse and representative enough. The problem of choice of the test problems has been considered in a lot of works (see, for example, [5–8]). Unfortunately, in many cases, the proposed sets contain a small number of test problems, and it is difficult to obtain the problems with desired properties. The most important drawback consists of the fact that the constraints are absent in the proposed test problems as a rule (or the constraints are relatively simple: linear, convex, etc.).

A novel approach to the generation of any number of the global optimization problems with non-convex constraints for performing multiple computational experiments in order to obtain a reliable evaluation of the efficiency of the developed optimization algorithms has been proposed. When generating the test problems, the necessary number of constraints and desired fraction of the feasible domain relative to the whole search domain can be specified. In addition, the locations of the global minimizers in the generated problems are known a priori that simplifies the evaluation of the results of the computational experiments essentially.

2   Test Problem Classes

A well-known approach to investigating and comparing the multiextremal optimization algorithms is based on testing these methods by solving a set of problems, chosen randomly from some specially designed class.

One generator for random samples of two-dimensional test functions has been described in [9,10]. This generator produces two-dimensional functions according to the formula

$$\varphi(y) = - \left\{ \left(\sum_{i=1}^{7} \sum_{j=1}^{7} A_{ij} g_{ij}(y) + B_{ij} h_{ij}(y) \right)^2 + \left(\sum_{i=1}^{7} \sum_{j=1}^{7} C_{ij} g_{ij}(y) + D_{ij} h_{ij}(y) \right)^2 \right\}^{1/2}, \tag{3}$$

where $g_{ij}(y) = \sin(i\pi y_1)\sin(j\pi y_2)$, $h_{ij}(y) = \cos(i\pi y_1)\cos(j\pi y_2)$, $y = (y_1, y_2) \in R^2$, $0 \le y_1, y_2 \le 1$, and coefficients $A_{ij}, B_{ij}, C_{ij}, D_{ij}$ are taken uniformly in the interval $[-1, 1]$.

Let us consider a scheme for constructing the generator GCGen (Global Constrained optimization problem Generator) which allows to generate the test global optimization problems with m constraints. Obviously, one can generate $m + 1$ functions, the first m of these ones can be considered as the constraints and the $(m + 1)$-th function – as the objective one. However, in this case, the conditional global minimizer of the objective function is unknown, and the preliminary estimate of this one (for example, by scanning over a uniform grid) will be time-consuming. At the same time, one could not control the size of the feasible domain. In particular, the constraints might be incompatible, and the feasible domain might be empty.

Below, the rules, which allow formulating the constrained global optimization problems so that:

- one could control the size of feasible domain with respect to the whole domain of the parameters' variation;
- the global minimizer of the objective function would be known a priori taking into account the constraints;
- the global minimizer of the objective function without accounting for the constraints would be out of the feasible domain (with the purpose of simulating the behavior of the constraints and the objective function in the applied constrained optimization problems)

are proposed.

The rules defining the operation of the generator of the constrained global optimization with the properties listed above consist in the following.

1. Let us generate $m + 1$ functions $f_j(y), y \in D, 1 \le j \le m + 1$, by some generating scheme (for instance, by using the formula (3)). The constraints will be constructed on the base of the first m functions, the $(m+1)$-th function will serve for the construction of the objective function.
2. In order to know the global minimizer in the constrained problem a priori, let us make it to be the same to the global minimizer in the unconstrained problem. To do so, let us perform a linear transformation of coordinates so that the global minimizers of the constraint functions y_j^*, $1 \le j \le m$, would transit into the minimizer of the objective function y_{m+1}^*. This way, the functions $\overline{f_j}(y)$, $1 \le j \le m$, with the same point of extremum will be constructed.
3. In order to control the size of the feasible domain, let us construct an auxiliary function (a combined constraint)

$$H(y) = \max_{1 \le j \le m} \overline{f_j}(y)$$

and compute its values in the nodes of a uniform grid in the domain D; the number of the grid nodes in the conducted experiments should be big enough (in our experiments it was $\min\{10^7, 10^{2N}\}$). Then, let us find the maximum and minimum values of the function $H(y)$ in the grid nodes, H_{max} and H_{min}, respectively, and construct a characteristic $s(i)$ – the number of points, in which the values of $H(y)$ fall into the range

$$\left[H_{min}, H_{min} + i\frac{H_{max} - H_{min}}{100} \right], \ 1 \le i \le 100.$$

Then, the functions

$$\overline{f_j}(y) \le q = H_{min} + i\frac{H_{max} - H_{min}}{100}, \ 1 \le j \le m,$$

where i is selected to be the minimal one satisfying the inequality

$$\Delta \leq \frac{s(i)}{s(100)}$$

will construct a problem with the feasible domain occupying the fraction Δ, $0 < \Delta < 1$, of the whole search domain.

4. The test problem of global constrained optimization can be stated as follows

$$\min \left\{ \varphi(y) : y \in D, \ g_j(y) = \overline{f_j}(y) - q \leq 0, \ 1 \leq j \leq m \right\}$$

where

$$\varphi(y) = f_{m+1}(y) - \beta \sum_{j=1}^{m} \max \left\{ 0, \overline{f_j}(y) - q \right\}^\alpha,$$

where α is a positive integer number and $\beta > 0$ is selected in such a way as to provide the global minimum location of the function $\varphi(y)$ in the infeasible part of the search domain D. For instance, to guarantee this property the value of β can be set as follows

$$\beta > (h_{max} - h_{min})/(H_{max} - q)^\alpha,$$

where h_{max} and h_{min} are the maximum and minimum values of the function $f_{m+1}(y)$ respectively.

3 Some Numerical Results

As an illustration, the level lines of the objective functions and the zero-level lines of the constraints for problems constructed on the base of functions (3) with $\alpha = 3, \beta = 1$ and the volume fractions of the feasible domains $\Delta = 0.4, 0.6$ are shown in Fig. 1. The feasible domains are highlighted by green. The change of volume and, at the same time, the increase of complexity of the feasible domains are seen clearly. Figure 1(a,b) also shows the points of 628 and 764 trials, correspondingly, performed by the *index method* for solving constrained global optimization problems until the required accuracy $\epsilon = 10^{-2}$ was achieved. The conditional optimizer is shown as a red point and the best estimation of the optimizer is shown as a blue point.

The index method has been proposed and developed in [11–13]. The approach is based on a separate accounting for each constraint of the problem and is not related to the use of the penalty functions. According to the rules of the index method, every iteration includes a sequential checking of fulfillment of the problem constraints at this point. The first occurrence of violation of any constraint terminates the trial and initiates the transition to the next iteration. This allows: (i) accounting for the information on each constraint separately and (ii) solving the problems, in which the function values may be undefined out of the feasible domain. It should be noted, that the index method can be efficiently parallelized for accelerators [14,15].

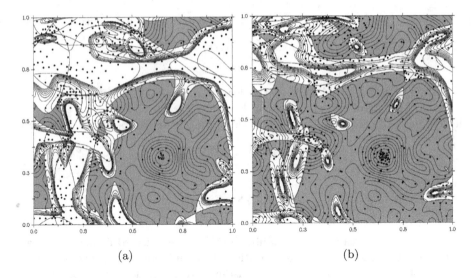

(a)   (b)

Fig. 1. The problems based on functions (3)

4   Conclusion

This paper considers a method for generating global optimization test problems with non-convex constraints that allows:

- to control the size of feasible domain with respect to the whole domain of the parameters' variation;
- to know a priori the conditional global minimizer of the objective function;
- to generate the unconditional global minimizer of the objective function out of the feasible domain (to simulate the constraints and objective function in the applied optimization problems).

The demonstration of the developed approach in application to well-known index method for solving complex multiextremal optimization problems with non-convex constraints is considered.

The developed approach allows generating any number of test global optimization problems with non-convex constraints for performing multiple computational experiments in order to obtain a reliable evaluation of the efficiency of the developed optimization algorithms. To develop the proposed approach further, the development of new test classes for the optimization problems of various dimensionalities is planned.

Acknowledgments. This study was supported by the Russian Science Foundation, project No 16-11-10150.

References

1. Famularo, D., Pugliese, P., Sergeyev, Y.D.: A global optimization technique for checking parametric robustness. Automatica **35**, 1605–1611 (1999)
2. Kvasov, D.E., Menniti, D., Pinnarelli, A., Sergeyev, Y.D., Sorrentino, N.: Tuning fuzzy power-system stabilizers in multi-machine systems by global optimization algorithms based on efficient domain partitions. Electr. Power Syst. Res. **78**(7), 1217–1229 (2008)
3. Kvasov, D.E., Sergeyev, Y.D.: Deterministic approaches for solving practical black-box global optimization problems. Adv. Eng. Softw. **80**, 58–66 (2015)
4. Modorskii, V.Y., Gaynutdinova, D.F., Gergel, V.P., Barkalov, K.A.: Optimization in design of scientific products for purposes of cavitation problems. Solving global optimization problems on GPU cluster. In: Simos, T.E. (ed.) ICNAAM 2015, AIP Conference Proceedings, 1738, art. no. 400013 (2016)
5. Floudas, C.A., et al.: Handbook of Test Problems in Local and Global Optimization. Kluwer Academic Publishers, Dordrecht (1999)
6. Gaviano, M., Kvasov, D.E., Lera, D., Sergeyev, Y.D.: Software for generation of classes of test functions with known local and global minima for global optimization. ACM TOMS **29**(4), 469–480 (2003)
7. Ali, M.M., Khompatraporn, C., Zabinsky, Z.B.: A numerical evaluation of several stochastic algorithms on selected continuous global optimization test problems. J. Glob. Optim. **31**(4), 635–672 (2005)
8. Addis, B., Locatelli, M.: A new class of test functions for global optimization. J. Glob. Optim. **38**(3), 479–501 (2007)
9. Grishagin, V.A.: Operating characteristics of some global search algorithms. Probl. Stat. Optim. **7**, 198–206 (1978). [in Russian]
10. Gergel, V., Grishagin, V., Gergel, A.: Adaptive nested optimization scheme for multidimensional global search. J. Glob. Optim. **66**(1), 35–51 (2016)
11. Strongin, R.G., Sergeyev, Y.D.: Global Optimization with Non-Convex Constraints. Sequential and Parallel Algorithms. Kluwer Academic Publishers, Dordrecht (2000)
12. Sergeyev, Y.D., Famularo, D., Pugliese, P.: Index branch-and-bound algorithm for Lipschitz univariate global optimization with multiextremal constraints. J. Glob. Optim. **21**(3), 317–341 (2001)
13. Barkalov, K.A., Strongin, R.G.: A global optimization technique with an adaptive order of checking for constraints. Comput. Math. Math. Phys. **42**(9), 1289–1300 (2002)
14. Barkalov, K., Gergel, V., Lebedev, I.: Use of Xeon Phi coprocessor for solving global optimization problems. In: Malyshkin, V. (ed.) PaCT 2015. LNCS, vol. 9251, pp. 307–318. Springer, Cham (2015). doi:10.1007/978-3-319-21909-7_31
15. Barkalov, K., Gergel, V.: Parallel global optimization on GPU. J. Glob. Optim. **66**(1), 3–20 (2016)

Global Optimization Using Numerical Approximations of Derivatives

Victor Gergel and Alexey Goryachih[✉]

Lobachevsky State University of Nizhni Novgorod,
Gagarin Prospect, 23, Nizhni Novgorod 603600, Russian Federation
gergel@unn.ru, a_goryachih@mail.ru
http://www.unn.ru

Abstract. This paper presents an efficient method for solving global optimization problems. The new method unlike previous methods, was developed, based on numerical estimations of derivative values. The effect of using numerical estimations of derivative values was studied and the results of computational experiments prove the potential of such approach.

Keywords: Multiextremal optimization · Global search algorithm · Lipschitz condition · Numerical derivatives · Computational experiments

1 Introduction

The global optimization problem is a problem of finding the minimum value of a function $\varphi(x)$

$$\varphi(x^*) = \min\{\varphi(x) : x \in [a, b]\}. \tag{1}$$

For numerical solving of the problem (1) optimization methods usually generate a sequence of points y_k which converges to the global optimum x^*.

Suppose that the optimized function $\varphi(x)$ is multiextremal. Also assume that the optimized function $\varphi(x)$ satisfies the Lipschitz condition

$$|\varphi(x_2) - \varphi(x_1)| \le L |x_2 - x_1|, x_1, x_2 \in [a, b], \tag{2}$$

where $L > 0$ is the Lipschitz constant. In addition to (2) also assume that the first derivative of the optimized function $\varphi'(x)$ satisfies the Lipschitz condition

$$\left|\varphi'(x_2) - \varphi'(x_1)\right| \le L_1 \left\| x_2 - x_1 \right\|, x_1, x_2 \in [a, b]. \tag{3}$$

2 One-Dimensional Global Optimization Algorithm Using Numerical Estimations of Derivatives

Conditions (2) and (3) allow estimating possible values of the optimized function $\varphi(x)$ more accurately, improving efficiency of global optimization methods. This paper presents an improvement of global optimization algorithms that uses derivatives of the optimized function $\varphi(x)$.

© Springer International Publishing AG 2017
R. Battiti et al. (Eds.): LION 2017, LNCS 10556, pp. 320–325, 2017.
https://doi.org/10.1007/978-3-319-69404-7_25

2.1   Core One-Dimensional Global Search Algorithm Using Derivatives

The new method that is proposed in this paper based on the Adaptive Global Method with Derivatives (AGMD). This method was developed in the framework of information-statistical theory of multiextremal optimization [4,5]. This method was proposed in [1] and a similar approach was described in [3].

The computational scheme of the AGMD can be given as follows.

The first two iterations are executed at the boundary points of the interval $[a, b]$. Assume that k ($k > 1$) iterations of global search have been performed, each iteration includes a calculation of a value of the optimized function $\varphi(x)$ and its derivative $\varphi'(x)$, this calculation will be further referred to as trial.

Next point of trial of $(k+1)$ iteration is calculated according to the following rules:

Rule 1. Renumber the points of previous trials by subscripts in increasing order

$$a = x_0 < x_1 < \cdots < x_i < \cdots < x_k = b.$$

Rule 2. Compute the estimation of the Lipschitz constant from (3) for the first derivative of the optimized function

$$M_i = max \begin{cases} |\varphi'(x_i) - \varphi'(x_{i-1})|/|x_i - x_{i-1}|, \\ -2\left[\varphi(x_i) - \varphi(x_{i-1}) - \varphi'(x_{i-1})(x_i - x_{i-1})\right]/(x_i - x_{i-1})^2, \\ 2\left[\varphi(x_i) - \varphi(x_{i-1}) - \varphi'(x_i)(x_i - x_{i-1})\right]/(x_i - x_{i-1})^2, \end{cases}$$

$$(4)$$

$$M = max(M_i), 1 \leq i \leq k,$$

$$m = \begin{cases} rM, & M > 0, \\ 1, & M = 0, \end{cases}$$

where $r > 1$ is the reliability parameter of the algorithm.

Rule 3. Construct a minorant for the optimized function on each interval (x_i, x_{i+1}), $0 \leq i \leq k - 1$:

$$\varphi^{min}(x) = \begin{cases} \varphi(x_i) + \varphi'(x_i)(x - x_i) - 0.5\,m(x - x_i)^2, & x \in (x_i, x_i'), \\ A_i(x - x_i') + 0.5\,m(x - x_i')^2 + B_i, & x \in [x_i', x_i''], \\ \varphi(x_{i+1}) + \varphi'(x_{i+1})(x - x_{i+1}) - 0.5\,m(x - x_{i+1})^2, & x \in (x_i'', x_{i+1}), \end{cases}$$

$$(5)$$

in order to make the minorant and its first derivative continuous the intervals (x_i', x_i'') and the values A_i and B_i are defined by the relation

$$x_i' = \frac{[\varphi(x_i) - \varphi(x_{i+1})x_{i+1}] + m(x_{i+1}^2 - x_i^2)/2 - md_x^2}{m(x_{i+1} - x_i) + (\varphi'(x_{i+1}) - \varphi'(x_i))},$$

$$x_i'' = \frac{[\varphi(x_i) - \varphi(x_{i+1})x_{i+1}] + m(x_{i+1}^2 - x_i^2)/2 + md_x^2}{m(x_{i+1} - x_i) + (\varphi'(x_{i+1}) - \varphi'(x_i))},$$

where the intervals (x_i', x_i'') and the values A_i and B_i are defined in order to make the minorant and its first derivative continuous (see [1]).

Rule 4. For each interval (x_i, x_{i+1}) $(0 \leq i \leq k-1)$ the characteristic

$$R(i) = \begin{cases} \varphi^{min}(x_i^{min}), & x_i^{min} \in [x_i', x_i''], \\ min(\varphi^{min}(x_i'), \varphi^{min}(x_i'')), & x_i^{min} \notin [x_i', x_i''], \end{cases}$$

where

$$x_i^{min} = \frac{-\varphi'(x_{i-1}) + m(x_i - x_{i-1}) + mx_i}{m}.$$

Rule 5. Find the interval (x_{t-1}, x_t) with the minimal characteristic $R(t)$

$$R(t) = min\{R(i) : 1 \leq i \leq k\}. \tag{6}$$

In the case when there exist several intervals satisfying (6) the interval with the minimal number t is taken for certainty.

Rule 6. Compute the next point of the next trial x^{k+1} accordingly

$$x^{k+1} = \begin{cases} x_t^{min}, & x_t^{min} \in [x_t', x_t''], \\ x_t', & \varphi^{min}(x_t') \leq \varphi^{min}(x_t''), \\ x_t'', & \varphi^{min}(x_t') > \varphi^{min}(x_t''). \end{cases}$$

The stopping condition is defined by the following relation

$$|x_t - x_{t-1}| \leq \varepsilon, \tag{7}$$

where ε is the accuracy, $\varepsilon > 0$.

Convergence conditions of the algorithm are described in [1].

2.2 One-Dimensional Global Search Algorithm Using Numerical Derivatives

As it was mentioned above, the derivative may be unknown or its values are time-consuming to compute. In this paper the modification of the AGMD based on numerical differentiation is proposed.

The following relations are used for numerical estimations of values of the first derivative:

– left-hand and right-hand approximation of values of the first derivatives by two values of the function

$$\overline{\varphi}_i' = \frac{\varphi(x_{i+1}) - \varphi(x_i)}{x_{i+1} - x_i}, 0 \leq i \leq k - 1,$$

$$\overline{\varphi}_i' = \frac{\varphi(x_k) - \varphi(x_{k-1})}{x_k - x_{k-1}},$$

– approximation of values of the first derivatives at left, right and center points
by three values of the function

$$\overline{\varphi}_0' = \frac{1}{H_1^2}\left(-(2+\delta_2)\varphi(x_0) + \frac{(1+\delta_2)^2}{\delta_2}\varphi(x_1) - \frac{1}{\delta_2}\varphi(x_2)\right),$$

$$\overline{\varphi}_k' = \frac{1}{H_{k-1}^k}\left(\delta_k\varphi(x_{k-2}) - \frac{(1+\delta_k)^2}{\delta_k}\varphi(x_{k-1}) + \frac{(2+\delta_k)}{\delta_k}\varphi(x_k)\right),$$

for $1 \leq i \leq k-1$

$$\overline{\varphi}_i' = \frac{1}{H_i^{i+1}}\left(-\delta_{i+1}\varphi(x_{i-1}) - \frac{(\delta_{i+1}^2-1)}{\delta_{i+1}}\varphi(x_i) + \frac{1}{\delta_{i+1}}\varphi(x_{i+1})\right),$$

where $H_i^{i+1} = h_i + h_{i+1}$, $\delta_{i+1} = \frac{h_{i+1}}{h_i}$ and $h_i = x_i - x_{i-1}$. Additional details can
be found in [11].

The algorithm of AGMD with numerical estimations of values of the first
derivative will be further referred to as the Adaptive Global Method with Numerical Derivatives (AGMND). In the scheme of AGMND we just replace $\varphi'(x_i)$ by
$\overline{\varphi}_i'$ $(0 \leq i \leq k)$, it forms two modifications of AGMND: AGMND-2 and AGMND-
3 with approximation of derivative values by 2 or 3 values of the optimized
functions correspondingly.

3 Results of Computational Experiments

For the first series of experiments five well known methods of global optimization
were compared: Galperin's algorithm (GA), Piyavskii's algorithm (PA), Global
Search Algorithm (GSA) proposed by Strongin, AGMD and AGMND with three
different scheme of approximation of derivative values by 2 and 3 values of the
function AGMND-2, AGMND-3 correspondingly.

Within the executed experiments a set of multiextremal functions from [1]
(it is also used in [2]) was chosen.

The accuracy of global search is equal to $\varepsilon = 10^{-4}(b-a)$. Galperin's algorithm
and Piyavskii's algorithm use the exact Lipschitz constant. For GSA parameter
of methods r is equal to 2. For AGMD and AGMND (AGMND-2, AGMND-3)
parameter of methods r is equal to 1.1. All results (except AGMND) were taken
from [1].

Table 1 contains the results of methods in term of the numbers of trials that
had to be performed before the stopping condition (7) (empty cells mean that
exact solutions were not found). The last row in the table contains the average
number of trials from nonempty cells.

As shown in the results of AGMND-3 and AGMD are quite close. It is important to note that the method with numerical derivatives looks even more effective
since each trial in AGMD includes calculation of the function and its derivative
in the relations (4) and (5) from the scheme. But in some cases AGMND-3 did
not find the global minimum before the stopping condition (7) was satisfied.

Table 1. The results of comparison of one-dimensional methods of global optimization

	GA	PA	GSA	AGMD	AGMND-2	AGMND-3
1	377	149	127	16	25	17
2	308	155	135	13	30	14
3	581	195	224	50	136	67
4	923	413	379	15	27	12
5	326	151	126	14	40	18
6	263	129	112	22	51	28
7	383	153	115	13	32	14
8	530	185	188	47	147	42
9	314	119	125	12	33	16
10	416	203	157	12	30	-
11	779	373	405	29	58	29
12	746	327	271	23	48	16
13	1829	993	472	59	26	20
14	290	145	108	15	48	16
15	1613	629	471	41	71	63
16	992	497	557	49	27	25
17	1412	549	470	44	90	42
18	620	303	243	10	32	11
19	302	131	117	12	32	15
20	1412	493	81	24	51	24
Average	720.8	314.6	244.15	26	51.7	24.83

In the next experiment different values of the reliability parameter r were compared with the following dynamic schemes for setting r: $r = r_\infty + d/k$, where k was a number of trials, $r_\infty = 1.1$, $d = 10$.

Results of the experiment with the constant parameters $r = 1.1, 1.5, 2.0$ and with variable parameters for AGMND-3 are presented in Table 2. These results show that the proposed dynamic scheme are the most efficient.

Table 2. The results of comparison of different schemes for setting the reliability parameter

	1	2	3	4	5	6	7	8	9	10	11	12	13	14	15	16	17	18	19	20	Average
1.0	17	14	67	12	18	28	14	43	16	-	29	16	20	16	63	25	-	11	15	24	24.89
1.5	21	15	66	16	22	27	16	64	20	-	28	25	2	17	38	32	-	14	17	28	27.17
2.0	23	18	77	20	24	28	20	72	22	21	30	30	29	28	35	38	47	16	18	28	31.2
$r_\infty + d/k$	18	14	49	16	18	24	17	47	16	16	30	21	25	17	34	30	32	15	15	25	23.95

4 Conclusion

In the framework of the proposed approach to solving global optimization problems, the algorithm of AGMND-3 showed results close to results of AGMD. The method with numerical derivatives looks even more effective since each trial in AGMD includes calculation of the function and its derivative.

For further research it is necessary to continue computational experiments on higher-dimensional optimization problems, as well as to provide some theoretical basis for AGMND.

Acknowledgments. This research was supported by the Russian Science Foundation, project No 16₋11-10150" Novel efficient methods and software tools for time-consuming decision making problems using supercomputers of superior performance."

References

1. Gergel, V.P.: A method of using derivatives in the minimization of multiextremum functions. Comput. Math. Math. Phys. **36**(6), 729–742 (1996)
2. Sergeyev, Y.D., Mukhametzhanov, M.S., Kvasov, D.E., Lera, D.: Derivative-free local tuning and local improvement techniques embedded in the univariate global optimization. J. Optim. Theor. Appl. **171**(1), 186–208 (2016)
3. Sergeyev, Y.D.: Global one-dimensional optimization using smooth auxiliary functions. Math. Program. **81**(1), 127–146 (1998)
4. Strongin, R.G., Sergeyev, Y.D.: Global Optimization with Non-Convex Constraints: Sequential and Parallel Algorithms. Kluwer Academic Publishers, Dordrecht (2000)
5. Strongin, R.G.: Numerical methods in multiextremal problems: information-statistical algorithms. Nauka, Moscow (1978). (in Russian)
6. Barkalov, K., Gergel, V.P.: Parallel global optimization on GPU. J. Global Optim. **66**(1), 3–20 (2016)
7. Gergel, V.P., Kuzmin, M.I., Solovyov, N.A., Grishagin, V.A.: Recognition of surface defects of cold-rolling sheets based on method of localities. Int. Rev. Automat. Control **8**(1), 51–55 (2015)
8. Barkalov, K., Gergel, V., Lebedev, I.: Use of xeon phi coprocessor for solving global optimization problems. In: Malyshkin, V. (ed.) PaCT 2015. LNCS, vol. 9251, pp. 307–318. Springer, Cham (2015). doi:10.1007/978-3-319-21909-7_31
9. Paulavicius, R., Zilinskas, J.: Advantages of simplicial partitioning for Lipschitz optimization problems with linear constraints. Optim. Lett. **10**(2), 237–246 (2016)
10. Paulavicius, R., Sergeyev, Y.D., Kvasov, D.E., Zilinskas, J.: Globally-biased DIS-IMPL algorithm for expensive global optimization. J. Global Optim. **59**(2–3), 545–567 (2014)
11. Griewank, A., Walther, A.: Evaluating Derivatives: Principles and Techniques of Algorithmic Differentiation, 2nd edn. Society for Industrial and Applied Mathematics, Philadelphia (2008)

Global Optimization Challenges in Structured Low Rank Approximation

Jonathan Gillard[1(✉)] and Anatoly Zhigljavsky[1,2]

[1] School of Mathematics, Cardiff University, Cardiff CF24 4AG, UK
{gillardjw,zhigljavskyaa}@cardiff.ac.uk
[2] Lobachevsky Nizhny Novgorod State University,
23 Prospekt Gagarina, 603950 Nizhny Novgorod, Russia

Abstract. In this paper, we investigate the complexity of the numerical construction of the so-called Hankel structured low-rank approximation (HSLRA). Briefly, HSLRA is the problem of finding a rank r approximation of a given Hankel matrix, which is also of Hankel structure.

Keywords: Structured low rank approximation · Hankel matrices · Time series analysis

1 Statement of the Problem

The aim of low-rank approximation methods is to approximate a matrix containing observed data, by a matrix of pre-specified lower rank r. The rank of the matrix containing the original data can be viewed as the order of complexity required to fit to the data exactly, and a matrix of lower complexity (lower rank) 'close' to the original matrix is often required. A further requirement is that if the original matrix of the observed data is of a particular structure, then the approximation should also have this structure. Let L, K and r be given positive integers such that $1 \leq r < L \leq K$. Denote the set of all real-valued $L \times K$ matrices by $\mathbb{R}^{L \times K}$. Let $\mathcal{M}_r = \mathcal{M}_r^{L \times K} \subset \mathbb{R}^{L \times K}$ be the subset of $\mathbb{R}^{L \times K}$ containing all matrices with rank $\leq r$, and $\mathcal{H} = \mathcal{H}^{L \times K} \subset \mathbb{R}^{L \times K}$ be the subset of $\mathbb{R}^{L \times K}$ containing matrices of some known structure. The set of structured $L \times K$ matrices of rank $\leq r$ is $\mathcal{A} = \mathcal{M}_r \cap \mathcal{H}$. Assume we are given a matrix $\mathbf{X}_* \in \mathcal{H}$. The problem of structured low rank approximation (SLRA) is:

$$f(\mathbf{X}) \to \min_{\mathbf{X} \in \mathcal{A}} \tag{1}$$

where $f(\mathbf{X}) = d(\mathbf{X}, \mathbf{X}_*)$ is a squared distance on $\mathbb{R}^{L \times K} \times \mathbb{R}^{L \times K}$. In this paper we only consider the case where $\mathcal{H}$ is the set of Hankel matrices and thus refer to (1) as HSLRA.

One distance to define f in (1) is given by

$$||\mathbf{X}||_{\mathbf{Q},\mathbf{R}}^2 = \text{Trace } \mathbf{Q} \mathbf{X} \mathbf{R} \mathbf{X}^T, \tag{2}$$

where $\mathbf{Q}$ and $\mathbf{R}$ are positive definite matrices. We will call this norm the $(\mathbf{Q}, \mathbf{R})$-norm. Note that if $\mathbf{Q}$ and $\mathbf{R}$ are identity matrices then (2) defines the Frobenius norm (which is the most popular choice).

R. Battiti et al. (Eds.): LION 2017, LNCS 10556, pp. 326–330, 2017.
https://doi.org/10.1007/978-3-319-69404-7_26

2   Optimization Challenges

2.1   Challenge 1: Selecting f

The selection of the distance f represents one of the challenges in HSLRA. To motivate this challenge, consider a problem of modelling a time series. Assume we are given a vector $Y = (y_1, \ldots, y_N)^T \in \mathbb{R}^N$. A common problem of time series analysis is

$$\rho(S, Y) \to \min_{S \in \mathbb{L}_{\leq r}} \tag{3}$$

where $\rho(\cdot, \cdot)$ is a distance on $\mathbb{R}^N \times \mathbb{R}^N$ and $\mathbb{L}_{\leq r} \subset \mathbb{R}^N$ is the set of vectors in $\mathbb{R}^N$ which satisfy a linear recurrence relation (LRR) of order $\leq r$; we say that a vector $S = (s_1, \ldots, s_N)^T$ satisfies an LRR of order $\leq r$ if

$$s_n = a_1 s_{n-1} + \ldots + a_r s_{n-r}, \quad \text{for all } n = r + 1, \ldots, N, \tag{4}$$

where $a_1, \ldots, a_r$ are some real numbers with $a_r \neq 0$. The model (4) includes, as a special case, the model of a sum of exponentially damped sinusoids:

$$s_n = \sum_{\ell=1}^{q} a_\ell \exp(d_\ell n) \sin(2\pi \omega_\ell n + \phi_\ell), \, n = 1, \ldots, N, \tag{5}$$

where typically $q = r/2$.

The optimization problem (3) can be equivalently formulated as a matrix optimization problem, where vectors in (3) are represented by $L \times K$ Hankel matrices. With a vector $Z = (z_1, \ldots, z_N)^T$ of size N and given $L < N$ we associate an $L \times K$ Hankel matrix

$$\mathbf{X}_Z = \begin{pmatrix} z_1 & z_2 & \cdots & z_K \\ z_2 & z_3 & \cdots & z_{K+1} \\ \vdots & \vdots & \vdots & \vdots \\ z_L & z_{L+1} & \cdots & z_N \end{pmatrix} \in \mathcal{H},$$

where $K = N - L + 1$. We also write this matrix as $\mathbf{X}_Z = \mathbb{H}(Z)$ and note that $\mathbb{H}$ makes a one-to-one correspondence between the spaces $\mathbb{R}^{N+1}$ and $\mathcal{H}$ so that for any matrix $\mathbf{X} \in \mathcal{H}$ we may uniquely define $Z = \mathbb{H}^{-1}(\mathbf{X})$ with $\mathbf{X} = \mathbb{H}(Z)$. The matrix version of the optimization problem (3) can now be written as

$$d(\mathbf{X}, \mathbf{X}_Y) \to \min_{\mathbf{X} \in \mathcal{A}} \tag{6}$$

where $d(\cdot, \cdot)$ is a distance on $\mathbb{R}^{L \times K} \times \mathbb{R}^{L \times K}$ and $\mathcal{A}$ is the set of all $L \times K$ Hankel matrices of rank $\leq r$. This is the optimization problem (1).

The optimization problems (3) and (1) are equivalent if the distance functions $\rho(\cdot, \cdot)$ in (3) and $d(\cdot, \cdot)$ in (1) are such that

$$\rho(Z, Z') = c \cdot d(\mathbb{H}(Z), \mathbb{H}(Z')) \tag{7}$$

for $Z, Z' \in \mathbb{R}^N$, where $c > 0$ is arbitrary. It can be assumed that $c = 1$ without loss of generality. The standard choice of the distance $d(\cdot, \cdot)$ in the HSLRA problem (1) is $d(\mathbf{X}, \mathbf{X}') = \|\mathbf{X} - \mathbf{X}'\|_F$, where $\| \cdot \|_F$ is the Frobenius norm. The primary reason for this choice is the availability of the singular value decomposition (SVD) which constitutes the essential part of many algorithms attempting to solve the HSLRA problem (1).

However, if the distance $d(\mathbf{X}, \mathbf{X}')$ in (1) is $d(\mathbf{X}, \mathbf{X}') = \|\mathbf{X} - \mathbf{X}'\|_F$ then the distance ρ in (3) takes a particular form. One would prefer to define the distance function $\rho(\cdot, \cdot)$ in (3) and acquire the distance $d(\cdot, \cdot)$ for (1) from (7), rather than vice-versa, which is a common practice. Different distances $\rho(\cdot, \cdot)$ can be used and may be desired. There is one serious problem, however, related to the complexity of the resulting HSLRA problem (1). If $d(\cdot, \cdot)$ is defined by the Frobenius norm then the HSLRA problem (1), despite being difficult, is still considered as solvable since there is a very special tool available at intermediate stages, the SVD. If $d(\cdot, \cdot)$ does not allow the use of SVD or similar tools then the HSLRA problem (1) becomes practically unsolvable (except, of course, for some very simple cases).

Work in the paper [6] extends the choice of the norms that define distances in (1) and (3) preserving the availability of the SVD. These norms allow the construction of exactly the same algorithms that can be constructed for the Frobenius norm and, since the family of the norms considered is rather wide, we are able to exactly or approximately match any given distance in (3). More precisely, in [6] we consider the class of distances in (3) of the form $\rho(Z, Z') = \|Z - Z'\|_{\mathbf{W}}$ with

$$\|Z\|_{\mathbf{W}}^2 = Z^T \mathbf{W} Z = \sum_{n=1}^{N} w_n z_n^2. \tag{8}$$

Further challenges arise when observations of the time series Y are classified as 'exact' or 'missing'. In the former case the observation would require infinite weight, whilst in the latter the observation would require zero weight. Both cases give rise to difficulties computing with infinite and infinitesimals, and there is need for methodology that allows one to represent infinite and infinitesimal numbers by a finite number of symbols to execute arithmetical operations. There is great potential for the use of grossone and the infinity computer (see [11] for more details on these topics).

2.2   Challenge 2: Complexity of the Optimization Problem

Natural approaches for solving the initial optimization problem (1) would use global optimization techniques for optimizing parameters in either representation (4) or (5). In the case of (4), the parameters are the coefficients of the LRR: $a_1, \ldots, a_r$, and the initial values $s_1, \ldots, s_{r-1}$. If we were to use (5) then the set of parameters is $\{(a_\ell, d_\ell, \omega_\ell, \phi_\ell), \ell = 1, \ldots, q\}$. In both cases, the parametric optimization problem is extremely difficult with multi-extremality and large Lipschitz constants of the objective functions [4]. The number of local minima is known to increase linearly with the number of observations. Many of the existing algorithms depend on local optimization based algorithms and do not move significantly from a starting point, see [2,3], and [5]. The difficulty of solving parametric versions of (3) is well understood and that is the reason why HSLRA described by (1), the equivalent matrix formulation of (3), is almost always considered instead of (3). As already stated, there is little work in the literature describing how to use weights in HSLRA. However, some recent work has commented that even unstructured weighted low-rank approximation is difficult, see [7]. Increasing N leads to more erratic cost function; as N increases, the number of local minima also increases. In many examples the number of local minimizers increases linearly in N.

2.3   Challenge 3: Construction of Numerical Methods

One of the earliest known approaches to obtain a solution of (1) is the so-called Cadzow iterations which are the alternating projections of the matrices, starting at a given structured matrix, to the set of matrices of rank $\leq r$ (by performing a singular value decomposition) and to the set of Hankel matrices (by diagonal averaging). Despite the fact that Cadzow iterations guarantee convergence to the set $\mathcal{A}$, they can easily be shown to be sub-optimal (see [1]). They remain popular due to their simplicity. One Cadzow iteration corresponds to the technique known as singular spectrum analysis, which has been an area of research developed by the PI (see for example [1]). The main recent contributions to finding a solution of (1) are described below.

1. *Structured total least norm.* Proposed by Park et al. [9], this class of methods is aimed at rank reduction of a given Hankel matrix by 1 (that is, $r = L - 1$).
2. *Fitting a sum of damped sinusoids* [10]. Methods in this category parameterize the vector of observations as a sum of damped sinusoids and use the set of unknown parameters as a feasible domain. This has been discussed earlier.
3. *Local optimization methods starting at an existing approximation.* Markovsky and co-authors [8] have developed methodology and software to locally improve an existing solution (or approximate solution) of (1).

In summary all of these methods suffer from a number of flaws [3] (i) the rank of the matrix can only be reduced by one, (ii) they are based on local optimizations and may not move significantly from this initial approximation and (iii) none have guaranteed convergence to the global optimum. Additionally (and importantly) the focus in the literature has been on the case when the distance f in (1) is taken to be the Frobenius norm (that is $\mathbf{Q}$ and $\mathbf{R}$ being the identity matrices).

Acknowledgements. This work was supported by the project No. 15-11-30022 "Global optimization, supercomputing computations, and applications" of the Russian Science Foundation.

References

1. Gillard, J.: Cadzow's basic algorithm, alternating projections and singular spectrum analysis. Stat. Interface **3**(3), 335–343 (2010)
2. Gillard, J., Zhigljavsky, A.: Analysis of structured low rank approximation as an optimization problem. Informatica **22**(4), 489–505 (2011)
3. Gillard, J., Zhigljavsky, A.: Optimization challenges in the structured low rank approximation problem. J. Global Optim. **57**(3), 733–751 (2013)
4. Gillard, J.W., Kvasov, D.: Lipschitz optimization methods for fitting a sum of damped sinusoids to a series of observations. Stat. Interface **10**(1), 59–70 (2017)
5. Gillard, J., Zhigljavsky, A.: Stochastic algorithms for solving structured low-rank matrix approximation problems. Commun. Nonlinear Sci. Numer. Simul. **21**(1), 70–88 (2015)
6. Gillard, J., Zhigljavsky, A.: Weighted norms in subspace-based methods for time series analysis. Numer. Linear Algebra Appl. **23**(5), 947–967 (2016)
7. Gillis, N., Glineur, F.: Low-rank matrix approximation with weights or missing data is NP-hard. SIAM J. Matrix Anal. Appl. **32**(4), 1149–1165 (2011)
8. Markovsky, I.: Low Rank Approximation: Algorithms, Implementation, Applications. Springer, London (2012)
9. Park, H., Zhang, L., Rosen, J.B.: Low rank approximation of a Hankel matrix by structured total least norm. BIT Numer. Math. **39**(4), 757–779 (1999)
10. Sergeyev, Y.D., Kvasov, D.E., Mukhametzhanov, M.S.: On the least-squares fitting of data by sinusoids. In: Pardalos, P.M., Zhigljavsky, A., Žilinskas, J. (eds.) Advances in Stochastic and Deterministic Global Optimization, Chap. 11. SOIA, vol. 107, pp. 209–226. Springer, Cham (2016). doi:10.1007/978-3-319-29975-4_11
11. Sergeyev, Y.D.: Numerical computations and mathematical modelling with infinite and infinitesimal numbers. J. Appl. Math. Comput. **29**(1–2), 177–195 (2009)

A D.C. Programming Approach to Fractional Problems

Tatiana Gruzdeva$^{(\boxtimes)}$ and Alexander Strekalovsky

Matrosov Institute for System Dynamics and Control Theory of SB RAS,
Lermontov Str., 134, 664033 Irkutsk, Russia
{gruzdeva,strekal}@icc.ru

Abstract. This paper addresses a rather general fractional optimization problem. There are two ways to reduce the original problem. The first one is a solution of an equation with the optimal value of an auxiliary d.c. optimization problem with a vector parameter. The second one is to solve the second auxiliary problem with nonlinear inequality constraints. Both auxiliary problems turn out to be d.c. optimization problems, which allows to apply Global Optimization Theory [11, 12] and develop two corresponding global search algorithms that have been tested on a number of test problems from the recent publications.

Keywords: Fractional optimization · Nonconvex problem · Difference of two convex functions · Equation with vector parameter · Global search algorithm

1 Introduction

We consider the following problem of the fractional optimization [1,10]

$$(\mathcal{P}_f) \qquad f(x) := \sum_{i=1}^{m} \frac{\psi_i(x)}{\varphi_i(x)} \downarrow \min_x, \quad x \in S,$$

where $S \subset \mathbb{R}^n$ is a convex set and $\psi_i : \mathbb{R}^n \to \mathbb{R}, \quad \varphi_i : \mathbb{R}^n \to \mathbb{R},$

$$(\mathcal{H}_0) \qquad \psi_i(x) > 0, \; \varphi_i(x) > 0 \; \forall x \in S, \; i = 1, \ldots, m.$$

The fractional programming problems arise in various economic applications and real-life problems [10]. However, it is well-known that the sum-of-ratios program is NP-complete [4]. Surveys on methods for solving this problem can be found in [1,10], but the development of new efficient methods for a fractional program still remains an important field of research in mathematical optimization.

In the recent two decades, we have succeeded in developing the Global Search Theory, which perfectly fits optimization theory and proved to be rather efficient in terms of computations [11,12]. Now we are going to apply this theory to solving fractional programming problems.

© Springer International Publishing AG 2017
R. Battiti et al. (Eds.): LION 2017, LNCS 10556, pp. 331–337, 2017.
https://doi.org/10.1007/978-3-319-69404-7_27

Generalizing the Dinkelbach's idea [3], we propose to attack such problems with the help of reduction of the fractional program $(\mathcal{P}_f)$ to solving an equation with the optimal value function of a d.c. minimization problem and the vector parameter that satisfies the nonnegativity assumption. To this end, we have to use the solution to the following d.c. minimization problem, treated here as an auxiliary one:

$$(\mathcal{DC}) \qquad f(x) = g(x) - h(x) \downarrow \min, \quad x \in D,$$

where $g(\cdot)$, $h(\cdot)$ are convex functions, and D is a convex set, $D \subset \mathbb{R}^n$.

Next, we also propose reduction of the sum-of-ratios problem to the problem of minimizing a linear function on the nonconvex feasible set given by d.c. functions. In this case, we need to solve the following nonconvex problems

$$(\mathcal{DCC}) \qquad f_0(x) \downarrow \min_x, \quad x \in S, \quad f_i(x) = g_i(x) - h_i(x) \leq 0, \quad i = 1, \ldots n,$$

where $g_i(\cdot)$, $h_i(\cdot)$ $i = 1, \ldots n$, are convex functions, $S \subset \mathbb{R}^n$, $f_0(\cdot)$ is a continuous function.

Thus, based on the solution of these two classes of d.c. optimization problems we developed two new methods for solving a general fractional program.

The outline of the paper is as follows. In Sect. 2, instead of considering a fractional program directly, we propose to combine solving of the corresponding d.c. minimization problem $(\mathcal{DC})$ with a search with respect to the vector parameter. In Sect. 3, we substantiate the reduction of the sum-of-ratios fractional problem to the optimization problem with nonconvex constraints $(\mathcal{DCC})$. Finally, in Sect. 4, we show some comparative computational testing of two approaches on instances with a small number of variables and terms in the sum.

2   Reduction to the D.C. Minimization Problem

Let us consider the following parametric optimization problem

$$(\mathcal{P}_\alpha) \qquad \Phi(x, \alpha) := \sum_{i=1}^{m} [\psi_i(x) - \alpha_i \varphi_i(x)] \downarrow \min_x, \quad x \in S,$$

where $\alpha = (\alpha_1, \ldots, \alpha_m)^\top \in \mathbb{R}^m$ is the vector parameter.

Further, let us introduce the function $\mathcal{V}(\alpha)$ of the optimal value to Problem $(\mathcal{P}_\alpha)$: $\mathcal{V}(\alpha) := \inf_x \{\Phi(x, \alpha) \mid x \in S\} = \inf_x \left\{ \sum_{i=1}^{m} [\psi_i(x) - \alpha_i \varphi_i(x)] : x \in S \right\}$.

In addition, suppose that the following assumptions are fulfilled:

$(\mathcal{H}_1)$ $\quad$ (a) $\mathcal{V}(\alpha) > -\infty$ $\forall \alpha \in \mathcal{K}$, where $\mathcal{K}$ is a convex set from $\mathbb{R}^m$;
$\qquad$ (b) $\forall \alpha \in \mathcal{K} \subset \mathbb{R}^m$ there exists a solution $z = z(\alpha)$ to Problem $(\mathcal{P}_\alpha)$

In what follows, we say that the data of Problem $(\mathcal{P}_f)$ satisfy "the nonnegativity condition", if the following inequalities hold

$$(\mathcal{H}(\alpha)) \qquad \psi_i(x) - \alpha_i \varphi_i(x) \geq 0 \quad \forall x \in S, \quad i = 1, \ldots, m.$$

Theorem 1. [5] *Suppose that in Problem* $(\mathcal{P}_f)$ $\psi_i(x) > 0$, $\varphi_i(x) > 0$ *and the assumption* $(\mathcal{H}_1)$ *is satisfied. In addition, let there exist a vector* $\alpha_0 = (\alpha_{01}, \dots, \alpha_{0m})^\top \in \mathcal{K} \subset \mathbb{R}^m$ *at which "the nonnegativity condition"* $(\mathcal{H}(\alpha_0))$ *holds. Besides, suppose that in Problem* $(\mathcal{P}_{\alpha_0})$ *the following equality takes place:*

$$\mathcal{V}(\alpha_0) := \inf_x \left\{ \sum_{i=1}^m [\psi_i(x) - \alpha_{0i}\varphi_i(x)] \ : \ x \in S \right\} = 0. \tag{1}$$

Then, any solution $z = z(\alpha_0)$ *to Problem* $(\mathcal{P}_{\alpha_0})$ *is a solution to Problem* $(\mathcal{P}_f)$, *so that* $z \in Sol(\mathcal{P}_{\alpha_0}) \subset Sol(\mathcal{P}_f)$.

According to Theorem 1, in order to verify the equality (1), we should be able to find a global solution to Problem $(\mathcal{P}_\alpha)$ for every $\alpha \in \mathbb{R}_+^m$. Since $\psi_i(\cdot)$, $\varphi_i(\cdot)$, $i = 1, \dots, m$, are d.c. functions it can be readily seen that Problem $(\mathcal{P}_\alpha)$ belongs to the class of d.c. minimization. As a consequence, in order to solve Problem $(\mathcal{P}_\alpha)$, we can apply the Global Search Theory [11,12].

Hence, instead of solving Problem $(\mathcal{P}_f)$, we propose to combine solving of Problem $(\mathcal{P}_\alpha)$ with a search with respect to the parameter $\alpha \in \mathbb{R}^m$ in order to find the vector $\alpha_0 \in \mathbb{R}_+^m$ such that $\mathcal{V}(\alpha_0) = 0$.

Denote $\Phi_i(x) := \psi_i(x) - \alpha_i^k \varphi_i(x)$, $i = 1, \dots, m$. Let $[0, \alpha^+]$ be a segment for varying α. To choose α^+ we should take into account that due to $(\mathcal{H}(\alpha))$ and

$(\mathcal{H}_0)$, we have $\forall i = 1, \dots, m : \alpha_i \leq f_i(x) \triangleq \dfrac{\psi_i(x)}{\phi_i(x)} \leq \sum_{i=1}^m \dfrac{\psi_i(x)}{\phi_i(x)} = f(x) \ \forall x \in S$,

so, for example, α^+ can be chosen as $\alpha_i^+ = f_i(x^0)$, $i = 1, \dots, m$.

Method for solving the equation $\mathcal{V}(\alpha) = 0$

Step 0. (Initialization) $k = 0$, $v^k = 0$, $u^k = \alpha^+$, $\alpha^k = \frac{\alpha^+}{2} \in [v^k, u^k]$.
Step 1. Find a solution $z(\alpha^k)$ to Problem $(\mathcal{P}_{\alpha^k})$ using the global search strategy for d.c. minimization [11,12].
Step 2. (Stopping criterion) If $\mathcal{V}_k := \mathcal{V}(\alpha^k) = 0$ and $\min_i \Phi_i(z(\alpha^k)) \geq 0$, then

STOP: $z(\alpha^k) \in Sol(\mathcal{P}_f)$ (in virtue of Theorem 1).
Step 3. If $\mathcal{V}_k > 0$, then set $v^{k+1} = \alpha^k$, $\alpha^{k+1} = \frac{1}{2}(u^k + \alpha^k)$, $k = k + 1$ and go to Step 1.
Step 4. If $\mathcal{V}_k < 0$, then set $u^{k+1} = \alpha^k$, $\alpha^{k+1} = \frac{1}{2}(v^k + \alpha^k)$, $k = k + 1$ and go to Step 1.
Step 5. If $\min_i \Phi_i(z(\alpha^k)) < 0$, then set $\alpha_i^{k+1} = \frac{\psi_i(z(\alpha^k))}{\varphi_i(z(\alpha^k))} \ \forall i : \Phi_i(z(\alpha^k)) < 0$,

$\alpha_i^{k+1} = \alpha_i^k \ \forall i : \Phi_i(z(\alpha^k)) \geq 0$. In addition, set $v^{k+1} = 0$, $u^{k+1} = t_{k+1}\alpha^{k+1}$,

where $t_{k+1} = \dfrac{\alpha^+}{\max_i \alpha_i}$, $k = k + 1$, and return to Step 1.

Let us emphasize the fact that the algorithm for solving Problem $(\mathcal{P}_f)$ of fractional optimization consists of 3 basic stages: the (a) local and (b) global searches in Problem $(\mathcal{P}_\alpha)$ with a fixed vector parameter α and (c) the method for finding the vector parameter α at which the optimal value of Problem $(\mathcal{P}_\alpha)$ is zero.

3   Reduction to the Problem with D.C. Constraints

In this section we reduce the fractional program to the optimization problem with a nonconvex feasible set.

Proposition 1. [6] *Let the pair* $(x_*, \alpha_*) \in \mathbb{R}^n \times \mathbb{R}^m$ *be a solution to the following problem:*

$$\sum_{i=1}^{m} \alpha_i \downarrow \min_{(x,\alpha)}, \quad x \in S, \quad \frac{\psi_i(x)}{\varphi_i(x)} \le \alpha_i, \quad i = 1, \ldots, m. \tag{2}$$

Then $\dfrac{\psi_i(x_*)}{\varphi_i(x_*)} = \alpha_{*i}, \quad i = 1, \ldots, m.$

Corollary 1. *For any solution* $(x_*, \alpha_*) \in \mathbb{R}^n \times \mathbb{R}^m$ *to the problem* (2), *the point* x_* *will be a solution to Problem* $(\mathcal{P}_f)$.

The inequality constraints in the problem (2) can be replaced by the equivalent constraints $\psi_i(x) - \alpha_i \varphi_i(x) \le 0$, $i = 1, \ldots, m$, since $\varphi_i(x) > 0$ $\forall x \in S$. This yields the following problem with m nonconvex constraints:

$$(\mathcal{P}) \quad f_0 := \sum_{i=1}^{m} \alpha_i \downarrow \min_{(x,\alpha)}, \quad x \in S, \quad f_i := \psi_i(x) - \alpha_i \varphi_i(x) \le 0, \quad i = 1, \ldots, m.$$

We intend to solve this problem using the exact penalization approach for d.c. optimization developed in [13]. Therefore, we introduce the penalized problem

$$(\mathcal{P}_\sigma) \quad \theta_\sigma(x) = f_0(x) + \sigma \max\{0, f_i(x), i \in I\} \downarrow \min, \quad x \in S.$$

It can be readily seen that the penalized function $\theta_\sigma(\cdot)$ is a d.c. function. The theory enables us to construct an algorithm which consists of two principal stages: (a) local search, which provides an approximately critical point; (b) procedures of escaping from critical points.

Actually, since $\sigma > 0$, $\theta_\sigma(x) = G_\sigma(x) - H_\sigma(x)$, $H_\sigma(x) := h_0(x) + \sigma \sum_{i \in I} h_i(x)$,

$$G_\sigma(x) := \theta_\sigma(x) + H_\sigma(x) = g_0(x) + \sigma \max \left\{ \sum_{i=1}^{m} h_i(x); \max_{i \in I} [g_i(x) + \sum_{j \in I,\, j \neq i} h_j(x)] \right\},$$

it is clear that $G_\sigma(\cdot)$ and $H_\sigma(\cdot)$ are convex functions.

Let the Lagrange multipliers, associated with the constraints and corresponding to the point z^k, $k \in \{1, 2, \ldots\}$, be denoted by $\lambda := (\lambda_1, \ldots, \lambda_m) \in \mathbb{R}^m$.

Global search scheme

Step 1. Using the local search method from [14], find a critical point z^k in $(\mathcal{P})$.

Step 2. Set $\sigma_k := \sum_{i=1}^{m} \lambda_i$. Choose a number $\beta : \inf(G_\sigma, S) \le \beta \le \sup(G_\sigma, S)$.

Choose an initial $\beta_0 = G_\sigma(z^k)$, $\zeta_k = \theta_\sigma(z^k)$.

Step 3. Construct a finite approximation

$$R_k(\beta) = \{v^1, \ldots, v^{N_k} \mid H_\sigma(v^i) = \beta + \zeta_k, \ i = 1, \ldots, N_k, \ N_k = N_k(\beta)\}$$

Step 4. Find a δ_k-solution $\bar{u}^i$ of the following Linearized Problem:

$$(\mathcal{P}_\sigma L_i) \quad G_\sigma(x) - \langle \nabla H_\sigma(v^i), x \rangle \downarrow \min_x, \quad x \in S.$$

Step 5. Starting from the point $\bar{u}^i$, find a KKT-point u^i by the local search method from [14].

Step 6. Choose the point u^j : $f(u^j) \leq \min\{f_0(u^i), \ i = 1, ..., N\}$.

Step 7. If $f_0(u^j) < f_0(z^k)$, then set $z_{k+1} = u^j$, $k = k+1$ and go to Step 2.

Step 8. Otherwise, choose a new value of β and go to Step 3.

According to Corollary 1, the point z^* resulting from the global search strategy will be a solution to the original fraction program. It should be noted that, in contrast to the approach from Sect. 2, α_i will be found simultaneously with the solution vector x.

4 Computational Simulations

Two approaches from above for solving the fractional programs $(\mathcal{P}_f)$ via d.c. optimization problems were successfully tested. The algorithm based on the method for solving the equation $\mathcal{V}(\alpha) = 0$ from Sect. 2 (F1-algorithm) and the algorithm based on the global search scheme from Sect. 3 (F2-algorithm) were applied to an extended set of test examples for the various starting points. Several instances of fractional problems from [2, 7–9] with a small number of variables and a small number of terms in the sum were used for computational experiments. Additionally, randomly generated fractional problems with linear or quadratic functions in the numerators and the denominators of ratios with up to 200 variables and 200 terms in the sum were successfully solved. All computational experiments were performed on the Intel Core i7-4790K CPU 4.0 GHz. All convex auxiliary problems (linearized problems) on the steps of F1-, F2-algorithms were solved by the software package IBM ILOG CPLEX 12.6.2.

Table 1 presents results of some comparative computational testing of two approaches (F1-, F2-algorithms) and employs the following designations: *name* is the test example name; n is the number of variables (problem's dimension); m is the number of terms in the sum; $f(x^0)$ is the value of the goal function to Problem $(\mathcal{P}_f)$ at the starting point; $f(z)$ is the value of the function at the solution provided by the algorithms; *it* is the number of iterations of F1- or F2-algorithms; *Time* stands for the CPU time of computing (seconds).

Observe that one iteration of F1-algorithm and one iteration of F2-algorithm differ in processing time and, therefore, cannot be compared. In the F1-algorithm it denotes the number of times that we varied the parameter α, while in the F2-algorithm it stands for the number of iterations of the global search in solving the nonconvex Problem $(\mathcal{P}_\sigma)$.

Computational experiments showed that solving of the fraction program should combine the two approaches. For example, we can use the solution to Problem $(\mathcal{P}_\sigma)$ to search for the parameter α that reduces the optimal value function of Problem $(\mathcal{P}_\alpha)$ to zero.

Table 1. Performance of two algorithms on test fractional program.

name	n	m	$f(x_0)$	$f(z)$	F1-algorithm		F2-algorithm	
					it	Time	it	Time
Problem [9]	2	1	0.333333	0.333333	19	0.02	1	0.01
			0.750000	0.333333	18	0.02	2	0.01
Problem [8]	3	1	0.943038	0.931298	22	0.03	4	0.03
			1.054217	0.931298	23	0.04	5	0.03
Problem [2]	2	2	4.293433	1.428571	213	0.16	2	9.82
			1.627273	1.428571	108	0.08	3	0.07
Problem 1 [7]	2	2	3.524024	2.829684	191	0.26	3	0.02
			2.829684	2.829684	19	0.05	1	0.02
Problem 2 [7]	3	3	3.000000	3.000000	18	0.04	1	0.04
			3.136952	3.000000	321	0.24	7	0.11
Problem 3 [7]	3	3	3.000000	2.889069	333	0.34	2	0.05
			2.904068	2.889069	259	0.26	2	0.05

5   Conclusions

In this paper, we showed how fractional programs can be solved by applying the Global Search Theory of d.c. optimization. The methods developed were justified and tested on an extended set of problems with linear or quadratic functions in the numerators and denominators of the ratios.

Acknowledgments. This work has been supported by the Russian Science Foundation, Project No. 15-11-20015.

References

1. Bugarin, F., Henrion, D., Lasserre, J.-B.: Minimizing the sum of many rational functions. Math. Prog. Comput. **8**, 83–111 (2016)
2. Chun-feng, W., San-yang, L.: New method for solving nonlinear sum of ratios problem based on simplicial bisection. Syst. Eng. Theory Pract. **33**(3), 742–747 (2013)
3. Dinkelbach, W.: On nonlinear fractional programming. Manage. Sci. **13**, 492–498 (1967)
4. Freund, R.W., Jarre, F.: Solving the sum-of-ratios problem by an interior-point method. J. Global Optim. **19**(1), 83–102 (2001)
5. Gruzdeva, T.V., Strekalovsky, A.S.: An approach to fractional programming via d.c. optimization. AIP Conf. Proc. **1776**, 090010 (2016)
6. Gruzdeva, T.V., Strekalovsky, A.S.: An approach to fractional programming via d.c. constraints problem: local search. In: Kochetov, Y., Khachay, M., Beresnev, V., Nurminski, E., Pardalos, P. (eds.) DOOR 2016. LNCS, vol. 9869, pp. 404–417. Springer, Cham (2016). doi:10.1007/978-3-319-44914-2_32

7. Ma, B., Geng, L., Yin, J., Fan, L.: An effective algorithm for globally solving a class of linear fractional programming problem. J. Softw. **8**(1), 118–125 (2013)

8. Pandey, P., Punnen, A.P.: A simplex algorithm for piecewise-linear fractional programming problems. European J. of Oper. Res. **178**, 343–358 (2007)

9. Raouf, O.A., Hezam, I.M.: Solving fractional programming problems based on swarm intelligence. J. Ind. Eng. Int. **10**, 56–66 (2014)

10. Schaible, S., Shi, J.: Fractional programming: the sum-of-ratios case. Optim. Methods Softw. **18**, 219–229 (2003)

11. Strekalovsky, A.S.: On solving optimization problems with hidden nonconvex structures. In: Rassias, T.M., Floudas, C.A., Butenko, S. (eds.) Optimization in Science and Engineering, pp. 465–502. Springer, New York (2014). doi:10.1007/978-1-4939-0808-0_23

12. Strekalovsky, A.S.: Elements of nonconvex optimization. Nauka, Novosibirsk (2003). [in Russian]

13. Strekalovsky, A.S.: On the merit and penalty functions for the d.c. optimization. In: Kochetov, Y., Khachay, M., Beresnev, V., Nurminski, E., Pardalos, P. (eds.) DOOR 2016. LNCS, vol. 9869, pp. 452–466. Springer, Cham (2016). doi:10.1007/978-3-319-44914-2_36

14. Strekalovsky, A.S.: On local search in d.c. optimization problems. Appl. Math. Comput. **255**, 73–83 (2015)

Objective Function Decomposition in Global Optimization

Oleg V. Khamisov[✉]

Melentiev Energy Systems Institute SB RAS, Lermontov, Russia
khamisov@isem.irk.ru

Abstract. In this paper we consider global optimization problems in which objective functions are explicitly given and can be represented as compositions of some other functions. We discuss an approach of reducing the complexity of the objective by introducing new variables and adding new constraints.

Keywords: Global optimization · Decomposition · Induced constraint · d.c. function

1 Introduction

In this paper we consider global optimization problems in which objective functions are explicitly given and can be represented as compositions of some other functions. Many practical problems can be formulated in a such form [8,9,11]. An approach similar to the described below was suggested earlier in [4,7,13]. In [5] an equivalent approach was suggested for utility function, i.e. for the case when objective composite function has some monotonicity properties.

2 Objective Function Decomposition and the Induced Constraint

Consider the following mathematical programming problem

$$\min g(x), x \in X, \tag{1}$$

where g is a continuous composite function $g(x) = F(f_1(x), \ldots, f_p(x))$, $F : R^p \to R$, $f_i : X \to R$ are continuous functions, $X \subset R^n$ is a compact set.

Introducing new variables $y_i = f_i(x)$, $i = 1, \ldots, p$ we formulate the following equivalent problem

$$\min F(y), \tag{2}$$

$$y_i = f_i(x), \ i = 1, \ldots, p, \ x \in X. \tag{3}$$

Assume that $p < n$ or function F is "less complicated" than f. In such case we can obtain a reduction in difficulty of the initial problem providing that Eq. (3)

© Springer International Publishing AG 2017
R. Battiti et al. (Eds.): LION 2017, LNCS 10556, pp. 338–344, 2017.
https://doi.org/10.1007/978-3-319-69404-7_28

are practically tractable. The latter means, for example, a possibility to solve Eq. (3) in x for given y subject to inclusion $x \in X$ by an efficient algorithm. Solving problem (1) corresponds to finding optimal and feasible point simultaneously. In problem (2)–(3) optimality (i.e. minimization in (2)) and feasibility (i.e. determining x for a given y in (3)) stages are separated: they are performed in different spaces. What is exactly done in the reduction of problem (1) to problem (2)–(3) and what is understood under objective function decomposition in this paper is deleting some complexity from the objective function to the constraints, i.e. moving a part of difficulty from the optimality stage the feasibility stage. The motivation of a such decomposition is a desire to distribute difficulty of the initial problem between objective and constraints more or less uniformly. It is necessary to mention that structure of the objective function is given, we just use it. In this case we perform explicit decomposition. There are many cases when we need to discover good (or efficient) decomposition. In the latter case the decomposition is implicit.

In practical minimization of F in (2) it is quite often necessary to localize a global minimum in some compact subset of R^p. Define values

$$\underline{f}_i = \min_{x \in X} f_i(x), \ \overline{f}_i = \max_{x \in X} f_i(x), \ i = 1, \dots, p. \tag{4}$$

Instead of exact calculation of $\underline{f}_i$ and $\overline{f}_i$ we quite often have to use approximate values $\underline{y}_i \leq \underline{f}_i$, $\overline{y}_i \geq \overline{f}_i$, $i = 1, \dots, p$. Let us define the set $Y \subset R^p$ as the image of X under nonlinear continuous mapping (or transformation) $f : X \to Y$, $f(x) = (f_1(x), \dots, f_p(x))$,

$$Y = \{y \in R^p : y = f(x) \text{ for some } x \in X\}. \tag{5}$$

Then the initial problem can be rewritten in the following way

$$\min F(y), \tag{6}$$

$$\underline{y}_i \leq y_i \leq \overline{y}_i, \ i = 1, \dots, p, \tag{7}$$

$$y \in Y. \tag{8}$$

Since (8) is a reformulation of the feasibility stage constraint (3) the inclusion $y \in Y$ will be referred to as induced constraint.

3   Agreed Decomposition

We will say that the composite objective function g has agreed variable decomposition

$$g(x) = F(f_1(x^1), \dots, f_p(x^p)), \tag{9}$$

where $x^i \in X^i \subset R^{n_i}$, $i = 1, \dots, p$, $X^1 \times \dots \times X^p = X$ and $n_1 + \dots + n_p = n$. Conversely, we will say that the function g has disagreed variable decomposition if g is still representable in the form (9) and $x^i \in R^{n_i}$, $n_i < n$, $i = 1, \dots, p$, but $n_1 + \dots + n_p > n$.

In the case of agreed variable decomposition we rewrite the problem (6)–(8) in the following way

$$\min F(y), \tag{10}$$

$$\underline{f}_i \leq y_i \leq \overline{f}_i, \ i = 1, \ldots, p, \tag{11}$$

$$y_i = f_i(x^i), \ x^i \in X^i, \ i = 1, \ldots, p. \tag{12}$$

Note, that in (11) we use exact lower and upper bounds on y_i since otherwise the inclusion $x^i \in X^i$ can be violated.

Due to variable decomposition property (12) we can use the following three-stage approach.

I. Variable bounding stage. Determine values $\underline{f}_i$ and $\overline{f}_i$ through (4).
II. Optimal solution stage. Solve the problem (10)–(11) and find optimal point y^*.
III. Feasibility stage. Solve the feasibility problem (12) for $y = y^*$ and obtain optimal point x^* for the initial problem. It easy to see that if $F(y) = \sum y_i$ we have the well-known separable problem.

Example 1. Consider the following three dimensional problem from [2]

$$g(x) = -(x_1 - 1)(x_1 + 2)(x_2 + 1)(x_2 - 2)x_3^2,$$

$$X = [-2, 2] \times [-2, 2] \times [-2, 2].$$

Let $F(y_1, y_2, y_3) = -y_1 y_2 y_3$, $f_1(x_1) = (x_1 - 1)(x_1 + 2)$, $f_2(x_2) = (x_2 + 1)(x_2 - 2)$, $f_3(x_3) = x_3^2$, $X^i = [-2, 2]$, $i = 1, 2, 3$.

I. Variable bounding stage. Each function f_i is convex, so it easy to determine

$$\underline{f}_1 = \underline{f}_2 = -2.25, \ \underline{f}_3 = 0, \ \overline{f}_1 = \overline{f}_2 = \overline{f}_3 = 4.$$

II. Optimal solution stage. Since $y^0 = 0$ is feasible and $F(y^0) = 0$ then the global minimum of F must be nonpositive. Lower bound $\underline{f}_3 = 0$, hence in order to minimize F variables y_1 and y_2 should be positive or negative simultaneously. In the first case the minimal value is attained at point $y^1 = (-2.25, -2.25, 4)$ with $F(y^1) = -25$. In the second case the minimal value is attained at point $y^2 = (4, 4, 4)$ with $F(y^2) = -36$. Therefore, $y^* = y^2$ is the unique global minimum of the optimal value stage.
III. Feasibility stage. Solving the feasibility problem for $y = y^* = (4, 4, 4)$ we obtain two solutions $x^{*1} = (2, -2, 2)$ and $x^{*2} = (2, -2, -2)$.
This example shows us that in some cases the problem can be solved analytically.

Example 2. Consider now the well-known Shubert function [10]

$$g(x_1, x_2) = f_1(x_1) \cdot f_2(x_2) = \left(\sum_{i=1}^{5} i \cos\left[(i+1)x_1 + i\right] \right) \cdot \left(\sum_{i=1}^{5} i \cos\left[(i+1)x_2 + i\right] \right),$$

$$X = X_1 \times X_2, \ X_1 = X_2 = [-10, 10].$$

It is obvious to set $F(y_1, y_2) = y_1 y_2$.

I. Variable bounding stage. It is not difficult to find out that

$$\underline{f}_1 = \underline{f}_2 = -12.87088549, \ \overline{f}_1 = \overline{f}_2 = 14.50800793.$$

To find $\underline{f}_i$ and $\overline{f}_i$, $i = 1, 2$ it is necessary to solve global univariate optimization problems. At present time there exist efficient approaches for global univariate optimization (see, for example, [12]), so we assume that such problems are computationally tractable.

II. Optimal solution stage. Global minimum value of bilinear function F is attained at two points $y^{*,1} = (y_1^{*,1}, y_2^{*,1}) = (-12.87088549, 14.50800793)$ and $y^{*,2} = (y_2^{*,1}, y_1^{*,1})$ with $F(y^{*,1}) = F(y^{*,2}) = -186.7309088$.

III. Feasibility stage. For every $i = 1, 2$ each equation $y_j^{*,i} = f_j(x_j)$, $j = 1, 2$ has three solutions in the corresponding x_j. Hence, for every $i = 1, 2$ we have 9 solutions that gives 18 global minimum points in total.

It is worthwhile to note that in the Example 2 we first found global minimum value and then found all global minimum points, i.e. these two problems are separated and their separate solution turned out to be easier than solution of the initial global optimization problem.

4   Reducing the Induced Constraint to a d.c. Inequality

Recall [16] that a function h is called d.c. function if it can be represented as $h(x) = r(x) - q(x)$, where functions r and q are convex. Let us introduced the function

$$\Phi(y) = \min_{x \in X} \left\{ \sum_{i=1}^{p} (y_i - f_i(x))^2 \right\}. \tag{13}$$

It is well known that function Φ is a d.c. function:

$$\Phi(y) = \sum_{i=1}^{p} y_i^2 - \max_{x \in X} \left\{ \sum_{i=1}^{p} (2f_i(x)y_i - f_i^2(x)) \right\} = \Phi_1(y) - \Phi_2(y)$$

with convex functions Φ_1 and (implicit) Φ_2. Therefore, the induced constraint can be rewritten as the d.c. inequality $\Phi(y) \leq 0$. The final reduced form of the initial problem is the following

$$\min F(y), \tag{14}$$

$$\Phi(y) \leq 0, \tag{15}$$

$$\underline{y}_i \leq y_i \leq \overline{y}_i, \ i = 1, \ldots, p. \tag{16}$$

Such reduction is effective when $p < n$ (or even $p \ll n$) and inner optimization problem for calculating values of Φ_1 can be effectively solved. The most appropriate example here is given by linear functions f_i and small p, say, $p \leq 10$. Then the complexity is formed by nonconvex problem in p variables and it can be solved by different kinds of branch and bounds methods. If F is a d.c. function then problem (14)–(16) is a d.c. optimization problem and different d.c. optimization methods (see [6,14,16]) can be used.

Comment. Assume that functions f_i are Lipschitz continuous with constants L_i, $i = 1, \ldots, p$. For given y define $\psi_i(x) = y_i - f_i(x)$, $i = 1, \ldots, p$. Let a point $\hat{x} \in X$ be given then

$$|\psi_i^2(x) - \psi_i^2(\hat{x})| = |(\psi_i(x) - \psi_i(\hat{x}))(\psi_i(x) + \psi_i(\hat{x}))|$$

$$\leq |(f_i(x) - f_i(\hat{x}))| \cdot |2\psi_i(\hat{x})| \leq 2|\psi_i(\hat{x})|L_i\|x - \hat{x}\| \quad \forall x : \psi_i(x) \leq \psi_i(\hat{x}).$$

Hence, the less value $|\psi_i(\hat{x})|$ the less is Lipschitz constant of ψ_i^2 at point $\hat{x}$, which in this case can be taken as $2|\psi_i(\hat{x})|L_i$. This property can essentially improve efficiency of Lipschitz optimization methods [11,15] for solving problem (13).

Example 3. Let us consider Goldstein-Price function [2,10]

$$g(x_1, x_2) = \left(1 + (x_1 + x_2 + 1)^2 \left(19 - 14x_1 + 3x_1^2 - 14x_2 + 6x_1x_2 + 3x_2^2\right)\right)$$
$$\times \left(30 + (2x_1 - 3x_2)^2 \left(18 - 32x_1 + 12x_1^2 + 48x_2 - 36x_1x_2 + 27x_2^2\right)\right),$$

$X = [-2, 2] \times [-2, 2]$. By linear transformation of variables g can be rewritten:

$$g(x_1, x_2) = \left(1 + (x_1 + x_2 + 1)^2 \left(3(x_1 + x_2 + 1)^2 - 20(x_1 + x_2 + 1) + 36\right)\right)$$
$$\times \left(30 + (2x_1 - 3x_2)^2 \left(3(2x_1 - 3x_2) - 16(2x_1 + 3x_2) + 18\right)\right).$$

Then, the introduction of new variables is obvious:

$$y_1 = x_1 + x_2 + 1, \ y_2 = 2x_1 - 3x_2. \tag{17}$$

For fixed (y_1, y_2) system (17) always has a unique solution in (x_1, x_2). Then

$$F(y_1, y_2) = \left(1 + y_1^2 \left(3y_1^2 - 20y_1 + 36\right)\right) \times \left(30 + y_2^2 \left(3y_2^2 - 16y_2 + 18\right)\right).$$

Bounds $\underline{y}_1 = -3, \underline{y}_2 = -10, \overline{y}_1 = 5, \overline{y}_2 = 10$ are easily calculated. Function F is the product of two univariate positive functions $F_1(y_1) = 1 + y_1^2 \left(3y_1^2 - 20y_1 + 36\right)$ and $F_2(y_2) = 30 + (2x_1 - 3x_2)^2 \left(3(2x_1 - 3x_2) - 16(2x_1 + 3x_2) + 18\right)$. This is the main advantage of the linear transformation. We minimize each function separately (as we mentioned above univariate global optimization assumed computationally tractable) and obtain global minimum point $y_1^* = 0, y_2^* = 3$. Substituting $y_1^* = 0, y_2^* = 3$ in system (17) instead of y_1, y_2 we obtain globally optimal solution $x_1^* = 0, x_2^* = -1$ for the initial problem.

5   Conclusion

An objective function decomposition in global optimization was discussed. Due to decomposition we can obtain reduction in solution difficulty. The suggested approach can be considered as a starting decomposition scheme depending on properties of F. Types of decomposition are generated by different classes of functions F. Among well-known classes we mention multiplicative functions, sum of ratios functions and so on. Other types of function F can be used.

Acknowledgments. This work is supported by the RFBR grant number 15-07-08986.

References

1. Bromberg, M., Chang, T.C.: A function embedding technique for a class of global optimization problems one-dimensional global optimization. In: Proceedings of the 28th IEEE Conference on Decision and Control, vol. 1–3, pp. 2451–2556 (1989)
2. Hansen, P., Jaumard, B.: Lipschitz optimization. In: Pardalos, P.M., Horst, R. (eds.) Handbook of Global Optimization, pp. 407–494. Kluwer Academic Publishers, Dordrecht (1995)
3. Hansen, P., Jaumard, B., Lu, S.H.: An analytical approach to global optimization. Math. Program. **52**(1), 227–254 (1991)
4. Hamed, A.S.E.-D., McCormick, G.P.: Calculations of bounds on variables satisfying nonlinear equality constraints. J. Glob. Optim. **3**, 25–48 (1993)
5. Horst, R., Thoai, N.V.: Utility functions programs and optimization over efficient set in multiple-objective decision making. JOTA **92**(3), 605–631 (1997)
6. Horst, R., Tuy, H.: Global Optimization: Deterministic Approaches. Springer, Heidelberg (1996). doi:10.1007/978-3-662-03199-5
7. McCormick, G.P.: Attempts to calculate global solution of problems that may have local minima. In: Lootsma, F. (ed.) Numerical Methods for Nonlinear Optimization, pp. 209–221. Academic Press, London, New York (1972)
8. Pardalos, P.M.: An open global optimization problem on the unit sphere. J. Glob. Optim. **6**, 213 (1995)
9. Pardalos, P.M., Shalloway, D., Xue, G.: Optimization methods for computing global minima of nonconvex potential energy functions. J. Glob. Optim. **4**, 117–133 (1994)
10. Paulavičius, R., Žilinskas, J.: Simplicial Global Optimization. Springer Briefs in Optimization. Springer, New York (2014). doi:10.1007/978-1-4614-9093-7
11. Pinter, J.: Global Optimization in Action. Kluwer Academic Publishers, Dordrecht (1996)
12. Sergeyev, Y.D., Strongin, R.G., Lera, D.: Introduction to Global Optimization Exploiting Space-Filling Curves. Springer Briefs in Optimization. Springer, New York (2013). doi:10.1007/978-1-4614-8042-6
13. Sniedovich, M., Macalalag, E., Findlay, S.: The simplex method as a global optimizer: a C-programming perspectuve. J. Glob. Optim. **4**, 89–109 (1994)
14. Strekalovsky, A.S.: On solving optimization problems with hidden nonconvex structures. In: Rassias, T.M., Floudas, C.A., Butenko, S. (eds.) Optimization in Science and Engineering, pp. 465–502. Springer, New York (2014). doi:10.1007/978-1-4939-0808-0_23

15. Strongin, R.G., Sergeev, Y.D.: Global Optimization with Non-convex Constraints: Sequential and Parallel Algorithms. Kluwer Academic Publishers, Dordrecht (2000)
16. Tuy, H.: D.C. optimization: theory, methods and algorithms. In: Pardalos, P.M., Horst, R. (eds.) Handbook of Global Optimization, pp. 149–216. Kluwer Academic Publishers, Dordrecht (1995)

Projection Approach Versus Gradient Descent for Network's Flows Assignment Problem

Alexander Yu. Krylatov[✉] and Anastasiya P. Shirokolobova

Saint Petersburg State University, Saint Petersburg, Russia
a.krylatov@spbu.ru

Abstract. The paper is devoted to comparison of two methodologically different types of mathematical techniques for coping with network's flows assignment problem. Gradient descent and projection approach are implemented to the simple network of parallel routes (there are no common arcs for any pair of routes). Gradient descent demonstrates zig-zagging behavior in some cases, while projection algorithm converge quadratically in the same conditions. Methodological interpretation of such phenomena is given.

Keywords: Network's flows assignment problem · Projection operator · Gradient descent

1 Introduction

Huge amount of different practical problems are solved due to models of network's flows assignment. The most remarkable among them are road networks, power grids and pipe networks [6,7]. The task it to estimate network's flows assignment *profile* according to demands between all source-sink pairs. Generally, there are many source-sink pairs in a network (multicommodity networks). In a multicommodity network flows from different commodities load common arcs simultaneously and influence on volume delays of each others.

In this paper we show that projection approach is more appropriate technique for coping with network's flows assignment problem than gradient descent. Moreover, zig-zagging bahavior of gradient descent in the neighborhood of the equilibrium solution is clarified.

2 Network's Flows Assignment Problem

Consider a network presented by connected directed graph $G = (N, A)$. Introduce notation: N – the set of sequentially numbered nodes of the graph G; A – the set of sequentially numbered arcs of the graph G; W – the set of nodes' pairs (source and sink) of the graph G, $w \in W$; K^w – the set of routes connecting source-sink pair $w \in W$; x_a – the flow on arc $a \in A$, $x = (\ldots, x_a, \ldots)$; c_a – capacity of arc $a \in A$, $c = (\ldots, c_a, \ldots)$; f_k^w – the flow on route $k \in K^w$;

© Springer International Publishing AG 2017
R. Battiti et al. (Eds.): LION 2017, LNCS 10556, pp. 345–350, 2017.
https://doi.org/10.1007/978-3-319-69404-7_29

F^w – the flow demand between source-sink pair $w \in W$; $t_a(x_a)$ – the link performance function (volume delay function) of arc $a \in A$; $\delta_{a,k}^w$ – indicator:

$$\delta_{a,k}^w = \begin{cases} 1, & \text{if the arc } a \in A \text{ lies along to route } k \in K^w; \\ 0, & \text{otherwise.} \end{cases}$$

Networks's flow assignment problem in a *link-route* formulation was first offered by Dafermos and Sparrow [1,2]. This formulation could be expressed in a form of the optimization program [6,8]:

$$\min_x \sum_{a \in A} \int_0^{x_a} t_a(u)du,$$

subject to $\sum_{k \in K^w} f_k^w = F^w$, $f_k^w \geq 0$ for any $k \in K^w$, $w \in W$ with definitional constraints $x_a = \sum_{w \in W} \sum_{k \in K^w} f_k^w \delta_{a,k}^w$ for any $a \in A$.

3   Simple Network of Parallel Routes

Network of parallel routes consists of two nodes (sources and sink) and n alternative arcs (routes). The demand between source and sink is F. The demand F is to be assigned among n routes: $F = \sum_{i=1}^n f_i$, $f_i \geq 0$, $i = \overline{1,n}$. Link performance function is smooth non-decreasing function: $t_i \in C^1(R^+)$, $t_i(x) - t_i(y) \geq 0$ when $x - y \geq 0$, $x, y \in R^+$, $i = \overline{1,n}$, where R^+ — non-negative orthant. Moreover, it is believed that $t_i(x) \geq 0$, $x \geq 0$ and $\partial t_i(x)/\partial x > 0$, $x > 0$, $i = \overline{1,n}$.

From mathematical perspective, link-route and link-node formulations are equivalent for the network of parallel routes. In such a case, network's flows assignment problem could be expressed as follows:

$$f^* = \arg\min_f \sum_{i=1}^n \int_0^{f_i} t_i(u)du, \tag{1}$$

subject to

$$\sum_{i=1}^n f_i = F, \tag{2}$$

$$f_i \geq 0 \quad \forall i = \overline{1,n}. \tag{3}$$

According to results obtained in [4], there exists an explicit projection operator to cope with the problem (1)–(3). For the sake of convenience let us introduce additional notations:

$$a_i(f_i) \overset{\text{def}}{=} t_i(f_i) - t_i'(f_i)f_i, \; b_i(f_i) \overset{\text{def}}{=} t_i'(f_i), \; i = \overline{1,n}.$$

Then, the projection operator Φ such as

$$f^* = \Phi(f^*)$$

could be expressed explicitly via $a_i(f_i)$ and $b_i(f_i)$, $i = \overline{1,n}$:

$$\Phi_i(f) = \begin{cases} \dfrac{1}{b_i(f_i)} \dfrac{F + \sum_{s=1}^{m} \frac{a_s(f_s)}{b_s(f_s)}}{\sum_{s=1}^{m} \frac{1}{b_s(f_s)}} - \dfrac{a_i(f_i)}{b_i(f_i)} & \text{for } i \le m, \\ 0 & \text{for } i > m, \end{cases} \tag{4}$$

when components f and $t(f)$ are indexed so that

$$a_1(f_1) \le a_2(f_2) \le \ldots \le a_m(f_m) \tag{5}$$

and m is defined from the condition

$$\sum_{i=1}^{m} \frac{a_m(f_m) - a_i(f_i)}{b_i(f_i)} \le F < \sum_{i=1}^{m} \frac{a_{m+1}(f_{m+1}) - a_i(f_i)}{b_i(f_i)}. \tag{6}$$

Due to projection operator Φ defined explicitly via (4)–(6) the corresponding projection algorithm was developed. At each iteration, the algorithm performed the following three steps.
$(k+1)$ iteration:

1. To index m_k components f^k and $t(f^k)$ so that

$$a_1(f_1^k) \le a_2(f_2^k) \le \ldots \le a_{m_k}(f_{m_k}^k).$$

2. To find $m_{k+1} \le m_k$ (amount of non-zero components f^{k+1}) from the condition

$$\sum_{i=1}^{m_{k+1}} \frac{a_{m_{k+1}}(f_{m_{k+1}}^k) - a_i(f_i^k)}{b_i(f_i^k)} \le F < \sum_{i=1}^{m_{k+1}} \frac{a_{m_{k+1}+1}(f_{m_{k+1}+1}^k) - a_i(f_i^k)}{b_i(f_i^k)}.$$

3. To compute f^{k+1}:

$$f_i^{k+1} = \frac{1}{b_i(f_i^k)} \frac{F + \sum_{s=1}^{m_{k+1}} \frac{a_s(f_s^k)}{b_s(f_s^k)}}{\sum_{s=1}^{m_{k+1}} \frac{1}{b_s(f_s^k)}} - \frac{a_i(f_i^k)}{b_i(f_i^k)}, \qquad i = \overline{1, m_{k+1}},$$

$$f_i^{k+1} = 0, \qquad i = \overline{m_{k+1}, n}.$$

4. Termination criterion

$$\sum_{i=1}^{m_{k+1}-1} \left| t_i(f_i^{k+1}) - t_{i+1}(f_{i+1}^{k+1}) \right| < \varepsilon.$$

Note that quadratic convergence of this algorithm is proved [4].

4   Simulation Results

Zig-zagging behavior of such widely used gradient descent as Frank-Wolfe algorithm became apparent in the 1970s [3,5]. That discussions were intuitive. Here we investigate it in detail on the example of simple network of parallel routes. Assume that volume delay functions for this network are defined as follows:

$$t_i(f_i) = c_i + d_i \cdot f_i^{\beta}, \quad i = \overline{1,n}.$$

Changing parameters c_i and d_i, $i = \overline{1,n}$ we could obtain different networks of paralle routes. Let's define five different patterns of such a network according to Table 1 with fixed demand $F = 20$.

Table 1. Five different patterns of the network topology

Pattern 1		Pattern 2		Pattern 3		Pattern 4		Pattern 5	
$n = 2$	$\beta = 2$	$n = 2$	$\beta = 3$	$n = 4$	$\beta = 3$	$n = 4$	$\beta = 3$	$n = 6$	$\beta = 2$
i	c_i	d_i	i	c_i	d_i	i	c_i	d_i	i	c_i	d_i	i	c_i	d_i

i	c_i	d_i	i	c_i	d_i	i	c_i	d_i	i	c_i	d_i	i	c_i	d_i
1	2	1	1	2	1	1	2	1	1	2	1	1	2	1
2	1	2	2	1	2	2	1	2	2	1	2	2	1	2
						3	1.25	2.5	3	200	200	3	200	200
						4	1	3	4	1	3	4	1	3
												5	300	300
												6	1.25	2.5

Network's flows assignment problem could be formulated for each pattern. Then corresponding problems could be solved. We test projection algorithm from [4] and Frank-Wolfe algorithm on the network of parallel routes. $(k+1)$ iteration of Frank-Wolfe algorithm:

1. Solve the linear programming subproblem

$$\min_f \sum_{i=1}^n t_i\left(f_i^k\right) f_i,$$

subject to (2) and (3). Let y^k be its solution, and $p^k = y^k - f^k$ the resulting search direction.
2. Find a step length l_k, which solves the problem $\min \left\{ T(f^k + lp^k) \,|\, 0 \le l \le 1 \right\}$, where T is the objective function (1).
3. Let $f^{k+1} = f^k + l_k p^k$ and $R^{k+1} = \left\{ f_i^{k+1} \,|\, f_i^{k+1} > 0.1, \, i = \overline{1,n} \right\}$ is the set of used routes.
4. If

$$\sum_{i,j \in R^{k+1}} \left| t_i(f_i^{k+1}) - t_j(f_j^{k+1}) \right| < \varepsilon,$$

then terminate, with f^{k+1} as the approximate solution. Otherwise, let $k :=$ $k + 1$, and go to Step 1.

Amount of iterations required by these two algorithms are available in Table 2. Infinite amount of iterations (∞) means zig-zagging behavior of algorithm.

Table 2. Projection algorithm versus Frank-Wolfe algorithm: amount of iterations

	Pattern 1	Pattern 2	Pattern 3	Pattern 4	Pattern 5
FW-algorithm	2	2	3	∞	∞
Projection algorithm	2	2	3	7	3

Highly remarkable that simple topology of the network allows us to draw revealing insights. Indeed, solutions of network's flows assignment problem corresponding to pattern 1, 2 and 3 have no zero components, i.e. there are no unused routes. However, solutions of network's flows assignment problem corresponding to pattern 4 and 5 have zero components, i.e. there are unused routes. Really, route 3 in pattern 4, and routes 3 and 5 in pattern 5 are obviously too "expensive" to use. Thus, it is quite clear in advance that $f_i^* = 0$ for $i = 3$ in pattern 4 and $f_i^* = 0$ for $i \in \{3, 5\}$ in pattern 5.

According to Table 2, Frank-Wolfe algorithm demonstrates zig-zagging behavior when there are zero components in equilibrium solution of network's flows assignment problem. Nevertheless, projection algorithm does not pay much attention to such a "trouble" (zero components in equilibrium solution) and demonstrates high convergence rate. Projection algorithm demonstrates such a performance primarily due to the third step of each iteration. Indeed, each iteration clarifies zero components in equilibrium solution and excludes them from consideration (7)-*II*. Eventually, projection algorithm seek solution in the space of non-zero components (red routes on Fig. 1b corresponds to zero components).

$$
\begin{pmatrix} f_1^k \\ f_2^k \\ f_3^k \\ \vdots \\ f_n^k \end{pmatrix} \rightarrow \begin{pmatrix} f_1^{k+1} \\ f_2^{k+1} \\ f_3^{k+1} \\ \vdots \\ f_n^{k+1} \end{pmatrix} \qquad \begin{pmatrix} f_1^k \\ f_2^k \\ \vdots \\ f_{m_k}^k \end{pmatrix} \rightarrow \begin{pmatrix} f_1^{k+1} \\ \vdots \\ f_{m_{k+1}}^{k+1} \end{pmatrix} \tag{7}
$$
$$
\qquad\quad I \qquad\qquad\qquad\qquad II
$$

In turn, Frank-Wolfe algorithm operate in the space of all components (7)-*I* (Fig. 1a). Then it experienced zig-zagging behavior in neighborhood of zero components. The larger the difference between alternative routes, the higher probability to experience zig-zagging behavior.

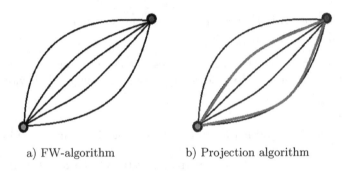

a) FW-algorithm   b) Projection algorithm

Fig. 1. Network of parallel routes

5   Conclusion and Future Work

Obtained results show obvious advantage of projection approach for network's flows assignment problem. The efficiency of the developed projection algorithm, first of all, caused by excluding of zero components during solving. Therefore, it is highly promising to develop projection algorithms. Generally, it is quite complicated task to find projection operator. Nevertheless, as it was shown in [4], for network's flows assignment problem projection operator could be obtained in an explicit form. Certainly, projection operator was obtained for the case of simple network topology. However, there is a good chance to apply methodology, implemented in paper [4], to a network of general topology via parallel decomposition algorithms [6].

Acknowledgement. The first author was jointly supported by a grant from the Russian Science Foundation (Project No. 17-71-10069).

References

1. Dafermos, S.C., Sparrow, F.T.: The traffic assignment problem for a general network. J. Res. Nat. Bur. Stan. **73B**, 91–118 (1969)
2. Dafermos, S.-S.C.: Traffic assignment and resource allocation in transportation networks. PhD thesis. Johns Hopkins University, Baltimore, MD (1968)
3. Holloway, C.A.: An extension of the Frank and Wolfe method of feasible directions. Math. Program. **6**, 14–27 (1973)
4. Krylatov, A.Y.: Network flow assignment as a fixed point problem. J. Appl. Ind. Math. **10**(2), 243–256 (2016)
5. Meyer, G.G.L.: Accelerated Frank-Wolfe algorithms. SIAM J. Control **12**, 655–663 (1974)
6. Patriksson, M.: The Traffic Assignment Problem: Models and Methods. Dover Publications, Inc., Mineola (2015)
7. Popov, I., Krylatov, A., Zakharov, V., Ivanov, D.: Competitive energy consumption under transmission constraints in a multi-supplier power grid system. Int. J. Syst. Sci. **48**(5), 994–1001 (2017)
8. Sheffi, Y.: Urban Transportation Networks: Equilibrium Analysis with Mathematical Programming Methods. Prentice-Hall, Inc., Englewood Cliffs (1985)

An Approximation Algorithm for Preemptive Speed Scaling Scheduling of Parallel Jobs with Migration

Alexander Kononov and Yulia Kovalenko$^{(\boxtimes)}$

Sobolev Institute of Mathematics,
4, Akad. Koptyug Avenue, 630090 Novosibirsk, Russia
alvenko@math.nsc.ru, julia.kovalenko.ya@yandex.ru

Abstract. In this paper we consider a problem of preemptive scheduling rigid parallel jobs on speed scalable processors. Each job is specified by its release date, its deadline, its processing volume and the number of processors, which are required for execution of the job. We propose a new strongly polynomial approximation algorithm for the energy minimization problem with migration of jobs.

Keywords: Parallel jobs · Speed scaling · Scheduling · Migration · Approximation algorithm

1 Introduction

Energy consumption of computing devices is an important issue in our days [9]. A popular technology to reduce energy usage is dynamic speed scaling, where a processor may vary its speed dynamically. Running a job at a slower speed is more energy efficient, however it takes longer time and may affect the performance. One of the algorithmic and complexity study of this area is devoted to revising classical scheduling problems with dynamic speed scaling (see e.g. [1,4,6,7,9,13,14] and others).

In our paper we consider a basic speed scaling scheduling of parallel jobs. Given a set $\mathcal{J} = \{1, \ldots, n\}$ of parallel jobs to be executed on m parallel speed-scalable processors. Each job $j \in \mathcal{J}$ is associated with a release date r_j, a deadline d_j and a processing volume (work) W_j. Moreover, job $j \in \mathcal{J}$ simultaneously requires exactly $size_j$ processors at each time point when it is in process. Such jobs are called rigid jobs [8].

We distinguish two variants of the problem. The first variant (*non-migratory variant*) allows the preemption of the jobs but not their migration. This means that a job may be interrupted and resumed later on the same subset of $size_j$ processors, but it is not allowed to continue its execution on a different subset of $size_j$ processors. In the second variant (*migratory variant*) both the preemption and the migration of jobs are allowed.

The standard homogeneous model in speed-scaling is considered. When a processor runs at a speed s, then the rate with which the energy is consumed

© Springer International Publishing AG 2017
R. Battiti et al. (Eds.): LION 2017, LNCS 10556, pp. 351–357, 2017.
https://doi.org/10.1007/978-3-319-69404-7_30

(the *power*) is s^α, where $\alpha > 1$ is a constant (usually, $\alpha \approx 3$ [9]). Each of m processors may operate at variable speed. However, we assume that if processors execute the same job simultaneously then all these processors run at the same speed. It is supposed that a continuous spectrum of processor speeds is available. The purpose is to find a feasible schedule, minimizing the total energy consumed on all the processors.

The idea of multiprocessor jobs receives growing attention in the scheduling theory [8]. Many computer systems offer some kinds of parallelism. The energy efficient scheduling of parallel jobs arises in testing and reliable computing, parallel applications on graphics cards, computer control systems and others.

2   Related Research

For the preemptive single-processor setting, Yao et al. [14] developed a polynomial time algorithm, that outputs a minimum energy schedule. The preemptive multiprocessor scheduling of single-processor jobs has been widely studied, see e.g. [1,4,6,13]. The authors proposed exact polynomial algorithms for this problem with migration. The works [1,4,13] are based on different reductions of the problem to maximum flow problems. As far as we know, the algorithm presented in [13] has the best running time among the above-mentioned algorithms.

Albers et al. [3] studied the preemptive problem on parallel processors where the migration of jobs among processors is disallowed. They showed that if all jobs have unit work and deadlines are agreeable, an optimal schedule can be computed in polynomial time. At the same time, the general speed scaling scheduling problem with unit-work jobs was proved to be NP-hard, even on two processors. A common rule to design algorithms for problems without migration is to first define some strategy that assigns jobs to processors, and then schedule the assigned jobs separately on each processor. Albers et al. [3] presented using this rule an $(\alpha^\alpha 2^{4\alpha})$-approximation algorithm for instances with agreeable deadlines, and an $2\left(2 - \frac{1}{m}\right)^\alpha$-approximation algorithm for instances with common release dates, or common deadlines. Greiner et al. [10] showed that any ρ-approximation algorithm for parallel processors with migration can be transformed into a $\rho B_{\lceil \alpha \rceil}$-approximation algorithm for parallel processors without migration, where $B_{\lceil \alpha \rceil}$ is the $\lceil \alpha \rceil$-th Bell number. The result holds when $\alpha \leq m$.

Bampis et al. [5] considered the problem on heterogeneous processors with preemption. They assume that each processor i has its own power function, $s^{\alpha(i)}$, and job's characteristics are processor dependent. For the case where job migrations are allowed, an algorithm has been proposed, that returns a solution within an additive error ε in time polynomial in the problem size and in $\frac{1}{\varepsilon}$. They also developed an approximation algorithm of ratio $(1 + \varepsilon)^\alpha \tilde{B}_\alpha$ for the problem without migration, where $\tilde{B}_\alpha$ is the generalized Bell number [5]. Recently, Albers et al. [2] proposed a faster combinatorial algorithm based on flows for preemptive scheduling of jobs whose density is lower bounded by a small constant, and the migration is allowed.

We extend the study of speed scaling scheduling to the case of parallel jobs. Recently, it has been proved that the speed scaling scheduling problem of rigid jobs is NP-hard even both the preemption and the migration of jobs are allowed [12]. An approximation algorithm has been proposed in [12] that returns a solution within an additive error $\varepsilon > 0$ and runs in time polynomial in m, $1/\varepsilon$ and the input size. Note that the algorithm is pseudopolynomial and it is based on solving a configuration linear program using the Ellipsoid method which is rather complicated and not efficient in practice. In the current paper we propose strongly polynomial $\left(2 - \frac{1}{m}\right)^{\alpha-1}$-approximation algorithm for scheduling of rigid jobs when the preemption and the job migrations are allowed.

3   Our Result

Here we consider the speed scaling scheduling problem of rigid jobs with migration and present $\left(2 - \frac{1}{m}\right)^{\alpha-1}$-approximation algorithm for this problem.

Our algorithm consists of two stages. At the first stage we solve an auxiliary min-cost max-flow problem in order to obtain a lower bound on the minimal energy consumption and an assignment of the jobs to time intervals. At this stage we follow the approach proposed in [13]. Then, at the second stage, we determine speeds of jobs and schedule them separately for each time interval.

The first stage. Due to the convexity of the speed-to-power function, the energy consumption is minimized if each job j is processed with a fixed speed s_j, which does not change during the processing of the job. Therefore, we can formulate the problem with the variables $p_j = W_j/s_j$, where p_j is treated as an actual processing time of job $j \in \mathcal{J}$. The objective function is written as follows:

$$F = \sum_{j=1}^{n} p_j size_j \left(\frac{W_j}{p_j}\right)^{\alpha}.$$

Let us divide the interval $[\min_j r_j, \max_j d_j]$ into subintervals $I_k = [t_{k-1}, t_k]$ by using the release dates r_j and the deadlines d_j for $j \in \mathcal{J}$ as break-points. Denote the length of interval I_k by Δ_k, $k = 1, \ldots, \gamma$, where $\gamma \leqslant 2n-1$. For a job j, denote the set of the available intervals by $\Gamma(j)$, where $\Gamma(j) = \{I_k : I_k \subseteq [r_j, d_j]\}$.

Let us construct a bipartite network $G = (V, A)$. The set of nodes is given by $V = \{s, t\} \cup \mathcal{J} \cup \mathcal{I}$, where $\mathcal{J}$ is the set of job nodes and $\mathcal{I} = \{I_1, \ldots, I_\gamma\}$ is the set of interval nodes. The set of arcs A is given as $A = A^s \cup A^0 \cup A^t$, where $A^s = \{(s, j) : j \in \mathcal{J}\}$, $A^0 = \{(j, I_k) : j \in \mathcal{J}, I_k \in \Gamma(j)\}$, $A^t = \{(I_k, t) : I_k \in \mathcal{I}\}$, so that the source s is connected to each job node, each interval node is connected to the sink t, and each job node j is connected to the nodes I_k associated with the available intervals $\Gamma(j)$. We define the arc capacities as follows:

$$\mu(s, j) = +\infty, \ (s, j) \in A^s,$$

$$\mu(j, I_k) = size_j \Delta_k, \ (j, I_k) \in A^0, \tag{1}$$

$$\mu(I_k, t) = m\Delta_k, \ (I_k, t) \in A^t. \tag{2}$$

We denote by $x(u, v)$ the amount of flow on an arc (u, v). Note that $p_j = \frac{x(s,j)}{size_j}$ defines a total duration of job j and $p_{j,I_k} = \frac{x(j,I_k)}{size_j}$ specifies a processing time of job j in the interval I_k. Let the cost of flow $x(s, j)$ is $x(s, j)\left(\frac{W_j size_j}{x(s,j)}\right)^\alpha$, which is a convex function with respect to $x(s, j)$. The cost of flow on all other arcs is set to be zero. Then the considered problem reduces to finding a maximum $s - t$ flow in $G = (V, A)$, that minimizes the total cost

$$\sum_{j=1}^{n} x(s, j)\left(\frac{W_j size_j}{x(s,j)}\right)^\alpha.$$

As shown in [13] the obtained min-cost max-flow problem can be solved in $O(n^3)$ time. It is not difficult to prove that any feasible schedule specifies a feasible flow with the same cost in the network G. The converse is not true, as we shall see in Example 1, below. At the second stage we use the "preemptive $size_j$-list-scheduling" algorithm [11] to construct a feasible schedule in each interval $I \in \mathcal{I}$.

The second stage. Let I be an arbitrary time interval and its length is equal to Δ. We denote by $\mathcal{J}'$ the subset of jobs, which are assigned to time interval I at the first stage, i.e. $\mathcal{J}' = \{j \in \mathcal{J} : x(j, I) > 0\}$. The capacity constraints (1) and (2) imply that $p_{j,I} = x(j, I)/size_j \leqslant \Delta$ and $\sum_{j \in \mathcal{J}'} p_{j,I} size_j \leqslant m\Delta$.

The "preemptive $size_j$-list-scheduling" algorithm for an interval I works as follows. At every decision point (i.e., the start time of the interval or completion time of a job) all currently running jobs are interrupted. Then not yet completed jobs are considered in order of nonincreasing $size_j$-values, and as many of them are greedily assigned to the processors as feasibly possible. The time complexity of the algorithm is $O(n^2)$.

We claim that the length of the constructed schedule is at most $\left(2 - \frac{1}{m}\right)\Delta$ (see Lemma 1 below). By increasing the speed of each job in $\left(2 - \frac{1}{m}\right)$ times we obtain a schedule of the length at most Δ. The total energy consumption for interval I is increased by a factor $\left(2 - \frac{1}{m}\right)^{\alpha-1}$. The final schedule is constructed by combining the schedules found for each individual interval, so we get $\left(2 - \frac{1}{m}\right)^{\alpha-1}$-approximate solution of the original speed scaling problem with rigid jobs. We note that the number of intervals does not exceed $2n - 1$ and, hence, the running time of the second stage is $O(n^3)$. As a result, we have

Theorem 1. *A $\left(2 - \frac{1}{m}\right)^{\alpha-1}$-approximate schedule can be found in $O(n^3)$ time for the preemptive speed scaling problem of rigid jobs with migration.*

Now we prove Lemma 1.

Lemma 1. *Given m processors, an interval I of duration Δ, and a set of jobs $\mathcal{J}'$ with processing times $p_{j,I} \leqslant \Delta$ and sizes $size_j$, where $\sum_{j \in \mathcal{J}'} p_{j,I} size_j \leqslant m\Delta$. The length of the schedule constructed by the "preemptive $size_j$-list-scheduling" algorithm is at most $\left(2 - \frac{1}{m}\right)\Delta$.*

Proof. Let l be the last job in the preemptive list-schedule (if there are several such jobs, we choose a job with the smallest value $size_j$), and let C_l be its completion time (the length of the schedule). We consider two cases: (I) $size_l > \frac{m}{2}$ and (II) $size_l \leqslant \frac{m}{2}$.

Case (I): $size_l > \frac{m}{2}$. According to the "preemptive $size_j$-list-scheduling" algorithm we obtain that exactly one job is executed at each time moment and $size_j \geqslant size_l \geqslant \frac{m+1}{2} = \frac{(m+0.5)(m-0.5)}{2(m-0.5)} + \frac{1}{4} \geqslant \frac{(m+0.5)(m-0.5)+0.25}{2(m-0.5)} = \frac{m^2}{2m-1}$ for all $j \in \mathcal{J}'$. It follows that $\sum_{j \in \mathcal{J}'} p_{j,I} size_j \geqslant \frac{m}{2-1/m} \sum_{j \in \mathcal{J}'} p_{j,I}$. From (2) we get $\frac{m}{2-1/m} \sum_{j \in \mathcal{J}'} p_{j,I} \leqslant m\Delta$ and $\sum_{j \in \mathcal{J}'} p_{j,I} \leqslant \Delta\left(2 - \frac{1}{m}\right)$.

Case (II): $size_l \leqslant \frac{m}{2}$. We claim that, at every point in time during the schedule, either job l is undergoing processing on some $size_l$ processors or job l is not executed and at least $(m - size_l + 1)$ processors are busy. Therefore, the total load of all processors $\sum_{j \in \mathcal{J}'} p_{j,I} size_j$ is at least $p_{l,I} size_l + (C_l - p_{l,I})(m - size_l + 1)$. Suppose that $C_l > \left(2 - \frac{1}{m}\right)\Delta$, then we get the inequality

$$
\sum_{j \in \mathcal{J}'} p_{j,I} size_j > p_{l,I} size_l + \left(\left(2 - \frac{1}{m}\right)\Delta - p_{l,I}\right)(m - size_l + 1)
$$

$$
= \Delta m + (\Delta - p_{l,I})(m - 2 size_l + 1) + \frac{\Delta}{m}(size_l - 1) \geqslant \Delta m,
$$

which leads to a contradiction. □

As shown in [11], the approximation ratio of $\left(2 - \frac{1}{m}\right)$ for the "preemptive $size_j$-list-scheduling" algorithm is tight even if $size_j = 1$ for all jobs. As a result, the energy consumption is increased in $\left(2 - \frac{1}{m}\right)^{\alpha-1}$ times when we put the resulting schedule inside the interval I. Now we show that the approximation ratio of our algorithm can not be improved even if we will use an exact algorithm for minimization of makespan at the second stage.

Example 1. Given m big jobs $j = 1, \ldots, m$ of work $W_j = m - 1$ and size $size_j = m$ and m small jobs $j = m + 1, \ldots, 2m$ with work $W_j = m$ and size $size_j = 1$. The release dates of all jobs are equal to 0. All small jobs have the common deadline m^2 and the deadline of big job j is $d_j = mj$, $j = 1, \ldots, m$.

Optimal solution has energy consumption m^3 and length m^2, by scheduling the big jobs from time 0 to time $m(m - 1)$ and using the last interval $[m(m - 1), m^2)$ for the small jobs. The speed of each processor is equal to 1.

Consider the following optimal solution of the min-cost max-flow problem. Let exactly one big job j and one small job $m + j$ be assigned to interval $[m(j - 1); mj)$, $j = 1, \ldots, m$. Though the total load of each pair jobs j and $m + j$ is m^2, it is required $2m - 1$ time units to execute these jobs with speed 1. Thus, after increasing the speed of each processor at each time point in $\left(2 - \frac{1}{m}\right)$ times, the total energy consumption is increased by a factor $\left(2 - \frac{1}{m}\right)^{\alpha-1}$.

4   Conclusion

We study the energy minimization problem of scheduling rigid jobs on m speed scalable processors. For migratory case of the problem we propose a strongly polynomial time approximation algorithm based on a reduction to the min-cost max-flow problem. The algorithm has approximation ratio $\left(2 - \frac{1}{m}\right)^{\alpha-1}$ and this bound is tight. Our result can be generalized to the case of job-dependent energy consumption when each job j has its own constant $\alpha_j > 1$. For this case our algorithm obtains the $\left(2 - \frac{1}{m}\right)^{\alpha-1}$-approximate solution where $\alpha = \max_{j \in \mathcal{J}} \alpha_j$.

Acknowledgements. This research is supported by the Russian Science Foundation grant 15-11-10009.

References

1. Albers, S., Antoniadis, A., Greiner, G.: On multi-processor speed scaling with migration. J. Comput. Syst. Sci. **81**, 1194–1209 (2015)
2. Albers, S., Bampis, E., Letsios, D., Lucarelli, G., Stotz, R.: Scheduling on power-heterogeneous processors. In: Kranakis, E., Navarro, G., Chávez, E. (eds.) LATIN 2016. LNCS, vol. 9644, pp. 41–54. Springer, Heidelberg (2016). doi:10.1007/978-3-662-49529-2_4
3. Albers, S., Müller, F., Schmelzer, S.: Speed scaling on parallel processors. In: 19th ACM Symposium on Parallelism in Algorithms and Architectures, SPAA 2007, pp. 289–298. ACM (2007)
4. Angel, E., Bampis, E., Kacem, F., Letsios, D.: Speed scaling on parallel processors with migration. In: Kaklamanis, C., Papatheodorou, T., Spirakis, P.G. (eds.) Euro-Par 2012. LNCS, vol. 7484, pp. 128–140. Springer, Heidelberg (2012). doi:10.1007/978-3-642-32820-6_15
5. Bampis, E., Kononov, A., Letsios, D., Lucarelli, G., Sviridenko, M.: Energy efficient scheduling and routing via randomized rounding. In: FSTTCS, pp. 449–460 (2013)
6. Bingham, B.D., Greenstreet, M.R.: Energy optimal scheduling on multiprocessors with migration. In: International Symposium on Parallel and Distributed Processing with Applications, ISPA 2008, pp. 153–161. IEEE, (2008)
7. Cohen-Addad, V., Li, Z., Mathieu, C., Milis, I.: Energy-efficient algorithms for non-preemptive speed-scaling. In: Bampis, E., Svensson, O. (eds.) WAOA 2014. LNCS, vol. 8952, pp. 107–118. Springer, Cham (2015). doi:10.1007/978-3-319-18263-6_10
8. Drozdowski, M.: Scheduling for Parallel Processing. Springer-Verlag, London (2009)
9. Gerards, M.E.T., Hurink, J.L., Hölzenspies, P.K.F.: A survey of offline algorithms for energy minimization under deadline constraints. Journ. Sched. **19**, 3–19 (2016)
10. Greiner, G., Nonner, T., Souza, A.: The bell is ringing in speed-scaled multiprocessor scheduling. In: 21st ACM Symposium on Parallelism in Algorithms and Architectures, SPAA 2009, pp. 11–18. ACM, (2009)
11. Johannes, B.: Scheduling parallel jobs to minimize the makespan. J. Sched. **9**, 433–452 (2006)
12. Kononov, A., Kovalenko, Y.: On speed scaling scheduling of parallel jobs with preemption. In: Kochetov, Y., Khachay, M., Beresnev, V., Nurminski, E., Pardalos, P. (eds.) DOOR 2016. LNCS, vol. 9869, pp. 309–321. Springer, Cham (2016). doi:10.1007/978-3-319-44914-2_25

13. Shioura, A., Shakhlevich, N., Strusevich, V.: Energy saving computational models with speed scaling via submodular optimization. In: Proceedings of Third International Conference on Green Computing, Technology and Innovation (ICGCTI2015), pp. 7–18 (2015)
14. Yao, F., Demers, A., Shenker, S.: A scheduling model for reduced CPU energy. In: 36th Annual Symposium on Foundation of Computer Science, FOCS 1995, pp. 374–382 (1995)

Learning and Intelligent Optimization for Material Design Innovation

Amir Mosavi[1,2(✉)] and Timon Rabczuk[1(✉)]

[1] Institute of Structural Mechanics, Bauhaus-Universitat Weimar,
Marienstr.15, 99423 Weimar, Germany
{amir.mosavi,timon.rabczuk}@uni-weimar.de
[2] Department of Computer and Information Science,
Norwegian University of Science and Technology,
Sem Saelandsvei 9, 7491 Trondheim, Norway

Abstract. Learning and intelligent optimization (LION) techniques enable problem-specific solvers with vast potential applications in industry and business. This paper explores such potentials for material design innovation and presents a review of the state of the art and a proposal of a method to use LION in this context. The research on material design innovation is crucial for the long-lasting success of any technological sector and industry and it is a rapidly evolving field of challenges and opportunities aiming at development and application of multi-scale methods to simulate, predict and select innovative materials with high accuracy. The LION way is proposed as an adaptive solver toolbox for the virtual optimal design and simulation of innovative materials to model the fundamental properties and behavior of a wide range of multi-scale materials design problems.

Keywords: Machine learning · Optimization · Material design

1 Introduction

Materials design is crucial for the long-lasting success of any technological sector, and yet every technology is founded upon a particular materials design set. This is why the pressure on development of new high-performance materials for use as high-tech structural and functional components has become greater than ever. Although the demand for materials is endlessly growing, experimental materials design is attached to high costs and time-consuming procedures of synthesis. Consequently simulation technologies have become completely essential for material design innovation [1]. Naturally the research community highly supports the advancement of simulation technologies as it represents a massive platform for further development of scientific methods and techniques. Yet computational material design innovation is a new paradigm in which the usual route of materials selection is enhanced by concurrent materials design simulations and computational applications [19].

Designing new materials is a multi-dimensional problem where multiple criteria of design need to be satisfied. Consequently material design innovation would require advanced multiobjective optimization (MOO) [13] and decision-support tools [12]. In addition the performance and behavior of new materials must be predicted in

© Springer International Publishing AG 2017
R. Battiti et al. (Eds.): LION 2017, LNCS 10556, pp. 358–363, 2017.
https://doi.org/10.1007/978-3-319-69404-7_31

different design scenarios and conditions [2]. In fact predictive analytics and MOO algorithms are the essential computation tools to tailor the atomic-scale structures, chemical compositions and microstructures of materials for desired mechanical properties such as high-strength, high-toughness, high thermal and ionic conductivity, high irradiation and corrosion resistance [7]. Via manipulating the atomic-scale dislocation, phase transformation, diffusion, and soft vibrational modes the material behavior on plasticity, fracture, thermal, and mass transport at the macroscopic level can be predicted and optimized accurately [17]. Therefore the framework of a predictive simulation-based optimization of advanced materials, which yet to be realized, represents a central challenge within material simulation technology [9]. Consequently material design innovation is facing the ever-growing need to provide a computational toolbox that allows the development of tailor-made molecules and materials through the optimization of materials behavior [10]. The goal of such toolbox is to provide insight over the property of materials associated with their design, synthesis, processing, characterization, and utilization [19].

2   Computational Materials Design Innovation

Computational materials design innovation aims at development and application of multiscale methods to simulate advanced materials with high accuracy [17]. A key to meet the ever-ongoing demand on increasing performance, quality, specialization, and price reduction of materials is the availability of simulation tools which are accurate enough to predict and optimize novel materials on a low computation cost [6]. A major challenge however would be the hierarchical nature inherent to all materials. Accordingly to understand a material property on a given length and time scale it is crucial to optimize and predict the mechanisms on shorter length and time scales all the way down to the most fundamental mechanisms describing the chemical bond. Consequently the materials systems are to be simultaneously studied under consideration of underlying nano-structures and Mesomanufacturing Scales. Such design process is highly nonlinear and requires an interactive MOO toolset [12].

2.1   Interdisciplinary Research and Research Gap

Structure calculations of materials [20], systematic storage of the information in database repositories [8], materials characterization and selection [18], and gaining new physical and environmental insights [9] account for big data technologies. In addition making decision for the optimal materials design needs MOO tools as well as an efficient decision-support system for post-processing [21]. This is considered as a design optimization process of the microstructure of materials with respect to desired properties and Mesoscale functionalities. Such process requires a smart agent which learns from dataset and makes optimal decisions. The solution of this inverse problem with the support of the virtual test laboratories and knowledge-based design would be the foundation of tailor-made molecules and materials toolbox. With such an integrated toolbox at hand the virtual testing concept and application is realized. This challenging task can only be accomplished through a variety of scale bridging methods which requires machine learning and optimization combined [4]. Furthermore a great deal of understanding on big data and prediction technologies for microstructure behavior of existing materials, as well as the ability to test the behavior of new materials at the atomic, microscopic and mesoscale is desired to confidently modifying the materials properties [7]. Numerical analysis further allows efficient experiments with entirely new materials and molecules [20]. Basic machine learning technologies such as artificial neural networks [21], and genetic algorithms [9], Bayesian probabilities and machine learning [8], data mining of spectral decompositions [7], refinement and optimization by cluster expansion [20], structure map analysis and neural networks [1], and support vector machines [19], have been recently used for this purpose.

Computational materials design innovation to perfect needs to dramatically improve and put crucial components in place. To be precise, data mining, efficient codes, Big data technologies, advanced machine learning techniques, intelligent and interactive MOO, open and distributed networks of repositories, fast and effective descriptors, and strategies to transfer knowledge to practical implementations are the research gaps to be addressed [6]. In fact the current solvers rely only on a single algorithm and address limited scales of the design problems [17]. In addition there is a lack of reliable visualization tools to better involve engineers into the design loop [11]. The absence of robust design, lack of the post-processing tools for multicriteria decision-making, lack of Big data tools for an effective consideration of huge materials database are further research gaps reported in literature [8]. To conclude, the process of computational material design innovation requires a set of up-to-date solvers to cover a wide range of problems. Further problem with the current open-source software toolboxes, reported in [6], is that they require a concrete specification on the mathematical model, and also the modeling solution is not flexible and adaptive. This has been a reason why the traditional computation tools for materials design have not been realistic and as effective. Consequently the vision of this work is to propose an interactive toolbox, where the solver determines the optimal choices via visualization tools as demonstrated in [5]. Ultimately the purpose is to construct a knowledge-based virtual test laboratory to simultaneously optimize the hybrid materials microstructure systems, e.g. textile composites. Whether building atomistic, continuum mechanics or multiscale models, the toolbox can provide a platform to rearrange the appropriate solver according to the problem at hand. Such platform contributes in advancement of innovative materials database leading to innovative materials design with the optimal functionality.

3   LION as a Solver

The complex body of information of computational materials design requires the most recent advancements in machine learning and MOO to scale to the complex and multiobjective nature of the optimal materials design problems [10]. From this perspective the materials design can be seen as a high potential research area and a continuous source of challenging problems for LION. In the LION way [3] every individual design task, according to the problem at hand, can be modeled on the basis of the solvers within the toolbox. To obtain a design model the methodology does not ask to specify a model, but it experiments with the current system. The appropriate model is created in the toolbox and further is used to identify a better solution in a learning cycle. The methodology is based on transferring data to knowledge to optimal decisions through LION way i.e. a workflow that is referred to as prescriptive analytics [4]. In addition an efficient Big data application [18] can be integrated to build models and extract knowledge. Consequently a large database containing the properties of the existing and hypothetical materials is interrogated in the search of materials with the desired properties. Knowledge exploits to automate the discovery of improving solutions i.e. connecting insight to decisions and actions [17]. As the result a massively parallelized multiscale materials modeling tools that expand atomistic-simulation-based predictive capability is established which leads to rational design of a variety of innovative materials and applications.

A variety of solvers integrated within the LION include several algorithms for data mining, machine learning, and predictive analytics which are tuned by cross-validation. These solvers provide the ability of learning from data, and are empowered by reactive search optimization (RSO) [4] i.e. the intelligent optimization tool that is integrated into the solver. The LION way fosters research and development for intelligent optimization and Reactive Search. Reactive Search stands for the integration of sub-symbolic machine learning techniques into local search heuristics for solving complex optimization problems via an internal online feedback loop for the self-tuning of critical parameters [3, 12]. In fact RSO is the effective building block for solving complex discrete and continuous optimization problems which can cure local minima traps. Further, cooperating RSO coordinates a collection of interacting solvers which is adapted in an online manner to the characteristics of the problem. LIONsolver [4], LIONoso (a non-profit version of LIONsolver), and Grapheur [5], are the software implementations of the LION way which can be customized for different usage contexts in materials design. These implementations have been used for solving a number of real-life problems including materials selection [18], engineering design [14, 15], computational mechanics [13], and Robotics [16].

4   Textile Composites Optimal Design

To evaluate the effectiveness of the LION way the case study of textile composites design with MOO, presented by Milani (2011), is reconsidered using *Grapheur*. This case study describes a novel application of LION way dealing with decision conflicts often seen among design criteria in composites materials design [18]. In this case study it is necessary to explore optimal design options by simultaneously analyzing materials properties in a multitude of disciplines, design objectives, and scales. The complexity increases with considering the fact that the design objective functions are not mathematically available and designer must be in the loop of optimization to evaluate the Mesomanufacturing Scales of the draping behavior of textile composites. The case study has a relatively large-scale decision space of electrical, mechanical, weight, cost, and environmental attributes.

To solve the problem an interactive MOO model is created with *Grapheur*. With the aid of the 7D visualization graph the designer in the loop formulates and systematically compares different alternatives against the large sets of design criteria to tackle complex decision-making task of exploring trade-offs and also designing break-even points. With the designer in the loop, interactive schemes are developed where *Grapheur* provides a versatile tool for stochastic local search optimization. Once the optimal candidates over the five design objectives preselected, screening the Mesomanufacturing Scales of draping figures takes place to identify the most suitable candidate. In this case study the interactive MOO toolset of *Grapheur* provides a strong user interface for visualizing the results, facilitating the solution analysis, and post-processing (Fig. 1).

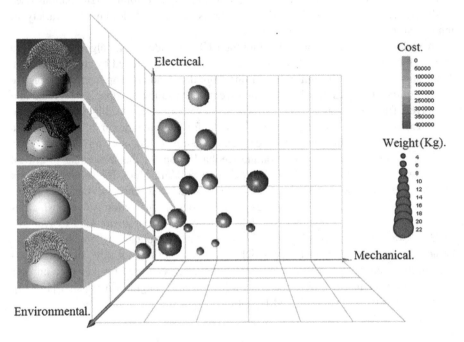

Fig. 1. 7D visualization graph for MOO and post-processing: Interactive MOO toolset of *Grapheur* on exploring trade-offs and simultaneous screening the Mesomanufacturing Scales: the multi-disciplinary property values of candidate materials are supplied from [12].

5   Conclusions

Computational material design innovation as an emerging area of materials science requires an adaptive solver to rule a wide range of materials design problems. The LION way provides a suitable platform for developing a computational toolbox for the virtual optimal design and simulation-based optimization of advanced materials to model, simulate, and predict the fundamental properties and behavior of multiscale materials. The proposed solver is a simple yet powerful concept presenting an integration of advanced machine learning and intelligent optimization techniques. With a strong interdisciplinary background the novel application of LION way connects computer science and engineering, and further strengthens the research direction of digital engineering.

References

1. Artrith, N.H., Alexander, U.: An implementation of artificial neural-network potentials for atomistic materials simulations. Comput. Mater. Sci. **114**, 135–150 (2016)
2. Bayer, F.A.: Robust economic Model Predictive Control using stochastic information. Automatica **74**, 151–161 (2016)
3. Battiti, R., Brunato, M.: The LION Way: Machine Learning plus Intelligent Optimization. Lionlab, University of Trento, Italy (2015)
4. Brunato, M., Battiti, R.: Learning and intelligent optimization: one ring to rule them all. Proc. VLDB Endow. **6**, 1176–1177 (2013)
5. Brunato, M., Battiti, R.: Grapheur: a software architecture for reactive and interactive optimization. In: Blum, C., Battiti, R. (eds.) LION 2010. LNCS, vol. 6073, pp. 232–246. Springer, Heidelberg (2010). doi:10.1007/978-3-642-13800-3_26
6. Ceder, G.: Opportunities and challenges for first-principles materials design and applications to Li battery materials. Mater. Res. Soc. Bull. **35**, 693–701 (2010)
7. Fischer, C.: Predicting crystal structure by merging data mining with quantum mechanics. Nat. Mater. **5**, 641–646 (2006)
8. Jain, A.: A high-throughput infrastructure for density functional theory calculations. Comput. Mater. Sci. **50**, 2295–2310 (2011)
9. Johannesson, G.H.: Combined electronic structure and evolutionary search approach to materials design. Phys. Rev. Lett. **88**, 255–268 (2002)
10. Lencer, D.: A map for phase-change materials. Nat. Mater. **7**, 972–977 (2008)
11. Mosavi, A.: Decision-making software architecture; the visualization and data mining assisted approach. Int. J. Inf. Comput. Sci **3**, 12–26 (2014)
12. Milani, A.: Multiple criteria decision making with life cycle assessment for material selection of composites. Express Polym. Lett. **5**, 1062–1074 (2011)
13. Mosavi, A., Vaezipour, A.: Reactive search optimization; application to multiobjective optimization problems. Appl. Math. **3**, 1572–1582 (2012)
14. Mosavi, A.: A multicriteria decision making environment for engineering design and production decision-making. Int. J. Comput. Appl. **69**, 26–38 (2013)
15. Mosavi, A.: Decision-making in complicated geometrical problems. Int. Comput. Appl. **87**, 22–25 (2014)
16. Mosavi, A., Varkonyi, A.: Learning in Robotics. Int. J. Comput. Appl. **157**, 8–11 (2017)
17. Mosavi, A., Rabczuk, T., Varkonyi-Koczy, A.R.: Reviewing the novel machine learning tools for materials design. In: Luca, D., Sirghi, L., Costin, C. (eds.) INTER-ACADEMIA 2017: Recent Advances in Technology Research and Education. Advances in Intelligent Systems and Computing, vol. 660, pp. 50–58. Springer, Cham (2018). doi:10.1007/978-3-319-67459-9_7
18. Mosavi, A., et al.: Multiple criteria decision making integrated with mechanical modeling of draping for material selection of textile composites. In Proceedings of 15th European Conference on Composite Materials, Venice, Italy (2012)
19. Saito, T.: Computational Materials Design, vol. 34. Springer Science & Business Media, Heidelberg (2013)
20. Stucke, D.P., Crespi, V.H.: Predictions of new crystalline states for assemblies of nanoparticles. Nano Lett. **3**, 1183–1186 (2003)
21. Sumpter, B.G., Noid, D.W.: On the design, analysis, and characterization of materials using computational neural networks. Annu. Rev. Mater. Sci. **26**, 223–277 (1996)

Statistical Estimation in Global Random Search Algorithms in Case of Large Dimensions

Andrey Pepelyshev[1,2(✉)], Vladimir Kornikov[2], and Anatoly Zhigljavsky[1,3]

[1] School of Mathematics, Cardiff University, Cardiff CF24 4AG, UK
{pepelyshevan,zhigljavskyaa}@cardiff.ac.uk
[2] Faculty of Applied Mathematics, St.Petersburg State University,
Saint Petersburg, Russia
vkornikov@mail.ru
[3] Lobachevsky Nizhny Novgorod State University,
23 Prospekt Gagarina, 603950 Nizhny Novgorod, Russia

Abstract. We study asymptotic properties of optimal statistical estimators in global random search algorithms when the dimension of the feasible domain is large. The results obtained can be helpful in deciding what sample size is required for achieving a given accuracy of estimation.

Keywords: Global optimization · Extreme value · Random search · Estimation of end-point

1 Introduction

We consider the problem of global minimization $f(x) \to \min_{x \in \mathbf{X}}$, where $f(\cdot)$ is the objective function and $\mathbf{X} \subset \mathbb{R}^d$ is a feasible domain. The set $\mathbf{X}$ is a compact set with non-empty interior and the objective function $f(\cdot)$ is assumed to satisfy some smoothness conditions which will be discussed below. Let $f_* = \min_{x \in \mathbf{X}} f(x)$ be the minimal value of $f(\cdot)$ and x_* be a global minimizer; that is, x_* is any point in $\mathbf{X}$ such that $f(x_*) = f_*$.

If the objective function is given as a 'black box' computer code and there is no information about this function available of Lipschitz type, then good stochastic approaches often perform better than deterministic algorithms, especially in large dimensions; see for example [3,4]. Moreover, stochastic algorithms are usually simpler than deterministic algorithms.

A general Global Random Search (GRS) algorithm constructs a sequence of random points $x_1, x_2, \ldots$ such that the point x_j has some probability distribution P_j, $j = 1, 2, \ldots$; we write this as $x_j \sim P_j$. For each $j \geqslant 2$, the distribution P_j may depend on the previous points $x_1, \ldots, x_{j-1}$ and on $f(x_1), \ldots, f(x_{j-1})$.

In the present paper, we will mostly concentrate on the so-called Pure Random Search (PRS) algorithm, where the points $x_1, x_2, \ldots$ are independent and have the same distribution $P = P_j$ for all j. Simplicity of PRS enables detailed examination of this algorithm.

© Springer International Publishing AG 2017
R. Battiti et al. (Eds.): LION 2017, LNCS 10556, pp. 364–369, 2017.
https://doi.org/10.1007/978-3-319-69404-7_32

2   Statistical Inference About f_* in Pure Random Search

Consider a PRS with $x_j \sim P$. Statistical inference about f_* can serve for the following purposes: (i) devising specific GRS algorithms like the branch and probability bounds methods, see [2, 6] and [4, Sect. 4.3], (ii) constructing stopping rules, see [5], and (iii) increasing efficiency of population-based GRS methods, see discussion in [3, Sect. 2.6.1]. Moreover, the use of statistical inferences in GRS algorithms can be very helpful in solving multi-objective optimization problems with non-convex objectives, see [6].

Since the points x_j in PRS are independent identically distributed (i.i.d.) with distribution P, the elements of the sample $Y = \{y_1, \ldots, y_n\}$ with $y_j = f(x_j)$ are i.i.d. with cumulative distribution function (c.d.f.) $F(t) = \Pr\{x \in \mathbf{X} : f(x) \leqslant t\} = \int_{f(x) \leqslant t} P(dx) = P(W(t - f_*))$, where $t \geqslant f_*$ and $W(\delta) = \{x \in \mathbf{X} : f(x) \leqslant f_* + \delta\}$, $\delta \geqslant 0$. Since the analytic form of $F(t)$ is either unknown or intractable (unless f is very simple), for making statistical inferences about f_* we need to use the asymptotic approach based the record values of the sample Y. It is known that (i) the asymptotic distribution of the order statistics is unambiguous, (ii) the conditions on $F(t)$ and $f(\cdot)$ when this asymptotic law works are very mild and typically hold in real-life problems, (iii) for a broad class of functions $f(\cdot)$ and distributions P, the c.d.f. $F(t)$ has the representation

$$F(t) = c_0(t - f_*)^\alpha + o((t - f_*)^\alpha), \ t \downarrow f_*, \tag{1}$$

where c_0 and α are some positive constants. The value of c_0 is not important but the value of α is essential. The coefficient α is called 'tail index' and its value is usually known, as discussed below.

Let η be a random variable which has c.d.f. $F(t)$ and $y_{1,n} \leqslant \ldots \leqslant y_{n,n}$ be the order statistics for the sample Y. By construction, f_* is the lower endpoint of the random variable η.

One of the most important result in the theory of extreme order statistics states (see e.g. [3, Sect. 2.3]) that if (1) holds then the c.d.f. $F(t)$ belongs to the domain of attraction of the Weibull distribution with density $\psi_\alpha(t) = \alpha t^{\alpha-1} \exp\{-t^\alpha\}$, $t > 0$. This distribution has only one parameter, the tail index α.

In PRS we can usually have enough knowledge about $f(\cdot)$ to get the exact value of the tail index α. Particularly, the following statement holds: if the global minimizer x_* of $f(\cdot)$ is unique and $f(\cdot)$ is locally quadratic around x_* then the representation (1) holds with $\alpha = d/2$. However, if the global minimizer x_* of $f(\cdot)$ is unique and $f(\cdot)$ is not locally quadratic around x_* then the representation (1) may hold with $\alpha = d$. See [4] for a comprehensive description of the related theory.

The result that α has the same order as d when d is large implies the phenomena called 'the curse of dimensionality'. Let us first illustrate this curse of dimensionality on a simple numerical example.

3   Numerical Examples

We investigate the minimization problem with the objective function $f(x) = e_1^T x$, where $e_1 = (1, 0, \ldots, 0)^T$, and the set $\mathbf{X}$ is the unit ball: $\mathbf{X} = \{x \in \mathbb{R}^d : ||x|| \leq 1\}$. The minimal value is $f_* = -1$ and the global minimizer $z_* = (-1, 0, \ldots, 0)^T$ is located at the boundary of $\mathbf{X}$. Consider the PRS algorithm with points x_j generated from the uniform distribution P_U on $\mathbf{X}$.

Let us give some numerical values. In a simulation with $n = 10^3$ and $d = 20$, we have received $y_{1,n} = -0.6435$, $y_{2,n} = -0.6107$, $y_{3,n} = -0.6048$ and $y_{4,n} = -0.6021$. In a simulation with $n = 10^5$ and $d = 20$, we have obtained $y_{1,n} = -0.7437$, $y_{2,n} = -0.7389$, $y_{3,n} = -0.7323$ and $y_{4,n} = -0.726$. In Fig. 1 we depict the differences $y_{k,n} - f_*$ for $k = 1, 4, 10$ and $n = 10^3, \ldots, 10^{13}$, where the horizontal axis has logarithmic scale. We can see that the difference $y_{k,n} - y_{1,n}$ is much smaller than the difference $y_{1,n} - f_*$; that demonstrates that the problem of estimating the minimal value of f_* is very hard.

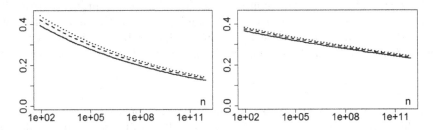

Fig. 1. Differences $y_{1,n} - f_*$ (solid), $y_{4,n} - f_*$ (dashed) and $y_{10,n} - f_*$ (dotted), where $y_{k,n}$, $k = 1, 4, 10$, are records of evaluations of the function $f(x) = e_1^T x$ at points $x_1, \ldots, x_n$ with uniform distribution in the unit hyperball in the dimension $d = 20$ (left) and $d = 50$ (right).

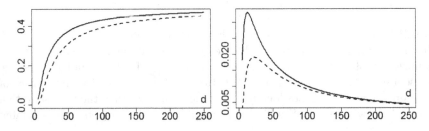

Fig. 2. The difference $y_{1,n} - f_*$ (left) and $y_{10,n} - y_{1,n}$ (right) for $n = 10^6$ (solid) and $n = 10^{10}$ (dashed), where $y_{j,n}$ is the j-th record of evaluations of the function $f(x) = e_1^T x$ at points $x_1, \ldots, x_n$ with uniform distribution in the unit hyperball in the dimension d; d varies in $[5, 250]$.

In Fig. 2 we observe that the difference $y_{1,n} - f_*$ increases as the dimension d grows, for fixed n. Thus, the minimization problem becomes more difficult in larger dimensions. Also, Fig. 2 shows that difference $y_{10,n} - y_{1,n}$ is much smaller than the difference $y_{1,n} - f_*$.

Consider now the optimal linear estimator based on the use of k order statistics; this estimator, as shown in [1,4], has the form

$$\hat{f}_{n,k} = \frac{1}{C_{k,\alpha}} \sum_{i=1}^{k} \frac{u_i}{\Gamma(i + 2/\alpha)} y_{i,n}, \tag{2}$$

where $\Gamma(\cdot)$ is the Gamma-function,

$$u_i = \begin{cases} \alpha + 1, & i = 1, \\ (\alpha - 1)\Gamma(i), & i = 2, \ldots, k - 1, \\ (\alpha - \alpha k - 1)\Gamma(k), & i = k, \end{cases}$$

$$C_{k,\alpha} = \begin{cases} \sum_{i=1}^{k} 1/i, & \alpha = 2, \\ \frac{1}{\alpha-2} (\alpha\Gamma(k+1)/\Gamma(k+2/\alpha) - 2/\Gamma(1+2/\alpha)), & \alpha \neq 2. \end{cases}$$

If the representation (1) holds, then for given k and α and as $n \to \infty$, the estimator $\hat{f}_{n,k}$ is a consistent and asymptotically unbiased estimator of f_* and its asymptotic mean squared error $E(\hat{f}_{n,k} - f_*)^2$ has maximum possible rate of convergence in the class of all consistent estimators including the maximum likelihood estimator of f_*, as shown in [4, Chap. 7]. This mean squared error has the following asymptotic form:

$$E(\hat{f}_{n,k} - f_*)^2 = C_{k,\alpha}(c_0 n)^{-2/\alpha} (1 + o(1)), \quad n \to \infty. \tag{3}$$

Using the Taylor series $\Gamma(k + 2/\alpha) = \Gamma(k) + \frac{2}{\alpha}\Gamma'(k) + O(1/\alpha^2)$ for large values of α, we obtain

$$C_{k,\alpha} \simeq \frac{1}{k} + \frac{2(\psi(k) - 1 + 1/k)}{\alpha k}, \tag{4}$$

for large α, where $\psi(\cdot) = \Gamma'(\cdot)/\Gamma(\cdot)$ is the psi-function. Quality of this approximation is illustrated on Figs. 3 and 4.

In practice of global optimization, the standard estimator of f_* is the current record $y_{1,n} = \min_{i=1,\ldots,n} f(x_i)$. Its asymptotic mean squared error is

$$E(\hat{f}_{n,k}(e_1) - f_*)^2 = \Gamma(1 + 2/\alpha)(c_0 n)^{-2/\alpha} (1 + o(1)), \quad n \to \infty.$$

Asymptotic efficiency of $y_{1,n}$ is therefore $\text{eff}(y_{1,n}) = C_{k,\alpha}/\Gamma(1 + 2/\alpha)$. This efficiency is illustrated on Fig. 5.

GRS algorithms have a very attractive feature in comparison with deterministic optimisation procedures. Specifically, in GRS algorithms we can use statistical procedures for increasing efficiency of the algorithms and devising stopping

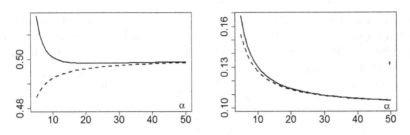

Fig. 3. The exact expression of $C_{k,\alpha}$ (solid) and the approximation (4) (dashed) for $k = 2$ (left) and $k = 10$ (right); α varies in $[5, 50]$.

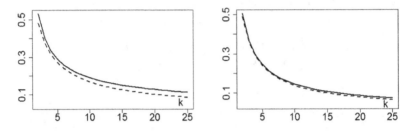

Fig. 4. The exact expression of $C_{k,\alpha}$ (solid) and the approximation (4) (dashed) for $\alpha = 4$ (left) and $\alpha = 7$ (right); as k varies in $[2, 25]$.

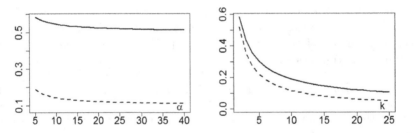

Fig. 5. Asymptotic efficiency $\mathrm{eff}(y_{1,n})$ of $y_{1,n}$. Left: $k = 2$ (solid) and $k = 10$ (dashed); as α varies in $[5, 40]$. Right: $\alpha = 5$ (solid) and $\alpha = 25$ (dashed); as k varies in $[2, 20]$.

rules. But do we lose much by choosing the points at random? We claim that if the dimension d is large then the use of quasi-random points instead of purely random does not bring any advantage. Let us try to illustrate this using some simulation experiments.

Using simulation studies we now investigate the performance of the PRS algorithm with $P = P_U$ and quasi-random points generated from the Sobol low-dispersion sequence. We examine the minimization problem with the objective function $f(x) = \sum_{s=1}^{d}(x_s - |\cos(s)|)^2$ and the set $\mathbf{X} = [0, 1]^d$ in the dimension $d = 15$. In this problem, the global minimum $f_* = 0$ is attained at the internal point $x_* = (|\cos(1)|, \ldots, |\cos(d)|)$. For each run of the PRS algorithm, we generate n points and compute the records $y_{1,n}$ and $y_{2,n}$, for $n = 10^3, 10^4, 10^5, 10^6$.

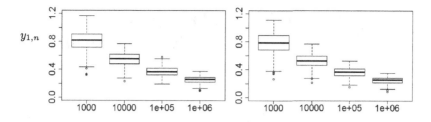

Fig. 6. Boxplot of records $y_{1,n}$ for 500 runs of the PRS algorithm with points generated from the Sobol low-dispersion sequence (left) and the uniform distribution (right), $d = 15$.

We repeat this procedure 500 times and show the obtained records as boxplots in Fig. 6.

We can see that the performance of the PRS algorithm with points generated from the Sobol low-dispersion sequence and the uniform distribution is very similar. We also note that the variability of $y_{1,n}$ is larger than variability of $y_{4,n}$ and the difference $y_{10,n} - y_{4,n}$ has a small variability.

Acknowledgements. The work of the first author was partially supported by the SPbSU project No. 6.38.435.2015 and the RFFI project No. 17-01-00161. The work of the third author was supported by the Russian Science Foundation, project No. 15-11-30022 'Global optimization, supercomputing computations, and applications'.

References

1. Zhigljavsky, A.: Mathematical Theory of Global Random Search. Leningrad University Press (1985). in Russian
2. Zhigljavsky, A.: Branch and probability bound methods for global optimization. Informatica **1**(1), 125–140 (1990)
3. Zhigljavsky, A., Žilinskas, A.: Stochastic Global Optimization. Springer, New York (2008)
4. Zhigljavsky, A.: Theory of Global Random Search. Kluwer Academic Publishers, Boston (1991)
5. Zhigljavsky, A., Hamilton, E.: Stopping rules in k-adaptive global random search algorithms. J. Global Optim. **48**(1), 87–97 (2010)
6. Zilinskas, A., Zhigljavsky, A.: Branch and probability bound methods in multi-objective optimization. Optim. Lett. **10**(2), 341–353 (2016). doi:10.1007/s11590-014-0777-z

A Model of FPGA Massively Parallel Calculations for Hard Problem of Scheduling in Transportation Systems

Mikhail Reznikov$^{(\boxtimes)}$ and Yuri Fedosenko$^{(\boxtimes)}$

Volga State University of Water Transport, Nizhny Novgorod, Russia
mirekez@gmail.com, fds@vgavt-nn.ru

Abstract. The FPGA calculation hardware was estimated in terms of performance of solving a hard nonlinear discrete optimization problem of scheduling in transportation systems. The dynamic programming algorithm for a large-scale problem is considered and a model of its decomposition into a set of smaller problems is investigated.

Keywords: Discrete optimization · Dynamic programming · Job-shop scheduling · FPGA calculations · Parallel computing

1 Introduction

A discrete nonlinear optimization problem that arises in real-life transportation systems of river multisectional trains while they pass servicing (loading, unloading or fueling) terminal (see, e.g., [1]) is considered. This problem is NP-hard and is characterized by an exponential time complexity depending on the schedule size n (see, e.g., [2, 3]). An executive planning requires the preparation of a servicing schedule that defines an order of the vessel processing. Such a schedule can be synthesized by an operator in a strongly limited time period before a servicing stage begins. From practical considerations, since one service cycle takes a full day, a one-hour limit was taken for a schedule creating time in this work. This time period should include dozens of runs of a computational algorithm with slightly different inputs. Since the resources consumption estimation for such an algorithm is $O(n2^nT)$, where T is the number of discrete time intervals, for even not high values of n and T as, for example, $n = 25$ and $T = 16$, the algorithm would require more than 13 billion of calculation operations and would take several minutes on a modern CPU. Thus, it is necessary to search for the best hardware approach and to investigate computational models providing a significant performance improvement.

The dynamic programming (DP) approach (see, e.g., [4]) which can be used for problem solving is memory-limited and its parallelization requires the development of special algorithms and architectures (see, e.g., [5, 6]). Some of perspective research directions in the field of specialized computational systems use multi-core graphical

The reported study was funded by RFBR according to the research project № 18-07-01078a.

R. Battiti et al. (Eds.): LION 2017, LNCS 10556, pp. 370–375, 2017.
https://doi.org/10.1007/978-3-319-69404-7_33

processing units (GPU), digital signal processors (DSP), neuro-computers, alternative physical imitation systems (see, e.g., [7]).

A separate direction of the investigation of massively parallel computing engines is based on the field programmable gate array (FPGA). The state-of-the-art in the field of FPGA-based discrete optimization is presented, for example, in [8, 9] and shows hardware approach benefits for suboptimal solution finding. In comparison with these works, in the present short paper the way to efficiently search for an optimal solution by using a hardware accelerator is suggested.

2   Service Scheduling Problem

The considered service scheduling problem consists of searching for an optimal schedule of jobs $\{z_1, z_2, \ldots, z_n\}$ from a set Z to be processed by a single machine P. The jobs are characterized by the following parameters $(i = 1, 2, \ldots, n)$:

t_i – the moment of readiness for processing of the job z_i,

τ_i – duration of processing of the job z_i,

$a_i(t' - t_i)$ – the linear penalty function (negative weight) for processing started at time t'.

The schedule of processing matches permutation $p = (p(1), p(2), \ldots, p(n))$ of the set of job's indexes. Let machine P be ready for processing from the time moment $t = 0$, processing of each job go without interruptions and simultaneous processing of several jobs be impossible. Let t'_k be a moment of processing start for the job with the index $p(k)$ $(k = 1, 2, \ldots, n)$. Then the correct schedule will be compact, i.e. $t'_1 = t_{p(1)}$ and $t'_k = \max\{t'_{k-1} + \tau_{p(k-1)}, t_{p(k)}\}$ for $k = 2, 3, \ldots, n$.

The problem is to find the schedule p with the minimal overall penalty over the whole set Z processed, i.e.

$$W(p) = \sum_{k=1}^{n} a_{p(k)}(t'_k - t_{p(k)}) \to \min. \tag{1}$$

As shown, e.g., in [1], problem (1) is strongly NP-hard.

2.1   Dynamic Programming Method

For an efficient problem solution the dynamic programming method is often used according to the Bellman's principle of optimality (see, e.g., [4]). Let $W_k^{min}(t, S)$ be the minimal overall penalty value for processing a set of jobs S by machine P, which becomes ready at the moment of discrete time t after the previously processed job $p(k)$. Let S be a subset of Z with jobs that have been already processed. Then, we can form the following recurrent expression (counting $W_n^{min}(t, Z) = 0$):

$$W_k^{min}(t, S) = \min_{\substack{i = \overline{1..n} \\ z_i \notin S}} (W_{k+1}^{min}(t' + \tau_i, S \cup z_i) + a_i(t' - t_i)), \tag{2}$$

where $t' = \max\{t, t_i\}$, and the minimal overall penalty is $W^* = W_0^{\min}(0, \emptyset)$. Thus, the optimization task can be represented by a recurrent hierarchy of subtasks. The solution to each task is based on all the solutions to all its subtasks.

The expression (t, S) is called the system state and each expression $W_k^{min}(t, S)$ can be calculated only once to be saved for a later usage.

The algorithm time costs are estimated by counting the overall number of unique states (t, S). The t value can be limited by some maximum moment T (some discretization can be chosen for the required accuracy), and the overall number of unique states will be $2^n T$. The requirements for the RAM usage will be exponential, too.

3   Parallel Implementation for FPGA

A heterogeneous approach which is considered in this work uses the classical von Neumann system connected to a special accelerator. Generally, it can be one of several types like GPU, FPGA or a specialized processor. In this work, we consider the possibility of using FPGA as a coprocessor to accelerate the optimization task solving. We suggest a model of massively parallel calculations for a part of the original problem.

For example, let us consider the original problem with $n = 22$ and $T = 32$. The process of the DP algorithm consists of a continuous calculation of $W_k^{min}(t, S)$ values for all possible system states (t, S) for the stage k from n to 0. Each W_k^{min} is calculated on the base of the previously calculated values. The structure of system states realizes the recurrent hierarchy. As shown on Fig. 1(a), the number of states for processing depends on its depth in hierarchy (S_h^k marks the set of processed jobs with the serial number h for the stage k). The CPU calculation time θ for each stage is shown on Fig. 1 (b). The most of time is spent for calculation of stages with the number k close to $n/2$. Hence, for a parallel realization we can calculate each stage separately and use all the resources.

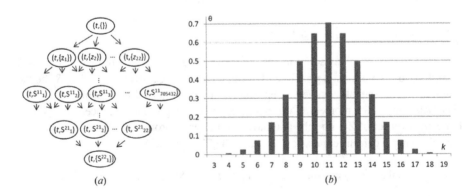

(a)   (b)

Fig. 1. (a) DP states hierarchy by stages; (b) time in seconds needed for each stage solving.

3.1   Decomposition of the Original Problem

Each system state (t, S) corresponds to some subtask with a lower size. The size of the subtask is equal to $(n - k)$ where k is the size of a set S. Each such subtask can be solved separately and without any connection with the original task. A local problem consists of a scheduling of the remaining $(n - k)$ jobs starting from the selected state.

This method is used to allow a larger problem to be solved by using solutions to smaller problems.

In general, a separate subtask calculation is harder because there is no secondary usage of the already calculated W_k^{min} values that can be stored in memory. So, this method will be efficient only in the case of ability to process much more states per second than a CPU does. For such a special computing, an FPGA implementation has been chosen.

The goal of the FPGA architecture is to continuously provide solutions to the small problem with schedule size m at a frequency of Integrated Circuit (IC), one solution per each clock cycle. Therefore, the calculation should use sufficient FPGA resources to process all calculations in parallel. The example of an IC for $m = 4$ and $T = 16$ is shown on Fig. 2(a) for demonstrative purposes. One ALU module is shown in Fig. 2 (b). The IC was built using hardware description language (HDL) by the following rules:

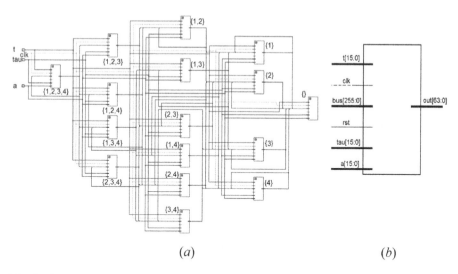

(a) (b)

Fig. 2. (a) ALU DP hierarchy example for $m = 4$, exact schedules are shown; (b) one ALU.

1. The first part generates ALU for all possible states (t, S) for t from 0 to T and $S \in Z$.
2. The second part generates busses $W_k^{min}(t, S)$ for each ALU and connects the result registers to all inputs of the other ALUs that require it.

This model for the size up to $m = 11$ has been synthesized for FPGA Virtex-7 565T (see [10]) with clock 300 MHz, providing one solution to task per clock cycle with pipelining.

The heterogeneous calculation process works in the following two cycles.

1. In the beginning of the solving process the FPGA calculates all W_{n-m}^{min} values for the stage $(n-m)$. Each value is provided as the solution to the subtask of size m.
2. The CPU gets these values from the FPGA, stores in memory tables and then starts the software-based dynamic programming calculation process from stage $(n-m)$.

As a result of this approach, the CPU skips the first m stages of the common DP algorithm calculations. In the case of ability of the FPGA coprocessor to solve a problem with the size equal to half of the original problem size n, the whole calculation time can be reduced almost by factor two. Increasing m will lead to the less calculation time. The problem solution time for a heterogeneous system containing FPGA-based solver for schedule size m is presented on Fig. 3. The time required for the problem solution using a 4-core modern CPU is shown as a reference. In particular, the time for $m = 11$ was taken from the performance estimation of the heterogeneous system prototype. The time for $m > 11$ is the result of mathematical modelling using a simulator. The original problem size is $n = 22$ and $T = 32$.

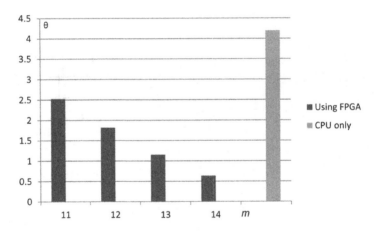

Fig. 3. Processing time θ of the original problem in seconds for different FPGA-based solver schedule size m compared with a 4-core CPU implementation.

The considered approach allows a linear scaling of DP algorithm for the original problem solving. However, the maximum size m of the problem which can be solved is significantly limited by the available FPGA resources. Taking into account the fact that the last generation of FPGA contains more resources than devices used in this work, we can conclude that coprocessors can provide a significant solving time reduction. By using several FPGA devices it is possible to provide a linear scalability of the solving algorithm.

From the other hand, this model demonstrates the way how FPGAs or, generally, application-specific integrated circuits (ASICs) can be used for a fast solving of NP-hard discrete optimization problems.

4   Conclusions

As the main result of this work, a special model of a heterogeneous architecture with the FPGA massively parallel calculations was suggested. The model is based on decomposition of the original problem using dynamic programming method and provides the following benefits:

- a 2–4 times synthesis time reduction with one FPGA architecture in comparison to the CPU-based realization,
- the linear scalability is possible if more FPGA devices are used.

References

1. Kogan, D.I., Fedosenko, Yu.S.: Optimal servicing strategy design problems for stationary objects in a one-dimensional working zone of a processor. Autom. Remote Control **71**(10), 2058–2069 (2010)
2. Kogan, D.I., Fedosenko, Yu.S.: The discretization problem: analysis of computational complexity, and polynomially solvable subclasses. Discrete Math. Appl. **6**(5), 435–447 (1996)
3. Sammarra, M., Cordeau, J.-F., Laporte, G., Monaco, M.F.: A tabu search heuristic for the quay crane scheduling problem. J. Sched. **10**(4), 327–336 (2007)
4. Bellman, R.E., Dreyfus, S.E.: Applied Dynamic Programming. Princeton Univ. Press, Princeton (1962)
5. Cuencaa, J., Gimnezb, D., Martnez, J.: Heuristics for work distribution of a homogeneous parallel dynamic programming scheme on heterogeneous systems. Parallel Comput. **31**(7), 711–735 (2005)
6. Gergel, V.P.: High-Performance Computing for Multi-Processor Multi-Core Systems. MGU Press, Moscow (2010). (In Russian)
7. Brodtkorb, A.R., Dyken, C., Hagen, T.R., Hjelmervik, J.M., Storaasli, O.O.: State-of-the-art in Heterogeneous Computing. Scientific Programming **18**(1), 1–33 (2010)
8. Madhavan, A., Sherwood, T., Strukov, D.: Race logic: a hardware acceleration for dynamic programming algorithms. In: ISCA 2014, Minnesota, USA, pp. 517–528 (2014)
9. Tao, F., Zhang, L., Laili, Y.: Job shop scheduling with FPGA-based F4SA. In: Tao, F., Zhang, L., Laili, Y. (eds.) Configurable Intelligent Optimization Algorithm. SSAM, pp. 333–347. Springer, Cham (2015). doi:10.1007/978-3-319-08840-2_11
10. 7 Series FPGAs Data Sheet: Overview, March 2017. https://www.xilinx.com/support/documenta-tion/data_sheets/ds180_7Series_Overview.pdf

Accelerating Gradient Descent with Projective Response Surface Methodology

Alexander Senov[✉]

Faculty of Mathematics and Mechanics, Saint Petersburg State University,
198504 Universitetsky Prospekt, 28, Peterhof, St. Petersburg, Russia
alexander.senov@gmail.com
http://math.spbu.ru/eng

Abstract. We present a new modification of gradient descent algorithm based on surrogate optimization with projection into low-dimensional space. It consequently builds an approximation of the target function in low-dimensional space and takes the approximation optimum point mapped back to original parameter space as the next parameter estimate. An additional projection step is used to fight the curse of dimensionality. Major advantage of the proposed modification is that it does not change gradient descent iterations, thus it may be used with almost any zero- or first-order iterative method. We give a theoretical motivation for the proposed algorithm and experimentally illustrate its properties on modelled data.

Keywords: Least-squares · Steepest descent · Quadratic programming · Projective methods

1 Introduction

Mathematical optimization is a very popular and widely used tool in today's control, machine learning and other applications since almost any process or procedure is essentially about maximising or minimising some quantity (see, for example [5]). In the case of high function computation complexity, unknown Hessian or high dimensional parameter space, one can choose among many zero- or first-order iterative optimization algorithms (see [2,6–8]), although some of them share one but yet important drawback: at each iteration they explicitly consider only the current point and function value estimates ignoring preceding history and, thus, loosing potentially highly important information.

In contrast, *surrogate optimization methods* use history of the past points for approximating the objective function by another function called *surrogate* and takes the next estimate based on it (see, e.g., [4]). For example, *quadratic response surface methodology* constructs surrogate as a second order polynomial constructed via polynomial regression and take its minimum as the next estimate (see, e.g., [1]). Despite many advantages, many surrogate models share

R. Battiti et al. (Eds.): LION 2017, LNCS 10556, pp. 376–382, 2017.
https://doi.org/10.1007/978-3-319-69404-7_34

two drawbacks: they are memory and time consuming and their performance depends on the chosen surrogate adequacy.

Additionally, the so-called *multi-step* optimization methods do use history too (see, e.g., [8]). Many of them do use only fixed amount of history, e.g. *two-step Heavy-ball method* use only two past steps. Others do use parametrized amount of history, like *multi-step quasi-Newton method* [3], but in slightly different way (e.g., not using the projection trick, or without explicit quadratic approximation).

In this paper we propose a new method which is essentially an incorporation of quadratic response surface methodology into the gradient descent algorithm. To neutralize memory footprint of quadratic polynomial we use a projection trick. The general idea of the proposed algorithm is to use a sequence of points obtained from the gradient descent iterations as follows:

1. K_1 consecutive points used to train incremental principal component analysis algorithm which produces orthogonal projection matrix $\mathbf{P}$.
2. K_2 consecutive points used to collect *training* set in low-dimensional space obtained with $\mathbf{P}$.
3. Quadratic polynomial fitted to collected training set and the argument of the polynomial minimum returned back to the original space used as the next point estimate.

These steps are executed iteratively producing an additional point every $K_1 + K_2$ gradient descent iterations. We consider *gradient descent* algorithm as an example, but it may be easily replaced with some other zero-order or first-order iterative optimization method.

The paper is organized as follows. In Sect. 2 we describe the proposed algorithm which improves gradient descent by using the projective quadratic response surface methodology. In Sect. 3 we provide theoretical motivation behind it. Further, in Sect. 4 we report a case study on modelled data and discuss its results. Finally, Sect. 5 concludes the paper.

2   Algorithm Description

A pseudo-code of the proposed algorithm (Algorithm 1) is given in Fig. 1. It has the following parameters: $f : \mathbb{R}^d \to \mathbb{R}$ — function to be optimized; $\nabla_{\mathbf{x}} f : \mathbb{R}^d \to \mathbb{R}$ — function gradient or its approximation (in case of zero-order method); $\lambda \in \mathbb{R}$ — step size; $d \in \mathbb{N}_+$ — original space dimensionality; $q \in \mathbb{N}_+$ — projective space dimensionality; $T \in \mathbb{N}_+$ — number of iterations; IncrPCA — incremental PCA algorithm; $K_1 \in \mathbb{N}_+$ — number of points for incremental PCA fitting; $K_2 \in \mathbb{N}_+$ — number of points for surrogate construction.

We use $\overline{\mathbf{x}}$ to denote sample mean vector. One might be confused by "backward projection" step of transforming $\widehat{\mathbf{z}}$ in low-dimensional space to $\widehat{\mathbf{x}}$ in high-dimensional space (line 18 in Algorithm 1). This transformation is motivated by Proposition 1 (Sect. 3). As for the choice of principal component analysis as a tool for the orthogonal projection construction it is rather motivated by practice

```
1:  t ← 1
2:  while t ≤ T do
3:      P ← IncrPCA.init(q, d)                          ▷ Initialize projection matrix
4:      for k ← 1 to K₁ do
5:          xₜ ← xₜ₋₁ − λ∇ₓf(xₜ₋₁)                      ▷ Gradient descent step
6:          P ← IncrPCA.update(q, P, xₜ)
7:          t ← t + 1
8:      x̄ ← 0_d
9:      for k ← 1 to K₂ do
10:         xₜ ← xₜ₋₁ − λ∇ₓf(xₜ₋₁)                      ▷ Gradient descent step
11:         x̄ ← x̄ + xₜ/K₂
12:         zₖ ← Pxₜ
13:         yₖ ← f(xₜ)
14:         t ← t + 1
15:     Â, b̂, ĉ ← QuadraticLeastSquares ({zₖ, yₖ}₁^{K₂})
16:     if Â positive definite then
17:         ẑ ← −½Â⁻¹b̂
18:         xₜ ← Pᵀẑ + (I − PᵀP)x̄                       ▷ "backward projection"
19:         t ← t + 1
```

$$1: \quad t \leftarrow 1$$
$$2: \quad \textbf{while } t \leq T \textbf{ do}$$
$$3: \qquad \mathbf{P} \leftarrow \text{IncrPCA.init}(q, d) \qquad\qquad \triangleright \text{ Initialize projection matrix}$$
$$4: \qquad \textbf{for } k \leftarrow 1 \text{ to } K_1 \textbf{ do}$$
$$5: \qquad\quad \mathbf{x}_t \leftarrow \mathbf{x}_{t-1} - \lambda \nabla_{\mathbf{x}} f(\mathbf{x}_{t-1}) \qquad\qquad \triangleright \text{ Gradient descent step}$$
$$6: \qquad\quad \mathbf{P} \leftarrow \text{IncrPCA.update}(q, \mathbf{P}, \mathbf{x}_t)$$
$$7: \qquad\quad t \leftarrow t + 1$$
$$8: \qquad \bar{\mathbf{x}} \leftarrow \mathbf{0}_d$$
$$9: \qquad \textbf{for } k \leftarrow 1 \text{ to } K_2 \textbf{ do}$$
$$10: \qquad\quad \mathbf{x}_t \leftarrow \mathbf{x}_{t-1} - \lambda \nabla_{\mathbf{x}} f(\mathbf{x}_{t-1}) \qquad\qquad \triangleright \text{ Gradient descent step}$$
$$11: \qquad\quad \bar{\mathbf{x}} \leftarrow \bar{\mathbf{x}} + \mathbf{x}_t / K_2$$
$$12: \qquad\quad \mathbf{z}_k \leftarrow \mathbf{P}\mathbf{x}_t$$
$$13: \qquad\quad y_k \leftarrow f(\mathbf{x}_t)$$
$$14: \qquad\quad t \leftarrow t + 1$$
$$15: \qquad \widehat{\mathbf{A}}, \widehat{\mathbf{b}}, \widehat{c} \leftarrow \text{QuadraticLeastSquares}\left(\{\mathbf{z}_k, y_k\}_1^{K_2}\right)$$
$$16: \qquad \textbf{if } \widehat{\mathbf{A}} \text{ positive definite } \textbf{then}$$
$$17: \qquad\quad \widehat{\mathbf{z}} \leftarrow -\tfrac{1}{2}\widehat{\mathbf{A}}^{-1}\widehat{\mathbf{b}}$$
$$18: \qquad\quad \mathbf{x}_t \leftarrow \mathbf{P}^\top \widehat{\mathbf{z}} + (\mathbf{I} - \mathbf{P}^\top \mathbf{P})\bar{\mathbf{x}} \qquad\qquad \triangleright \text{ "backward projection"}$$
$$19: \qquad\quad t \leftarrow t + 1$$

Fig. 1. Proposed gradient descent modification

and intuition: the first q principal components have the largest possible variance. The more information (in terms of variance) we keep from the original points the more accurate our approximation will be.

As one can see, the proposed modification utilizes: $\mathcal{O}(K_1 qd)$ operations and $\mathcal{O}(qd)$ memory at step (1), $\mathcal{O}(K_2 qd)$ operations and $\mathcal{O}(K_2 q)$ memory at step (2) and $\mathcal{O}(K_2 q^2 + q^3)$ operations and $\mathcal{O}(K_2 q + q^2)$ memory at step (3). Hence, this modification adds at most $\mathcal{O}(qd + q^3)$ in the number of operations and $\mathcal{O}(q^2 + qd + K_2 q)$ in the memory consumption per single gradient descent iteration.

3   Theoretical Background

In this section we provide theoretical motivation behind the proposed algorithm. Proofs sketches are given in Appendix A.

Proposition 1. *Consider* $\mathbf{P} \in \mathbb{R}^{q \times d}$, $\mathbf{P}\mathbf{P}^\top = \mathbf{I}_q$ *and* $\{\mathbf{x}_1, \ldots, \mathbf{x}_K\} \subset \mathbb{R}^d$. *Then*

$$\widehat{\mathbf{x}} = \underset{\{\mathbf{x} \in \mathbb{R}^d \,:\, \mathbf{P}\mathbf{x} = \widehat{\mathbf{z}}\}}{argmin} \sum_{t=1}^{K} \|\mathbf{x}_t - \mathbf{x}\|_2^2 = (\mathbf{I} - \mathbf{P}^\top \mathbf{P}) \frac{1}{K} \sum_1^K \mathbf{x}_t + \mathbf{P}^\top \widehat{\mathbf{z}}.$$

This proposition motivates the "backward projection" step in the proposed algorithm: from set $\{\mathbf{x} \in \mathbb{R}^d \,:\, \mathbf{P}\mathbf{x} = \widehat{\mathbf{z}}\}$ we pick the point closest to K previous gradient descent estimates. Further, the following propositions quantify the effect of the projection on the function optimum point.

Proposition 2. *Consider function* $f(\mathbf{x}) = \mathbf{x}^\top \mathbf{A}\mathbf{x} + \mathbf{b}^\top \mathbf{x} + c$, *where* $\mathbf{A} \in \mathbb{R}^{d \times d}$, $\mathbf{A} \succ 0$, $\mathbf{b} \in \mathbb{R}^d$, $c \in \mathbb{R}$, $\mathbf{P} \in \mathbb{R}^{q \times d}$, $\mathbf{P}\mathbf{P}^\top = \mathbf{I}_q$, $\mathbf{z} = \mathbf{P}\mathbf{x}$, $\mathbf{v} = (\mathbf{I} - \mathbf{P}^\top \mathbf{P})\mathbf{x}$. *Then*

$$\operatorname*{argmin}_{\mathbf{z}} f(\mathbf{P}^\top \mathbf{z} + \mathbf{v}) = -\frac{1}{2}\mathbf{P}\mathbf{A}^{-1}\mathbf{b}.$$

Proposition 3. *Consider function* $f(\mathbf{x}) = \mathbf{x}^\top \mathbf{A}\mathbf{x} + \mathbf{b}^\top \mathbf{x} + c$, *where* $\mathbf{A} \in \mathbb{R}^{d \times d}$, $\mathbf{b} \in \mathbb{R}^d$, $c \in \mathbb{R}$ *and* $\mathbf{A} \succ 0$, *orthogonal projection matrix* $\mathbf{P} \in \mathbb{R}^{q \times d}$, $q < d$, *sequence of gradient descent estimates* $\{\mathbf{x}_t\}_1^K \subset \mathbb{R}^d$, *their projections* $\{\mathbf{z}_t\}_1^K \subset \mathbb{R}^q$, $\mathbf{z}_t = \mathbf{P}\mathbf{x}_t$ *and corresponding function values* $\{y_t\}_1^K$, $y_t = f(\mathbf{x}_t)$. *Let* $\widehat{\mathbf{x}} = -\frac{1}{2}\mathbf{P}^\top \mathbf{P}\mathbf{A}^{-1}\mathbf{b} + (\mathbf{I} - \mathbf{P}^\top \mathbf{P})\overline{\mathbf{x}}$. *Then*

$$\| \operatorname{argmin} f - \widehat{\mathbf{x}} \|_2^2 = \left\| (\mathbf{I} - \mathbf{P}^\top \mathbf{P}) \left(\frac{1}{2}\mathbf{A}^{-1}\mathbf{b} - \overline{\mathbf{x}} \right) \right\|_2^2.$$

This proposition provides an estimate of the difference between the true function optimum point and its estimate obtained after each iteration of the proposed algorithm (lines 3–19) and implies the following two important facts.

1 The difference between the optimum point and obtained estimate lies in the kernel of orthogonal projection $\mathbf{P}$. Hence, estimate $P\widehat{x}$ is the best estimate in terms of this projection.
2 The closer the averaged gradient descent estimates to the optimum points, the smaller is the error. Hence, $\widehat{\mathbf{x}}$ benefits from precision of gradient descent estimates.

4   A Case Study

First, we describe the modelling strategy. For modelling purposes we use the following default values: $f(\mathbf{x}) = \mathbf{x}^\top \mathbf{I}_d \mathbf{x} - \mathbf{1}_d^T \mathbf{x}$, where d is a variable parameter; $d \leftarrow 10$, $T \leftarrow 50$, $x_0 \sim \mathrm{U}[0,1]^d$, $\lambda = 10^{-5}$; $q \leftarrow 1$, $K_1 \leftarrow 10$, $K_2 \leftarrow 5$. We vary parameters T, d, K_1 and K_2 independently (with other parameters fixed) in the following ranges: $T \in [25, 30, 50, 100]$; $d \in [5, 10, 20, 50, 100]$; $K_1 \in [2, 5, 10, 20]$; $K_2 \in [2, 5, 10, 20]$. For each parameters values combination we execute the gradient descent algorithm and the proposed algorithm with the same initial estimate x_0 for 10^3 times and calculate their errors as Euclidean distance from the real optimum point to the algorithm estimate. Then we calculate ξ — a ratio of times when the proposed algorithm error was less than the gradient descent error.

Table 1 contains the results of the experiment described above. ξ monotonically decreases with T increased since the proposed algorithm works well at the start but fails to build a surrogate when the gradient descent oscillates near the optimum point. Further, there are no evident dependency on q since for the particular function the gradient descent estimates lie on the straight line. Thus, they are perfectly described by a single dimension. Situation with parameter K_1

Table 1. Results of the performed numerical experiments. ξ is the ratio of times when the proposed algorithm error was less than the gradient descent error

Parameter	Parameter value	ξ
T	25	0.91
T	30	0.90
T	50	0.80
T	100	0.51
d	5	0.81
d	10	0.78
d	20	0.80
d	50	0.81
d	100	0.79
K_1	2	0.79
K_1	5	0.78
K_1	10	0.78
K_1	20	0.83
K_2	2	0.10
K_2	3	0.80
K_2	5	0.82
K_2	10	0.79

is similar to the parameter q and may be explained by the same reason. Finally, a huge difference in quality between $K_2 = 2$ and $K_2 = 3$ is explained by the fact that one needs at least three points in one-dimensional space to construct the second order polynomial.

5   Conclusion

We propose a modification to the gradient descent based on the quadratic response surface methodology with the projection trick. We show that the proposed modification can provide the best optimum approximation with respect to the considered projection. The synthetic case study shows that the modified gradient descent can be superior with respect to the original one in terms of the optimum point estimation error. This modification may be used with other zero- or first-order iterative optimization method thus improving their performance.

Acknowledgments. This work was supported by Russian Science Foundation (project 16-19-00057).

A Proofs

Proof. (of Proposition 1)

$$\partial_{\mathbf{x}} \underset{\{\mathbf{x}\in\mathbb{R}^d\,:\,\mathbf{Px}=\widehat{\mathbf{z}}\}}{\operatorname{argmin}} \sum_{t=1}^{K} \|\mathbf{x}_t - \mathbf{x}\|_2^2$$

$$= \partial_{\mathbf{x}} \sum_{t=1}^{K} \| \mathbf{P}^{\top}\mathbf{Px}_t + \left(\mathbf{I} - \mathbf{P}^{\top}\mathbf{P}\right)\mathbf{x}_t - \left(\mathbf{P}^{\top}\widehat{\mathbf{z}} + \left(\mathbf{I} - \mathbf{P}^{\top}\mathbf{P}\right)\mathbf{x}\right) \|_2^2$$

$$= 2(\mathbf{I} - \mathbf{P}^{\top}\mathbf{P}) \sum_{t=1}^{K} \mathbf{x}_t - 2K\left(\mathbf{I} - \mathbf{P}^{\top}\mathbf{P}\right)\mathbf{x} = 0.$$

Since $(\mathbf{I} - \mathbf{P}^{\top}\mathbf{P})$ is not invertible, the above equation has an infinite number of solutions. Hence, we are free to choose any one of them, e.g. $\mathbf{x} = \frac{1}{K}\sum_1^K \mathbf{x}_t$.

Proof. (of Proposition 2)

$$f(\mathbf{x}) = \left(\mathbf{P}^{\top}\mathbf{z} + \mathbf{v}\right)^{\top}\mathbf{A}\left(\mathbf{P}^{\top}\mathbf{z} + \mathbf{v}\right) + \mathbf{b}^{\top}\left(\mathbf{P}^{\top}\mathbf{z} + \mathbf{v}\right) + c$$
$$= \mathbf{z}^{\top}\mathbf{PAP}^{\top}\mathbf{z} + \left(\mathbf{b}^{\top} + \mathbf{v}^{\top}\mathbf{A}\right)\mathbf{P}^{\top}\mathbf{z} + \left(\mathbf{v}^{\top}\mathbf{Av} - \mathbf{b}^{\top}\mathbf{v} + c\right).$$

Substituting $\mathbf{v}$ back and taking derivative with respect to $\mathbf{z}$, we obtain:

$$0 = 2\mathbf{PAP}^{\top}\widehat{\mathbf{z}} + \left(\mathbf{b}^{\top} + \mathbf{x}^{\top}\left(\mathbf{I} - \mathbf{P}^{\top}\mathbf{P}\right)^{\top}\mathbf{A}\right)\mathbf{P}^{\top}$$

$$\rightsquigarrow \widehat{\mathbf{z}} = -\frac{1}{2}\left(\mathbf{PAP}^{\top}\right)^{-1}\left(\left(\mathbf{b}^{\top} + \mathbf{x}^{\top}\left(\mathbf{I} - \mathbf{P}^{\top}\mathbf{P}\right)^{\top}\mathbf{A}\right)\mathbf{P}^{\top}\right)^{\top}$$

$$= -\frac{1}{2}\mathbf{P}\left(\mathbf{A}^{-1}\mathbf{b} + \left(\mathbf{I} - \mathbf{P}^{\top}\mathbf{P}\right)\mathbf{x}\right) = -\frac{1}{2}\mathbf{PA}^{-1}\mathbf{b}.$$

Proof. (of Proposition 3) From Propositions 1 and 2: $\widehat{\mathbf{x}} = \left(\mathbf{I} - \mathbf{P}^{\top}\mathbf{P}\right)\overline{\mathbf{x}} - \frac{1}{2}\mathbf{P}^{\top}\mathbf{PA}^{-1}\mathbf{b}$. Hence

$$\|\operatorname{argmin} f - \widehat{\mathbf{x}}\|_2^2 = \left\| -\frac{1}{2}\mathbf{A}^{-1}\mathbf{b} - \widehat{\mathbf{x}} \right\|_2^2$$

$$= \left\| -\left(\mathbf{I} - \mathbf{P}^{\top}\mathbf{P}\right)\overline{\mathbf{x}} + \frac{1}{2}\mathbf{P}^{\top}\mathbf{PA}^{-1}\mathbf{b} - \frac{1}{2}\mathbf{A}^{-1}\mathbf{b} \right\|_2^2$$

$$= \left\| \left(\mathbf{I} - \mathbf{P}^{\top}\mathbf{P}\right)\left(\frac{1}{2}\mathbf{A}^{-1}\mathbf{b} - \overline{\mathbf{x}}\right) \right\|_2^2.$$

References

1. Box, G.E., Draper, N.R., et al.: Empirical Model-Building and Response Surfaces. John Wiley & Sons, New York (1987)
2. Boyd, S., Vandenberghe, L.: Convex Optimization. Cambridge University Press, New York (2004)

3. Ford, J., Moghrabi, I.: Multi-step quasi-Newton methods for optimization. J. Comput. Appl. Math. **50**(1–3), 305–323 (1994)
4. Forrester, A., Keane, A.: Recent advances in surrogate-based optimization. Prog. Aerosp. Sci. **45**(1), 50–79 (2009)
5. Granichin, O., Volkovich, V., Toledano-Kitai, D.: Randomized Algorithms in Automatic Control and Data Mining. Springer, Heidelberg (2015)
6. Granichin, O.N.: Stochastic approximation search algorithms with randomization at the input. Autom. Remote Control **76**(5), 762–775 (2015)
7. Nesterov, Y.: Introductory Lectures on Convex Optimization: A Basic Course. Springer, New York (2004)
8. Polyak, B.T.: Introduction to Optimization. Translations Series in Mathematics and Engineering. Optimization Software (1987)

Emmental-Type GKLS-Based Multiextremal Smooth Test Problems with Non-linear Constraints

Ya.D. Sergeyev[1,2], D.E. Kvasov[1,2], and M.S. Mukhametzhanov[1,2]([✉])

[1] University of Calabria, Rende (CS), Italy
{yaro,kvadim,m.mukhametzhanov}@dimes.unical.it
[2] Lobachevsky State University, Nizhni Novgorod, Russia

Abstract. In this paper, multidimensional test problems for methods solving constrained Lipschitz global optimization problems are proposed. A new class of GKLS-based multidimensional test problems with continuously differentiable multiextremal objective functions and non-linear constraints is described. In these constrained problems, the global minimizer does not coincide with the global minimizer of the respective unconstrained test problem, and is always located on the boundaries of the admissible region. Two types of constraints are introduced. The possibility to choose the difficulty of the admissible region is available.

Keywords: GKLS classes of test problems · Constrained optimization · Lipschitz global optimization · Nonlinear constraints

1 Introduction

Global optimization problems with and without constraints attract a great attention of researchers from both theoretical and practical viewpoints (see, e.g., [3,17,20], for the derivative-free global optimization, [5,9,13,18], for the real-life engineering problems, [2,17], for the parallel global optimization, etc.).

Let us consider the following constrained problem

$$f^* = f(x^*) = \min_{x \in D} f(x), \; D = \{(x_1, ..., x_N) : a_i \le x_i \le b_i, i = 1, ..., N\} \subset \mathbb{R}^N, \tag{1}$$

with p constraints

$$g_j(x) \le 0, \; j = 1, ..., p, \tag{2}$$

where the functions $f(x)$ and $g_j(x)$, $j = 1, ..., p$, satisfy the Lipschitz condition over the hyperinterval D

$$|f(x') - f(x'')| \le L||x' - x''||, \\ |g_j(x') - g_j(x'')| \le L_j||x' - x''||, \; j = 1, ..., p, \; x', x'' \in D, \tag{3}$$

where L and L_j, $0 < L < \infty$ and $0 < L_j < \infty$, $j = 1, ..., p$, are the Lipschitz constants for the functions $f(x)$ and $g_j(x)$ over the hyperinterval D, respectively (hereafter $|| \cdot ||$ denotes the Euclidean norm).

© Springer International Publishing AG 2017
R. Battiti et al. (Eds.): LION 2017, LNCS 10556, pp. 383–388, 2017.
https://doi.org/10.1007/978-3-319-69404-7_35

There exists a huge number of methods for solving (1)–(3) (see, e.g., [11, 16, 21, 22]). These methods often have a completely different nature and their numerical comparison can be very difficult (see, e.g., [14, 15] for a numerical comparison of metaheuristic and deterministic unconstrained global optimization algorithms). In [19], a new tool called "Operational zones" for an efficient numerical comparison of constrained and unconstrained global optimization algorithms of different nature has been proposed. To use it, classes of test problems are required. On the one hand, there exist many generators of test problems for global and local optimization (see, e.g., [23] for the landscape generators, [12] for the multidimensional assignment problem generator, [4, 7] for the wide analysis of different test classes and generators, [1, 10, 16] for different classes and generators of unconstrained test problems). On the other hand, collections of test problems are used usually in the framework of continuous constrained global optimization (see, e.g., [6]) due to absence of test classes and generators for such a type of problems. This paper introduces a new class of test problems with non-linear constraints, known minimizers, and parameterizable difficulty, where both the objective function and constraints are continuously differentiable.

2 GKLS Classes of Unconstrained Test Problems

Let us consider the unconstrained GKLS class of test problems with continuously differentiable objective function proposed in [8]. Test functions in this class are generated by defining a convex quadratic function systematically distorted by cubic polynomials in order to introduce local minima. The objective function $f(x)$ of the GKLS class is constructed by modifying a paraboloid Z

$$Z : z(x) = ||x - T||^2 + t, \ x \in D, \tag{4}$$

with the minimum value t at the point $T \in int(D)$, where $int(D)$ denotes the interior of D, in such a way that the resulting function $f(x)$ has m, $m \geq 2$, local minimizers: point T from (4) and points $M_i \in int(D)$, $M_i \neq T$, $M_i \neq M_j$, $i, j = 2, ..., m$, $i \neq j$. The paraboloid Z is modified by cubic polynomials $C_i(x)$ within balls $S_i \subset D$ (not necessarily entirely contained in D) around each point M_i, $i = 2, ..., m$ (with $M_1 = T$ being the vertex of the paraboloid and M_2 being the global minimizer of the problem), where

$$S_i = \{x \in \mathbb{R}^N : ||x - M_i|| \leq \rho_i, \ \rho_i > 0\}. \tag{5}$$

Each class contains 100 test problems and is defined by the following parameters: problem dimension, number of local minima, value of the global minimum, radius of the attraction region of the global minimizer, and distance from the global minimizer to the vertex of the paraboloid. An example of the test problem with 30 local minima is presented in Fig. 1a.

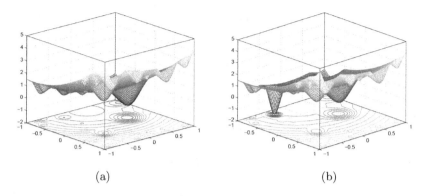

(a)   (b)

Fig. 1. Original GKLS test function (a) and Emmental-type GKLS-based test function (b) do not coincide even without constraints.

3   Constrained Emmental-Type Test Problems

In order to have a powerful tool for testing algorithms for constrained global optimization it is desirable to get test problems satisfying the following conditions:

- The global minimizer of the constrained problem is known, randomly generated, and is placed on the boundary of the feasible region;
- The global minimizer of the constrained problem differs from the global minimizer of the unconstrained problem.
- The constraints are non-linear and satisfy the Lipschitz condition over the hyperinterval D.
- The difficulty of the admissible region can be changed.

In order to satisfy the requirements given above, the Emmental-type GKLS-based class of test problems with smooth objective function is constructed by the following modifications of the GKLS class of unconstrained test problems.

Two types of constraints are proposed, where each constraint $g_i(x)$ is the exterior of a ball with a center G_i and a radius r_i, i.e.,

$$g_i(x) = r_i - ||x - G_i|| \le 0, \ i = 1, ..., p. \tag{6}$$

The first-type constraints are constructed as follows. First, the local minimizer (not the global one) M_{imin} nearest to the vertex of the paraboloid is taken as follows

$$imin = arg \min_{i=3,...,m} \{||M_i - M_1|| - \rho_i\}, \tag{7}$$

where ρ_i is the radius of the ball S_i (the index i starts from $i = 3$ since in original GKLS classes M_2 is the global minimizer). Then, the polynomial C_{imin} around M_{imin} is modified in a way, that its minimum value has been set to $-2 \cdot |f^*|$, where f^* is the optimum value of the original unconstrained GKLS problem. The global minimizer of the Emmental-type unconstrained test problem differs from the global minimizer of the original GKLS test problem due to this modification (see Fig. 1b). The ball with the radius $r_1 = ||M_1 - M_{imin}|| - \rho_{imin}$, and the center at the vertex of the paraboloid, i.e., $G_1 = M_1$, is taken as the first constraint.

Second, in order to guarantee that the global minimizer of the constrained test problem does not coincide with the global minimizer of the unconstrained modified problem, the ball with the radius $r_2 = \rho_{imin}$ and the center $G_2 = M_{imin}$ is taken as the second constraint.

Then, in order to guarantee that the global minimizer M_2 of the constrained test problem is placed on the boundary, the ball with the radius $r_3 = \frac{1}{2}\rho_2$ and the center at the point $G_3 = \frac{r}{2 \cdot ||M_1 - M_2||} M_1 + (1 - \frac{r}{2 \cdot ||M_1 - M_2||}) M_2$ is taken as the third constraint (see Fig. 2a).

If $p_1 > 3$, where p_1 is the number of the first-type constraints, $p_1 \leq p$, then the fourth constraint is constructed symmetrically to the third one with respect to the global minimizer M_2, i.e., $r_4 = r_3$ and $G_4 = M_2 + (M_2 - G_3)$ (see Fig. 2b).

The last $p_1 - 4$ first-type constraints are taken as $p_1 - 4$ different random balls S_j, $j \in \{3, ..., m\}$, $j \neq imin$, i.e., $r_i = \rho_{j(i)}$ and $G_i = M_{j(i)}$, $i = 5, ..., p_1$.

The $p_2 = p - p_1$ constraints of the second type, where $0 \leq p_2 \leq 2^N$, are built as follows. Random vertices $c_j = (c_1^j, ..., c_N^j)$, $c_i^j \in \{a_i, b_i\}$, $j = 1, ..., p_2$, of D are taken. Then, for each taken vertex c_j, the nearest local or global minimum $M_{i(j)}$ is found. The $(p_1 + j)$-th constraint is built as a ball with the center $G_{p_1+j} = c_j$ and the radius $r_{p_1+j} = ||c_j - M_{i(j)}||$, $j = 1, ..., p_2$ (see Fig. 2c).

The presented Emmental-GKLS class of test problems consists of 100 smooth objective functions with the same characteristics as the original GKLS-based test functions, with the global minimum located in a random point with a random radius of its region of attraction. Moreover, it is built using p constraints, including p_1 constraints of the first type, i.e., the constraints related to the local minima, with $3 \leq p_1 \leq m$, and $p_2 = p - p_1$ constraints of the second type, i.e., the constraints related to the vertices of D, with $0 \leq p_2 \leq 2^N$. The admissible regions for $(p_1, p_2) = (3, 0)$, $(4, 0)$, $(4, 2)$, and $(20, 2)$ are presented in Fig. 2.

We can conclude that in the obtained test class:

- The global minimizer x^* of the constrained problem is known, is placed on the boundaries of the admissible region in a point different w.r.t. global minimizer of the unconstrained Emmental-type GKLS problem.
- The global minimizer of the unconstrained Emmental-type GKLS problem is known. It coincides with the nearest local minimizer to the vertex of the paraboloid of the original unconstrained GKLS test problem, and differs from the global minimizer of the constrained test problem.
- The difficulty of the constrained problem can be easily changed. The simplest domain has $p_1 = 3$ and $p_2 = 0$ constraints and the hardest one has $p_1 = m + 1$ and $p_2 = 2^N$ constraints (notice that the search domain can be simply connected, biconnected or multiconnected). It should be noticed, that the global minimizer cannot be an isolated point and is always accessible from a feasible region having a positive volume.
- The constraints are non-linear and satisfy the Lipschitz condition over the hyperinterval D.

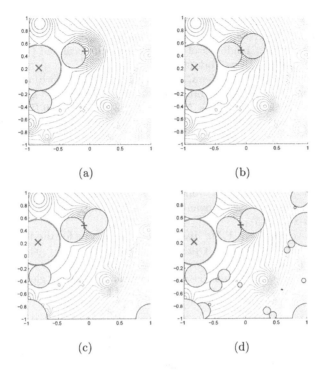

(a)   (b)

(c)   (d)

Fig. 2. The admissible region and the level curves of the objective function of the Emmental-type GKLS-based test problem with (a) $p_1 = 3$, $p_2 = 0$; (b) $p_1 = 4$, $p_2 = 0$; (c) $p_1 = 4$, $p_2 = 2$; (d) $p_1 = 20$, $p_2 = 2$. The admissible region in (d) contains 3 disjoint subregions. The vertex of the paraboloid is indicated by $\times$ and the global minimizer of the constrained problem is indicated by $+$.

Acknowledgements. This work was supported by the project No. 15-11-30022 "Global optimization, supercomputing computations, and applications" of the Russian Science Foundation.

References

1. Addis, B., Locatelli, M.: A new class of test functions for global optimization. J. Global Optim. **38**, 479–501 (2007)
2. Barkalov, K., Gergel, V., Lebedev, I.: Solving global optimization problems on GPU cluster. In: Simos, T.E. (ed.) ICNAAM 2015: 13th International Conference of Numerical Analysis and Applied Mathematics, vol. 1738, p. 400006. AIP Conference Proceedings (2016)
3. Barkalov, K.A., Strongin, R.G.: A global optimization technique with an adaptive order of checking for constraints. Comput. Math. Math. Phys. **42**(9), 1289–1300 (2002)
4. Beasley, J.E.: Obtaining test problems via internet. J. Global Optim. **8**(4), 429–433 (1996)
5. Famularo, D., Pugliese, P., Sergeyev, Y.D.: A global optimization technique for checking parametric robustness. Automatica **35**, 1605–1611 (1999)

6. Floudas, C.A., Pardalos, P.M.: A Collection of Test Problems for Constrained Global Optimization Algorithms. Springer-Verlag, Heidelberg (1990)
7. Floudas, C.A., et al.: Handbook of Test Problems in Local and Global Optimization, vol. 33. Springer, New York (1999)
8. Gaviano, M., Kvasov, D.E., Lera, D., Sergeyev, Y.D.: Algorithm 829: Software for generation of classes of test functions with known local and global minima for global optimization. ACM Trans. Math. Softw. **29**(4), 469–480 (2003)
9. Gergel, V.P., Kuzmin, M.I., Solovyov, N.A., Grishagin, V.A.: Recognition of surface defects of cold-rolling sheets based on method of localities. Int. Rev. Autom. Control **8**, 51–55 (2015)
10. Grishagin, V.A.: Operating characteristics of some global search algorithms. Probl. Stoch. Search **7**, 198–206 (1978). in Russian
11. Grishagin, V.A., Israfilov, R.: Multidimensional constrained global optimization in domains with computable boundaries. In: CEUR Workshop Proceedings, vol. 1513, pp. 75–84 (2015)
12. Grundel, D.A., Pardalos, P.M.: Test problem generator for the multidimensional assignment problem. Comput. Optim. Appl. **30**(2), 133–146 (2005)
13. Kvasov, D.E., Menniti, D., Pinnarelli, A., Sergeyev, Y.D., Sorrentino, N.: Tuning fuzzy power-system stabilizers in multi-machine systems by global optimization algorithms based on efficient domain partitions. Electr. Power Syst. Res. **78**, 1217–1229 (2008)
14. Kvasov, D.E., Mukhametzhanov, M.S.: One-dimensional global search: Nature-inspired vs. Lipschitz methods. In: Simos, T.E. (ed.) ICNAAM 2015: 13th International Conference of Numerical Analysis and Applied Mathematics, vol. 1738, p. 400012. AIP Conference Proceedings (2016)
15. Kvasov, D.E., Mukhametzhanov, M.S., Sergeyev, Y.D.: A numerical comparison of some deterministic and nature-inspired algorithms for black-box global optimization. In: Topping, B.H.V., Iványi, P. (eds.) Proceedings of the Twelfth International Conference on Computational Structures Technology, p. 169. Civil-Comp Press, UK (2014)
16. Pintér, J.D.: Global optimization: software, test problems, and applications. In: Pardalos, P.M., Romeijn, H.E. (eds.) Handbook of Global Optimization, vol. 2, pp. 515–569. Kluwer Academic Publishers, Dordrecht (2002)
17. Sergeyev, Y.D., Grishagin, V.A.: A parallel algorithm for finding the global minimum of univariate functions. J. Optim. Theory Appl. **80**, 513–536 (1994)
18. Sergeyev, Y.D., Kvasov, D.E., Mukhametzhanov, M.S.: On the least-squares fitting of data by sinusoids. In: Pardalos, P.M., Zhigljavsky, A., Žilinskas, J. (eds.) Advances in Stochastic and Deterministic Global Optimization, Chap. 11. SOIA, vol. 107, pp. 209–226. Springer, Cham (2016). doi:10.1007/978-3-319-29975-4_11
19. Sergeyev, Y.D., Kvasov, D.E., Mukhametzhanov, M.S.: Operational zones for comparing metaheuristic and deterministic one-dimensional global optimization algorithms. Math. Comput. Simul. **141**, 96–109 (2017)
20. Sergeyev, Y.D., Mukhametzhanov, M.S., Kvasov, D.E., Lera, D.: Derivative-free local tuning and local improvement techniques embedded in the univariate global optimization. J. Optim. Theory Appl. **171**, 186–208 (2016)
21. Sergeyev, Y.D., Strongin, R.G., Lera, D.: Introduction to Global Optimization Exploiting Space-Filling Curves. Springer, New York (2013)
22. Strongin, R.G., Sergeyev, Y.D.: Global Optimization with Non-convex Constraints: Sequential and parallel algorithms. Kluwer Academic Publishers, Dordrecht (2000)
23. Yuan, B., Gallagher, M.: On building a principled framework for evaluating and testing evolutionary algorithms: A continuous landscape generator. In: The 2003 Congress on Evolutionary Computation, pp. 451–458 (2003)

Author Index

Printed in the United States
By Bookmasters